ENERGY FROM BIOLOGICAL PROCESSES

Technical and Environmental Analyses

OFFICE OF TECHNOLOGY ASSESSMENT
BALLINGER ENERGY SERIES

BALLINGER PUBLISHING COMPANY
Cambridge, Massachusetts
A Subsidiary of Harper & Row, Publishers, Inc.

Library of Congress Catalog Card Number 80-600118

Foreword

In this volume of Energy From Biological Processes, OTA presents the technical and environmental analyses on which the conclusions in volume I are based. The "Part I: Biomass Resource Base" includes forestry, agriculture, processing wastes, and various unconventional sources including oil-bearing and aquatic plants. "Part II: Conversion Technologies and End Uses" considers thermochemical conversions, fermentation for ethanol production, anaerobic digestion, use of alcohol fuels, select energy balances, and a brief description of chemicals from biomass. In each case, appropriate technical, economic, and environmental details are presented and analyzed.

JOHN H. GIBBONS
Director

Energy From Biological Processes Advisory Panel

Thomas Ratchford, *Chairman*
Associate Executive Director, American Association for the Advancement of Science

Henry Art
  Center for Environmental Studies
  Williams College

Stanley Barber
  Department of Agronomy
  Purdue University

John Benemann
  Sanitary Engineering Laboratory
  University of California, Richmond

Paul F. Bente, Jr.
  Executive Director
  The Bio-Energy Council

Calvin Burwell
  Oak Ridge National Laboratory

Robert Hirsch
  EXXON Research and Engineering Co.

Robert Hodam
  California Energy Commission

Kip Howlett
  Georgia Pacific Corp.

Ralph Kienker
  Monsanto Co.

Dean Kleckner
  President
  Iowa Farm Bureau Federation

Kevin Markey
  Friends of the Earth

Jacques Maroni
  Energy Planning Manager
  Ford Motor Co.

Michael Neushul
  Marine Science Institute
  University of California, Santa Barbara

William Scheller
  Department of Chemical Engineering
  University of Nebraska

Kenneth Smith
  Office of Appropriate Technology
  State of California

Wallace Tyner
  Department of Agricultural Economics
  Purdue University

NOTE: The Advisory Panel provided advice and comment throughout the assessment, but the members do not necessarily approve, disapprove, or endorse the report for which OTA assumes full responsibility.

Working Group on Photosynthetic Efficiency and Plant Growth

Olle Bjorkman
  Carnegie Institution
  Stanford University

Glenn Burton
  Southern Region
  U.S. Department of Agriculture

Gary Heichel
  North Central Region
  U.S. Department of Agriculture
  University of Minnesota

Edgar Lemon
  Northeastern Region
  U.S. Department of Agriculture
  Cornell University

Richard Radmer
  Martin-Marietta Laboratory

Contractors and Consultants

The Baham Corp.
Charles Berg
California Energy Commission
Otto Doering III
Douglas Frederick
Charles Hewett
I. E. Associates
Larry Jahn
Robert Kellison
Neushul Mariculture, Inc.
Participation Publishers
Princeton University, Department of
  Aerospace and Mechanical Sciences

Purdue University, Departments of
  Agricultural Economics, Agricultural
  Engineering, and Agronomy
Santa Clara University, Department of
  Mechanical Engineering
State of California, Office of Appropriate
  Technology
T. B. Taylor, Associates
Texas A&M University, Departments of
  Agricultural Economics and Agricultural
  Engineering
Texas Tech University, Department of
  Chemical Engineering

University of California at Davis, College of
  Agricultural and Environmental Sciences
University of California, Richmond Field
  Station
University of Pennsylvania, Department of
  Chemical and Biochemical Engineering
University of Texas at Austin, Center for
  Energy Studies
University of Washington, College of
  Forest Resources
Ronald Zweig

Energy From Biological Processes Project Staff

Lionel S. Johns, *Assistant Director, OTA*
Energy, Materials, and International Security Division

Richard E. Rowberg, *Energy Program Manager*

Thomas E. Bull, *Project Director*

A. Jenifer Robison, *Assistant Project Director*
Audrey Buyrn*
Steven Plotkin, *Environmental Effects*
Richard Thoreson, *Economics*
Franklin Tugwell, *Policy Analysis*
Peter Johnson, *Ocean Kelp Farms*
Mark Gibson, *Federal Programs*

Administrative Staff

Marian Growchowski   Lisa Jacobson

Lillian Quigg   Yvonne White

Supplements to Staff

David Sheridan, *Editor*

Stanley Clark

OTA Publishing Staff

John C. Holmes, *Publishing Officer*

Kathie S. Boss   Debra M. Datcher   Joanne Mattingly

---

*Project director from April 1978 to December 1978.

Acknowledgments

OTA thanks the following people who took time to provide information or review part or all of the study.

Don Augenstein, Flow Laboratories
Edgar E. Bailey, Davy McKee Corp.
Richard Bailie, Environmental Energy Engineering, Inc.
Weldon Barton, U.S. Department of Agriculture
Charles Bendersky, Pyros, Inc.
Edward Bentz, National Alcohol Fuels Commission
Beverly Berger, U.S. Department of Energy
Brian Blythe, Davy McKee Corp.
Hugh Bollinger, Plant Resources Institute
Diane Bonnert, Soil Conservation Service
Carroll Bottum, Purdue University
Robert Buckman, U.S. Forest Service
Fred Buttel, Cornell University
James Childress, National Alcohol Fuels Commission
Raymond Costello, Mittlehauser Corp.
Gregory D'Allessio, U.S. Department of Energy
Ray Dideriksen, Soil Conservation Service
Richard Doctor, Argonne National Laboratory
James Dollard, U.S. Department of Energy
Warren Doolittle, U.S. Forest Service
Ernest Dunwoody, Mittlehauser Corp.
Ed Edelson, Pacific Northwest Laboratory
Eugene Eklund, U.S. Department of Energy
George Emert, University of Arkansas
John Erickson, U.S. Forest Service
William Farrell, EXXON Research and Engineering Co.
Winston C. Ferguson, Conservation Consultants of New England
Kenneth Foster, University of Arizona
Douglas Frederick, North Carolina State University
Ralph E. C. Fredrickson, Raphael Katzen Associates

Tim Glidden, Dartmouth College
Irving Goldstein, North Carolina State University
John Goss, University of California
Roy Gray, Soil Conservation Service
Loren Habegger, Argonne National Laboratory
John Harkness, Argonne National Laboratory
Sanford Harris, U.S. Department of Energy
Marilyn Herman, National Alcohol Fuels Commission
Edward Hiler, Texas A&M University
Dexter Hinckley, Flow Resources Corp.
Wally Hopp, Pacific Northwest Laboratory
John Hornick, U.S. Forest Service
William Jewell, Cornell University
Fred Kant, EXXON Research and Engineering Co.
J. L. Keller, Union Oil of California
Don Klass, Institute of Gas Technology
J. A. Klein, Oak Ridge National Laboratory
Al Kozinski, Amoco Oil Co.
Suk Moon Ko, Mitre Corp.
Kit Krickensberger, Mitre Corp.
Barbara Levi, Georgia Institute of Technology
Les Levine, U.S. Department of Energy
Edward Lipinsky, Battelle Columbus Laboratory
William Lockeretz, Northeast Solar Energy Center
Dwight Miller, U.S. Department of Agriculture
John Milliken, U.S. Environmental Protection Agency
Larry Newman, Mitre Corp.

Edward Nolan, General Electric Co.
John Nystrom, Arthur D. Little, Inc.
Ralph Overend, National Research Council of Canada
Billy Page, U.S. Forest Service
R. Max Peterson, U.S. Forest Service
David Pimentel, Cornell University
L. H. Pincen, U.S. Department of Agriculture
Harry Potter, Purdue University
T. B. Reed, Solar Energy Research Institute
Mark Rey, National Forest Products Association
Robert San Martin, U.S. Department of Energy
Kyosti Sarkanen, University of Washington
John Schaeffer, Schaeffer and Roland, Inc.
Rolf Skrinde, Olympic Associates
Thomas Sladek, Colorado School of Mines Research Institute
Frank Sprow, EXXON Research and Engineering Co.
George Staebler, Weyerhauser Corp.
Terry Surles, Argonne National Laboratory
Robert Tracy, U.S. Forest Service
R. Thomas Van Arsdall, U.S. Department of Agriculture
R. I. Van Hook, Oak Ridge National Laboratory
Thomas Weil, Amoco Chemicals
Donald Wise, Dynatech R&D
Robert Wolf, Congressional Research Service
Robert Yeck, U.S. Department of Agriculture
John Zerbe, U.S. Department of Agriculture

Contents

Part I.

Biomass Resource Base

Chapter 1

INTRODUCTION AND SUMMARY

INTRODUCTION AND SUMMARY

The biomass resource base potentially includes hundreds of thousands of different plant species and various animal wastes. In principle, plants can be cultivated anywhere there is a favorable climate with sufficient water. sunlight, and nutrients. In practice, there are numerous limitations, and the most important of these appears to be the soil type for land-based plants and cultivation and harvesting techniques for aquatic plants.

The largest *area* of underutilized land that is well suited to plant growth is the Nation's forestland. Through more intensive forest management—particularly on privately owned lands—the supply of wood for energy, as well as for traditional wood products, could be substantially increased. However, haphazard wood harvest could cause severe environmental damage and reduce the available supply of wood.

The highest *quality* land suitable for intensive cultivation of plants is the Nation's cropland. The best cropland is dedicated to food production, but there is some underutilized hayland and pastureland as well as land that can be converted to cropland. To a certain extent, grains—especially corn—can be grown for ethanol production and the distillery byproduct used as an animal feed to displace soybean production. More grain can then be grown on the former soybean land. As the ethanol production level grows, however, the animal feed market for the distillery byproduct will become saturated and grass production quickly will become a more effective energy option for the cropland. To a certain extent, environmental damage appears to be practically unavoidable with grain production, but grass cultivation is more environmentally benign.

The candidates for bioenergy crops are numerous, but crop development directed toward energy production is needed to compare the options and to establish cultivation requirements and yields.

In addition to energy crops, substantial quantities of crop residues can be collected and used for energy without exceeding cropland erosion standards.

The third major land category is rangeland, which vary from highly productive wetlands to deserts. Cultivation and harvesting techniques and plant growth are uncertain, and in drier regions the lack of water will limit yields unless the crops are irrigated. However, irrigation greatly increases the energy needed for farming and it is uncertain whether it will be socially acceptable to use the available water for energy production.

Aside from natural wetlands, there are other areas where freshwater plants might be grown and there are vast areas of ocean in which ocean farms might be built. Cultivation and harvesting techniques and crop yields are highly uncertain.

In addition to crop cultivation and residues there is biomass potential from processing and animal wastes.* Most processing wastes currently are used for energy, animal feed, or chemical production, but much of the remainder could be used for energy. Moreover, most of the manure from animals in confined livestock operations could be used for energy.

These biomass sources and various other aspects of the resource base are considered in the following chapters.

---

*Wastes are defined as byproducts of biomass processing that are not dispersed over a wide area and therefore need not be collected. Residues must be collected.

Chapter 2
FORESTRY

Chapter 2.—FORESTRY

Chapter 2
FORESTRY

Introduction

The use of wood for fuel is at least as old as civilization. Worldwide, wood is still a very important source of energy. The U.N. Food and Agricultural Organization estimates that the total annual world harvest of wood in 1975 was 90 billion ft[3] (about 25 Quads) of which nearly one-half was used directly for fuel.[1] Much of the wood that is processed into other products is available for fuel when the product is discarded from its original use, and indeed large but unknown quantities are used in this manner.

Wood has been a very important fuel in the United States, having been used for home heating and cooking, locomotive fuel, the generation of electricity for home, business, and industrial use, and for the generation of steam for industry. According to Reynolds and Pierson, more than half of the wood harvested from U.S. forests for the 300 years of American history preceding 1940 was used as fuel.[2] Consumption of wood fuel reached its peak in the United States in 1880 when 146 million cords (2.3 Quads) were used according to Panshin, et al.[3] The same authors report that per capita consumption of wood fuel peaked in 1860 at 4.5 cords/yr. During the past 100 years, the direct use of wood for fuel declined in the United States to about 30 million cords/yr (0.5 Quad/yr). It was used primarily as a fuel by the forest products industries, which used manufacturing residues, and for home fireplaces and outdoor cooking, which created demand for charcoal and hardwood roundwood.

There have, however, been periodic revivals of fuelwood use to replace conventional fuels in the United States. They have usually occurred during times of crises, such as World

Wars I and II, when conventional fuels became scarce. After the crises abated fuelwood use dwindled rapidly, even though the reemergence of the same conditions in the near future may have been expected.

During 1917-18, for example, the Eastern United States suffered a shortage of coal. Fuelwood was used whenever possible to replace coal, as were sawdust briquettes and other combustible biomass. Individual towns in New England organized "cutting bees" and "cut a cord" clubs for gathering wood fuel to offset the shortage of coal. Between 1916 and 1917 the price of fuelwood increased by about 20 to 30 percent.

The U.S. Forest Service prepared a publication explaining, among other things, how wood could be used as fuel to conserve coal.[4] It was thought at the time that coal reserves in the United States were dangerously low and that the war-induced shortage of 1917-18 had merely emphasized the inevitable need to conserve them. This publication advocated a broad Government policy for development of a fuelwood industry. The role that cutting fuelwood could play in forest management was considered, and an analysis of the economics of cutting and gathering, etc., was given. The report concluded that a fuelwood industry could be profitable and could benefit the forest in other ways as well. The document was published March 10, 1919, by which time the war had ended, and the Nation's fuel situation was already beginning to return to prewar conditions. There is no evidence that any of the recommendations were followed.

Since World War II, the major emphasis on wood use has been for lumber and paper pulp. The annual harvest of commercial wood (wood appropriate for the forest products industry) grew by 22 percent between 1952 and 1976. During this same period, the net growth of

---

[1]*Yearbook of Forest Products 1964-1975* (Rome: Food and Agriculture Organization of the United Nations, 1977).

[2]R. V. Reynolds and A. H. Pierson, "Fuelwood Used in the U.S. 1630-1930," USDA Cir. 641, 1942.

[3]A. J. Panshin, E. S. Harrar, J. S. Bethel, and W. J. Baker, *Forest Products* (New York: McGraw-Hill, 1962).

[4]USDA Bulletin 753, Forest Service, Mar. 10, 1919.

commercial wood (total growth of commercial timber less mortality of commercial timber) increased by 56 percent. In only one region in the country, the Pacific Coast, did the inventory of live commercial wood on commercial forestlands decline. In the Pacific Coast region, however, the growth, as a percentage of the standing inventory, is the lowest in the country due to the old age of the timber. Nationwide the inventory of commercial timber increased by 20 percent from 1952 to 1976. Thus, increased harvests of wood do not necessarily imply that the forests are being depleted.

The growth of wood depends not only on the climate and soil type, but also on the type and age of trees and the way the forest is managed. In this chapter, the potential for fuelwood production from the Nation's forests is examined.

Present Forestland

Forestland is defined as land that is at least 10-percent stocked with forest trees or has been in the recent past and is not permanently converted to other uses. The forestlands are divided into two categories: commercial and noncommercial. Commercial forestland is defined as forestland that is capable of producing at least 20 ft^3/acre-yr (0.3 dry ton/acre-yr) of commercial timber in naturally stocked stands and is *not* withheld from timber production (e.g., parks or wilderness areas). The rest is termed noncommercial.

The forest regions of the United States and the percentage of the total land area of each State that is forestland are shown in figure 1. Currently, there are 740 million acres of forestland in the United States, with about half in the East (i.e., North plus South) and half in the West. About 490 million acres are classified as commercial forestland and nearly three-quarters of this are in the East. The productive potential of commercial forestlands is shown in figure 2.

In addition, there are 205 million acres of noncommercial forestland, which are classified this way because of their low productive potential (i.e., less than 20 ft^3/acre-yr). Practically all of the noncommercial forestland is in the West. Despite the low-productivity classification, however, timber is harvested from many areas of land in this category.

Most of the forestland in the East is privately owned, while about 70 percent of the western forestland is publicly owned and managed by the Federal Government or State and local authorities.

The U.S. Department of Agriculture (USDA) projects that the forestland area will decrease by 3 percent by the year 2030 (about 0.4 million acres/yr or a total of 20 million acres).[5] In the 1980's, a significant portion of the decline will result from conversion to cropland, particularly in the Southeast. USDA projects that in the 1990's, most of the conversion will be to reservoirs, urban areas, highway and airport construction, and surface mining sites.

However, about 32 million acres of potential cropland are now classified as forestland (see ch. 3). Consequently, if a strong demand develops for cropland, then the decrease in forest area will be somewhat larger than USDA's projection.

---

[5]*An Assessment of the Forest and Range Land Situation in the United States,* review draft, USDA Forest Service, 1979.

Present Cutting of Wood

Forest wood is currently being cut for four purposes: 1) production of forest products industry roundwood, 2) production of household fuelwood, 3) timber stand improvements, and

Figure 1.—Forestland as a Percentage of Total Land Area

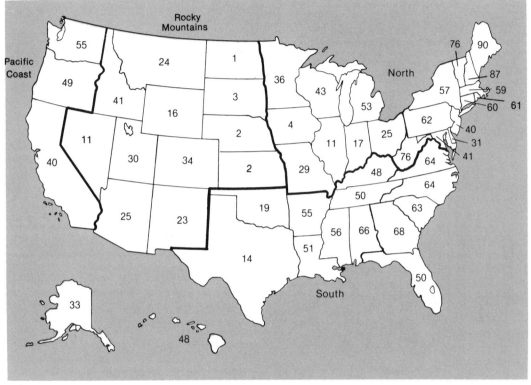

SOURCE: Forest Service, U.S. Department of Agriculture.

4) clearing of timberland for other uses. Each of these produces wood that can be or is used for energy.

Forest Products Industry Roundwood Harvesting

Currently, the forest products industry is harvesting 200 million dry ton/yr (3.1 Quads/yr) for lumber, plywood, pulp, round mine timber, etc.). During the processing of this wood, 90 million ton/yr of primary and secondary manufacturing wastes are produced. These wastes are discussed later under "Biomass Processing Wastes" in chapter 5.

In addition, the process of harvesting the wood generates considerable logging residue. The logging residue consists of the material left at the logging site after the commercial roundwood is removed. These residues are branches, small trees, rough and rotten wood, tops of harvested trees, etc.

The statistics on logging residues reported for 1970 and 1976 by the Forest Service underestimate the total quantity of residues generated by harvesting activities. The Forest Service data only include wood logging residues from growing stock trees.*

Not reported are:

1. bark—most studies of logging residue present volumes without bark;
2. residues from:
   - nongrowing stock trees on logged-over areas,

*Commercial stock trees that are 1) at least 5-inch diameter at breast height (dbh) and 2) not classified as rough or rotten.

Figure 2.—Area of Commercial Timberland by Region and Commercial Growth Capability as of January 1, 1977

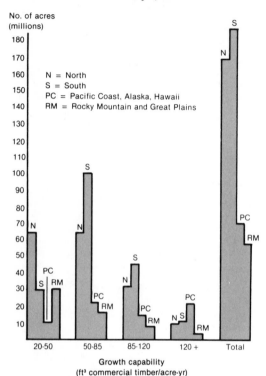

No. of acres (millions)

N = North
S = South
PC = Pacific Coast, Alaska, Hawaii
RM = Rocky Mountain and Great Plains

Growth capability
(ft³ commercial timber/acre-yr)

SOURCE: Data from *Forest Statistics, 1977*, Forest Service, U.S. Department of Agriculture, 1978.

- trees of growing stock species and quality, but less than 5-inch diameter at breast height,
- trees that would be growing stock trees except that they are classified as rough or rotten, and
- trees of noncommercial species;
3. tops and branches; and
4. stumps.

All of these logging residue components, as well as the residue presented in the aforementioned reports, are potentially usable as fuel.*

_____
*In this report, the stumpwood component is not considered.

From various sources[6][7][8] and OTA estimates, the ratios of growing stock residues to total biomass residues were derived.[9] Using these ratios and the Forest Service data for growing stock residues, the quantity of logging residues was estimated to be about 84 million dry tons (1.3 Quads) in 1976. The regional breakdown is shown in table 1, and a more detailed breakdown is shown in table 2.

Table 1.—Logging Residues Estimate[a]—Summary
(in million dry tons)

Region	Softwood	Hardwood	Total[b]
North	2.9	13.2	16.0
South	17.6	15.2	32.8
Rocky Mountain	7.0	0.02	7.0
Pacific Coast	27.1	1.1	28.2
Total[b]	54.5	29.5	84.1

[a]From a 1976 harvest of 130 million dry tons of softwood and 54 million dry tons of hardwood.
[b]Sums may not agree due to round off error.

SOURCE: J. S. Bethel, et al., "Energy From Wood," College of Forest Resources, University of Washington, Seattle, contractor report to OTA, April 1979.

There is some uncertainty as to whether various logging residue studies are in agreement as to what constitutes nongrowing stock logging residue. Loggers may avoid cutting nongrowing stock trees that hold little or no economic value. This practice would be common in selective logging. In many logging residue studies, it is unclear whether or not such uncut trees were considered residue. Some of the differences observed in logging residue factors reported by various authors in the same region may be due largely to these methodological differences. There is a danger that if uncut nongrowing stock is counted as a logging residue, it might again be counted as part of the biomass that should be removed by various silvicultural stand improvements. Every effort was made to avoid this type of double counting.

_____
[6]J. O. Howard, "Forest Residues—Their Volume, Value and Use," Part 2: *Volume of Residues From Logging Forest Industries*, 98 (12), 1971.
[7]R. L. Welch, "Predicting Logging Residues for the Southeast," USDA Forest Service Research Note SE-263, 1978.
[8]J. T. Bones, "Residues for Energy in New England," *Northern Logger and Timer Processor* 25 (12), 1977.
[9]J. S. Bethel, et al., "Energy From Wood," College of Forest Resources, University of Washington, Seattle, contractor report to OTA, April 1979.

Table 2.—Logging Residue Estimates (thousand dry tons)

	Harvest in 1976	From growing stock			From nongrowing stock			Tops and branches incl. bark	Total
		Wood	Bark	Total	Wood	Bark	Total		
Softwoods									
North............	7,448	823	85	908	597	64	661	1,323	2,892
South............	63 031	3,756	393	4,149	2,697	314	2,993	10,415	17,557
W. Pine	16,500	1,548	181	1,729	2,022	236	2,258	3,000	6,987
Coast............	43,190	7,496	876	8,372	8,117	949	9,066	9,668	27,106
Total..........	130,169	13,623	1,535	15,158	13,433	1,563	14,978	24,406	54,542
Hardwoods									
North............	24,546	4,214	313	4,527	1,410	100	1,510	7,147	13,184
South............	27,974	4,984	381	5,275	1,637	123	1,760	8,185	15,220
W. Pine	34	3	—	3	2	—	2	11	16
Coast............	1,094	345	36	381	255	27	282	458	1,121
Total..........	53,648	9,456	730	10,186	3,304	250	3,554	15,801	29,541

SOURCE: J. S. Bethel, et al., "Energy From Wood," College of Forest Resources, University of Washington, Seattle, contractor report to OTA, April 1979.

Household Fuelwood

The harvest of roundwood for use as household fuel was estimated in 1976 to be 657 million ft³, or approximately 10 million dry tons (0.16 Quad). These figures are similar to the results reported by Ellis, who found that 600 million ft³ of roundwood, excluding bark, were harvested for fuelwood.[10] Allowing a 10-percent increase for bark, this becomes 660 million ft³. The regional breakdown is shown in table 3. The quantity harvested in more recent years is considerably larger, however.

Table 3.—Fuelwood Harvests in 1976 (in million dry tons)

Region	Softwood	Hardwood	Total[a]
North...............	0.05	3.7	3.8
South	1.3	4.2	5.7
Rocky Mountains	0.43	0.01	0.44
Pacific Coast.........	0.33	0.11	0.39
Total	2.3	8.2	10.2

[a]Sums may not agree due to round off errors.

SOURCE: J. S. Bethel, et al., "Energy From Wood," College of Forest Resources, University of Washington, Seattle, contractor report to OTA, April 1979.

Stand Improvements

In normal forestry operations, there may be several times during the growth of a stand of trees that malformed, rough, or otherwise undesirable trees are cut to make more growing space for the higher quality trees. These cutting activities are generally referred to as stand improvements, and include stand conversions* and thinning operations. Wood from these activities or sources is suitable for fuel.

The data on the amount of current stand improvement activity are very limited and do not allow a detailed analysis. During the 1968-71 period, various practices, such as precommercial and commercial thinning, species conversion, weed control, and other stand improvements were carried out on a total of 1.4 million acres. This represents only 0.3 percent of the commercial timberland. Generally these practices are carried out irregularly, or on a when-and where-needed basis. Undoubtedly most of the activity is carried out on industry lands where intensive forest management is most advanced. A recent survey of forest industry firms that manage their own lands revealed the current level of these practices.[11] These are summarized in table 4.

In addition, there are timber stand improvements (excluding thinnings), species conversion, and weed control items, on about 1.7 million acres of low-quality stands per year. Yields would vary tremendously among these prac-

---

[10]T. H. Ellis, "Fuelwood," unpublished manuscript, 1978.

*Stand conversion is the practice of eliminating tree species currently occupying a stand and replacing them with other species.

[11]D. S. DeBell, A. P. Brunette, and D. C. Schweitzer, "Expectations From Intensive Culture on Industrial Forest Lands," *J. For.,* January 1977.

Table 4.—Current and Expected Annual Stand Improvements

Treatment	Percent of industry lands treated	Estimated acres treated[a]	Percent of firms expecting to maintain or increase level of treatment
Precommercial thinning ..	0.2	135,000	53
Timber stand improvement........	1.8	1,212,000	69
Commercial thinning	2.5	1,684,000	92
Species conversion	0.4	269,000	65
Weed control	0.3	202,000	50

[a]Percent of lands treated times total acreage owned by industry.

SOURCE: D. S. DeBell, A. P. Brunette, and D. C. Schweitzer, "Expectations From Intensive Culture on Industrial Forest Lands," *J. For.*, January 1977.

tices, but assuming 17 dry ton/acre (as derived by Bethel for rough, rotten, and salvageable trees in the South), this amounts to 29 million dry ton/yr.

Thinnings were also carried out on 1.8 million acres, but there is little information regarding the amounts of residue produced. Yields have been reported of 2.2 dry ton/acre in 4-year-old loblolly pine thinning,[12] and 17 to 28 dry ton/acre in pole timber hardwoods in the North.[13] If a national average of 10 ton/acre is assumed, thinning would provide 18 million dry ton/yr of residue.

---

[12]*Silviculture Biomass Farms* (McLean, Va.: The MITRE Corp., 1977).

[13]F. E. Biltoner, W. A. Hillstrom, H. M. Steinhill, and R. M. Gadmar, USDA Forest Service Research Paper NC-137, 1976.

Combining these two sources results in 47 million ton/yr (0.7 Quad/yr) of residues from stand improvements.

Clearing of Forestland

Clearing of forestland for other uses can provide a temporary, but potentially significant, local supply of wood. The yield per acre harvested varies widely with the locality. Assuming 30 ton/acre cleared, then USDA projections for forestland clearing would provide about 0.2 Quad/yr to 2030. If the forestland with a high and medium potential for conversion to cropland is all cleared over the next 15 years, then this would provide 1 Quad/yr of wood for these 15 years. Most of this would occur in the Southeast (see ch. 3).

Summary of Current Cutting of Wood

The forest products industry currently harvests about 200 million dry ton/yr (3.1 Quads/yr) of wood for lumber, plywood, paper pulp, and other products. The process generates an additional 84 million ton/yr (1.3 Quads/yr) of logging residues. Another 10 million dry tons (0.2 Quad/yr) are harvested for fuelwood, and about 47 dry ton/yr (0.7 Quad/yr) are cut during stand improvements. This results in a total harvest of about 340 million dry ton/yr or the eqivalent of 5.3 Quads/yr. Another 0.2 Quad/yr is obtained from clearing and converting forestlands to other uses.

Present Inventory of Forest Biomass

It is not a simple matter to derive the total forest biomass inventory from the Forest Service surveys. As noted earlier, this lack of an adequate census base stems from the traditional practice of evaluating the wood in a forest only in terms of what is assumed to be merchantable, rather than on a whole-tree or whole-biomass basis. Furthermore, the Forest Service does not survey noncommercial forestlands (about one-third of the total forest area). As a result of this inadequate information

base, the present inventory of forest biomass can only be estimated.

Noncommercial Forestland

As mentioned above, of the one-quarter billion acres of noncommercial land, 24 million acres (about 10 percent) are so classified because they are recreation or wilderness areas, or are being studied for these uses. These lands are not included in the inventory of standing timber. Approximately 205 million acres are classified as noncommercial because they are considered incapable of producing as much as 20 ft³ of commercial wood per acre-year. This criterion, is an arbitrary one, however, and timber is, in fact, harvested from many areas of land in this category. For this reason, the latter category of noncommercial forestlands is included in the inventory of standing timber.

Assuming that these 205 million acres produce an average of 10 ft³/acre-yr of commercial wood, that they are mature stands (80 years old or more), and that the aboveground biomass is 1.5 times the amount of commercial timber, the inventory of these noncommercial lands is 3.7 billion dry tons (57 Quads).

In addition, 23 million acres, mostly in Alaska, were classified in 1978 as noncommercial because they were considered inaccessible. Assuming a production capability of 35 ft³/acre-yr and the same assumptions as above, the inventory from these lands is 1.4 billion dry tons (22 Quads).

These two categories result in an inventory on 1978 noncommercial lands of about 5 billion dry tons (80 Quads).

Commercial Forestland

Approximately 488 million acres of forestland are classified by the Forest Service as commercial forestland for purposes of reporting a national forest survey. It is possible to estimate a fuel inventory from commercial forestland, using national forest survey data, with much more precision than was the case for noncommercial lands.

Two options were considered for developing estimates of total biomass on commercial forestland based on the national forest survey. One procedure involved the assumption of multipliers that would convert the basic product inventory data to whole-stem biomass estimates. A second method involved the use of stand tables from the national forest survey and allometric regression equations for estimating biomass for various tree components.[14]

[14]Bethel, op. cit.

For the purposes of this study, an estimate of total whole-stem biomass for the United States was developed, based on *Forest Statistics for the United States, 1977*.[15] Table 5 shows the result of this analysis for commercial forestland. The details of these computations and more extensive tables are given in OTA's contractor report "Energy From Wood."[16]

Table 5.—Estimated Aboveground Standing Biomass of Timber in U.S. Commercial Forestland
(excluding foliage and stumps, in billion dry tons[a])

Region	Hardwood	Softwood	Total
North.	5.2	1.3	6.5
South	4.6	2.3	6.9
Rocky Mountains	0.2	2.4	2.6
Pacific Coast.	0.6	4.2	4.8
Alaska.	0.08	1.3	1.4
Total	10.6	11.5	22.1

[a]Sums may not agree due to round off errors.

SOURCE: J. S. Bethel, et al., "Energy From Wood," College of Forest Resources, University of Washington, Seattle, contractor report to OTA, April 1979.

Adding commercial and noncommercial land inventories gives 27 billion tons (430 Quads), which is estimated to be the inventory of biomass in U.S. forests, excluding stumps, foliage, and roots and the biomass in parks and wildnerness areas, or areas being considered for these uses.*

Quantity Suitable for Stand Improvement

Of the 27 billion tons of standing biomass, some of the wood is of the type that would be removed in stand improvements. This would include brush, rough, rotten, salvageable dead wood, and low-quality hardwood stands occupying former conifer sites. In Alaska, there are roughly 330 million tons of this type of wood.[17] In the rest of the Pacific Coast region, there are 565 million tons, and in the Rocky Mountain region, 324 million tons. The North and South have 822 million and 978 million tons, respec-

[15]*Forest Statistics of the U.S., 1977*, USDA Forest Service, review draft.
[16]Bethel, op. cit.
*For the purposes of this report, stumps, roots, and foliage are excluded from whole-stem biomass.
[17]Ibid.

tively.[18] The total is 3.1 billion tons (49 Quads) of wood that would be appropriate for removal in stand improvements on commercial forestlands. This figure does not include foliage

[18] Ibid.

or all of the cuttings that could be used to convert stands of one kind of trees to a more productive type. Consequently, this is a conservative estimate of the biomass available from stand improvements.

Present and Potential Growth of Biomass in U.S. Commercial Forests

Current gross annual biomass growth in commercial U.S. forests has been estimated from Forest Service data to be 570 million dry ton/yr, of which 120 million ton/yr are mortality, and 450 million ton/yr net growth.[19] * The usual method of determining the productivity of a particular stand occupying a site is by reference to normal yield tables. These tables are models used to predict growth of active natural stands, and are based on stands of "full" or "normal " stock.

Because of the utilization assumptions built into normal yield tables, however, productivity may consistently be assigned a low, and misleading rating. For example, when the actual growth in 131 Douglas-fir plots scattered throughout western Washington and Oregon was compared with Forest Service Bulletin normal yield tables for Douglas fir, it was found that the yield tables consistently underestimated the actual growth. Actual growth in some age-site combinations was more than double the normal yield table value, and the overall average growth exceeded the yield table by nearly 40 percent.[20] Furthermore, in parts of the Rocky Mountains where Forest Service and industry lands are co-mingled, industry representatives report that measurements of actual growth are two to three times the productivity assigned by normal yield tables.[21] Because of the errors associated with estimating tree types, their number, and their size from normal yield tables, OTA estimates

that the actual current biomass growth on commercial forestland is one to two times the values derived from normal yield tables, or 570 million to 1,140 million dry ton/yr (9 to 18 Quads/yr). (See figure 3.)

These estimates do not take into account the productive potential of the forestland. Forest site productivity is estimated on the basis of the vegetation currently occupying the area at the time of the survey. But over 20 million acres of commercial forestland are unstocked, and much more land is stocked with species that are growing more slowly than could be achieved with species better suited to the site. The forest survey indicates that, due to these factors, current growth is about half the growth that could be achieved with full stocking of highly productive tree types (i.e., current growth is estimated by the Forest Service at 38 ft³/acre-yr while the land capability is estimated by USDA at 74 ft³/acre-yr). OTA therefore estimates the potential growth to be about two to four times that derived from normal yield tables, or 1.1 billion to 2.3 billion dry ton/yr (18 to 36 Quads/yr) with full stocking of productive tree species on commercial forestland. This corresponds to slightly more than 2 to 4 ton/acre-yr on the average.

Beyond the potential growth with unfertilized timber, studies in the Southeast indicate that fertilizers and genetic hybrids could increase the biomass growth by 30 percent.[22] However, not all of the potential growth is physically accessible or economically attractive as discussed below.

[19] H. Wahlgren and T. Ellis, "Potential Resource Availability With Whole Tree Utilization," TAPPI, vol. 61, No. 11, 1978.
*The 120 million tons of annual mortality are from growing stock trees only. Mortality from nongrowing stock trees is not known. Under intensive management, much of the mortality loss could potentially be captured for productive use.
[20] Bethel, op. cit.
[21] Ibid.

[22] Ibid.

Figure 3.—Forest Biomass Inventory, Growth, and Use (billion dry tons with equivalent values in Quads)

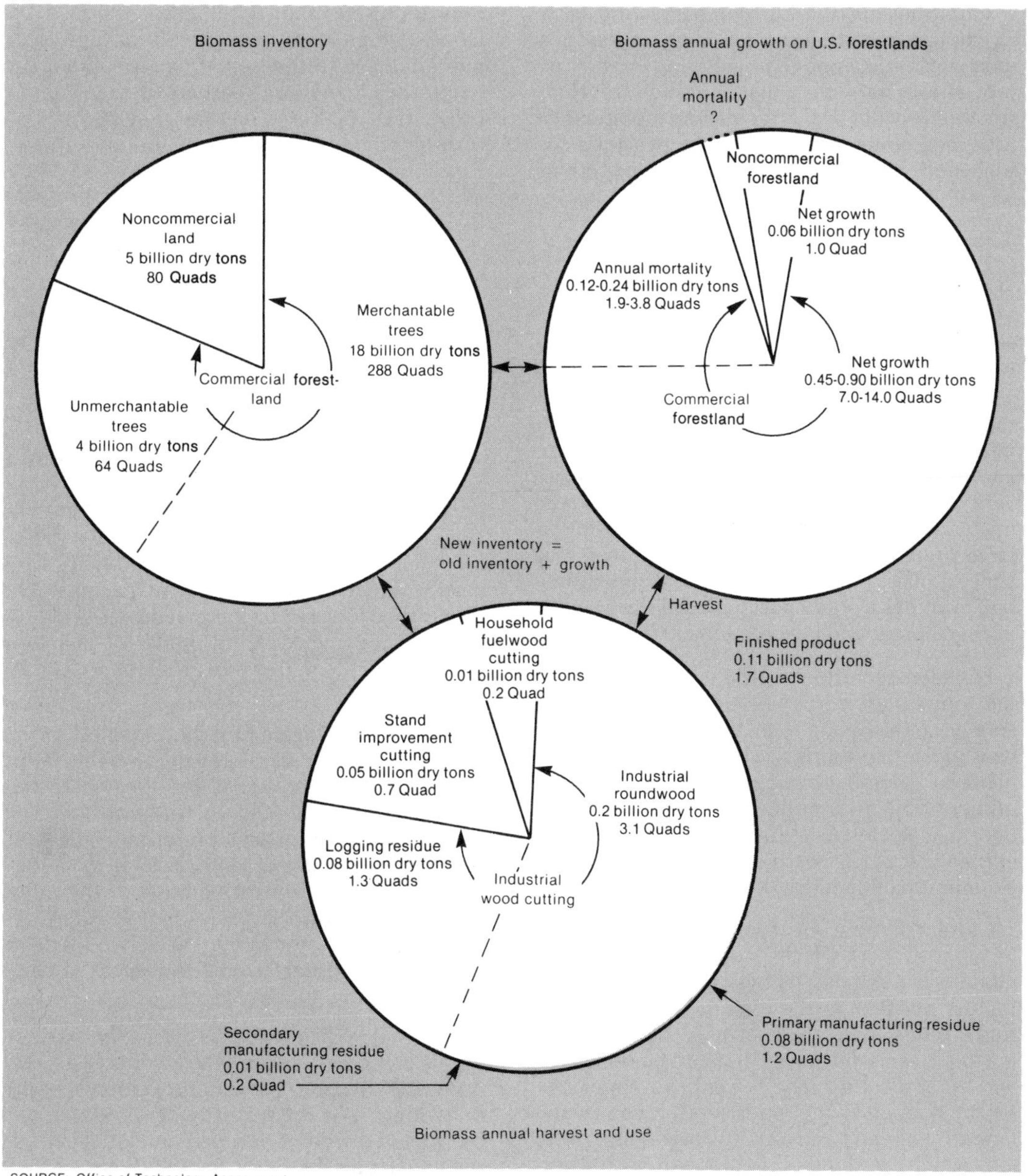

SOURCE: Office of Technology Assessment.

Forest Biomass Harvesting

Variations on the current harvesting techniques (described below) are likely to be common with fuelwood harvests and stand improvement activities that produce residues suitable for fuel. Nevertheless development of new techniques and equipment designed for fuelwood harvests and stand improvements could lower the cost.

Intensive forest management might typically consist of the following: The stand would be clearcut, and the slash (or logging residue) removed. The stand would then be replanted with the desired trees. After 5 to 20 years the stand would be thinned so as to provide more space for the remaining trees. The stand would then continue to be thinned at about 10-year intervals, by removing diseased, rough, rotten, and otherwise undesirable trees and brush. In very intensively managed stands, the trees might also be pruned to avoid the formation of large knots in the stem of the tree (e.g., for veneer). These periodic thinnings and (possible) prunings would continue until the stand is again clearcut and the entire cycle repeated.

For each operation mentioned above (except the replanting), some woodchips suitable for energy could be made available. The method chosen for harvesting the fuelwood would depend on a number of site-specific factors. The primary objective would be to fell and transport the selected trees or to transport the slash in the most cost-effective manner, while doing a minimum of damage to the remaining stand.

Currently there are four basic methods of logging, each of which is designed to accommodate a number of physical and economic factors peculiar to the logging site. Once the tree is felled: 1) it can be skidded (dragged) to a roadside as a *whole tree,* 2) it can be delimbed and the top cut off, and the entire stem or *tree length* skidded to the roadside, 3) it can be delimbed, topped, and cut (bucked) into *long logs* which are skidded, or 4) it can be cut into shorter logs or *short wood* which are skidded. The *whole-tree* skidding brings out the most biomass. However, if the limbs cannot be used

they represent a disposal problem. Also the *whole-tree* and *tree length* methods tend to do more damage to the timber being skidded and to the residual stand. If there is thick underbrush, the *whole-tree* method may be difficult or impossible. A weighing of the various factors appropriate to the site being logged results in the method used. If markets for the limbs develop, however, then more *whole-tree* skidding may be used than is now the case.

Once the wood is at the roadside, it can be cut and loaded or loaded directly into trucks for transport to the mill or conversion site. Alternatively, the wood can be chipped at the roadside with the chips being blown into a van for transport.

Two large-scale harvesting systems considered here are whole-tree harvesting and cable logging. In the whole-tree chip system, the trees are felled by a vehicle called a feller-buncher, which grabs the tree and uses a hydraulic shear to cut the tree at its base. The tree is then lowered to the ground for skidding. This method is most appropriate for relatively flat land and smaller trees (i.e., less than 20-inch diameter).

In the cable logging method, cables are extended from a central tower and the felled trees are dragged to a central point, where they are sorted and skidded to the roadside. This method is used primarily on terrain with steep slopes and large trees. Estimates for the equipment and annual operating costs of these two systems are shown in tables 6 to 9. There are other logging systems, but these two methods are fairly representative of the range of existing systems.

The major difference between the harvesting of various categories of wood (e.g., residues from logging, stand improvements, or primary logging) is the quantity of wood that can be removed from a site per unit of time, i.e., the logging productivity. Several factors affect the logging productivity, and the most important of these are shown in table 10. The production of the logging operations discussed above might range from 15,000 to 75,000 green ton/yr, leading to harvesting costs from about $5 to $30/green ton. In addition, transportation, possible roadbuilding, and stumpage fees (fees paid to the landowner for the right to harvest the wood) must be included. Transportation ranges from $0.06 to $0.20/ton-mile, and where roadbuilding is necessary, the costs will be considerably higher. Stumpage fees for fuelwood have been estimated at $0.40 to $1.00/green ton in New England,[23] but these will change with the market.

The costs of whole-tree chipping 33 stands in northern Wisconsin and the Michigan peninsula have been modeled by computer simulation.[24] In each case, the center of the county was assumed to be the destination for the wood. The supply curve for these stands is shown in figure 4, exclusive of stumpage fees. The cost average varies from $6 to $15/green ton ($12 to $30/dry ton) in 1978 dollars. The range of delivered costs included relogging of logging residues ($16.50 to $20.30/green ton), thinning ($10.00 to $13.80/green ton), and integrated logging for lumber and residue chipping ($9.75 to $12.30/green ton). An equalizing factor in the delivered cost is the stumpage fee.

[23]C. Hewett, School of Forestry, Yale University, private communication.
[24]J. A. Mattson, D. P. Bradley, and E. M. Carpenter, "Harvesting Forest Residues for Energy," *Proceedings of the Second Annual Fuels From Biomass Symposium* (Troy, N.Y.: Rensselaer Polytechnic Institute, June 20-22, 1978).

Table 6.—Assumptions for Whole-Tree Harvesting Equipment

Equipment	Initial cost (dollars)	Salvage value (percent)	Life years	Labor[a] $/hour
Whole-tree chipper				
380 hp	$115,000	20	5	$4.62
600 hp	132,000	20	5	4.62
Feller-buncher	100,000	20	5	4.62
Skidder (each)	55,000	20	4	4.20
Used skidder	10,000	10	3	
Lowboy trailer	10,000	10	10	
Used crawler	30,000	20	5	
Equipment moving truck	1,680/yr			
¾-ton crew cab pickup	8,400/yr			
½-ton pickup	7,862/yr			
Chain saws (3)	3,024/yr			
Other labor				
deck hands (2)				7.20
foreman				8.40
supervisor				2.81

[a]South, includes payroll benefits.

SOURCE: J. S. Bethel, et al., "Energy From Wood," College of Forest Resources, University of Washington, Seattle, contractor report to OTA, April 1979.

Where logging, transportation, and other costs are low, stumpage fees will be high and vice versa. The market will determine these fees, as well as the quantity and types of wood that can be economically harvested.

The 1979 delivered cost of fuel chips was about $12 to $18/green ton in New England.[25] A detailed national cost curve, however, would require a survey of all potential logging sites,

[25]Connecticut Valley Chipping, Plymouth, N.H., L. W. Hawhensen, president, letter to Conservation Consultants of New England, Dec. 20, 1979.

Table 7.—Annual Whole-Tree Chipping System Costs

Annual costs to pay all expenses and earn 15% aftertax ROI, shown in thousands of dollars (values in columns are shown only when a change occurs).

Region	Annual capital cost	Maintenance	Fuel, lube, etc.	Local taxes and insurance	Labor[a]	Miscellaneous equipment[b]	Total
System based on 380-hp chipper Initial investment: $375,000							
North	$221	$35	$44	$7	$ 83	$21	$442
South	221	—	—	—	62	—	390
West	221	—	—	—	104	—	432
System based on 600-hp chipper Initial investment: $447,000							
North	264	40	50	9	94	21	478
South	264	—	—	—	70	—	454
West	264	—	—	—	118	—	502

[a]Includes foreman and supervisor.
[b]Pickup trucks, chainsaws, etc.

SOURCE: J. S. Bethel, et al., "Energy From Wood," College of Forest Resources, University of Washington, Seattle, contractor report to OTA, April 1979.

Table 8.—Assumptions for Cable Yarding Equipment

Equipment	Initial cost (dollars)	Salvage value (percent)	Life years	Labor[a] $/hour
Yarder with 50-ft tower ..	$180,000	20	8	$10.29
Yarder with 90-ft tower ..	228,000	20	8	10.29
Radio and accessories ...	11,386	0	4	—
Whole-tree chipper				
380 hp	115,000	20	5	7.76
600 hp	132,000	20	5	7.76
Skidder (each)	55,000	20	4	7.06
Hydraulic loader	207,000	20	6	10.64
Used skidder	10,000	10	3	
Lowboy trailer.........	10,000	10	10	
Used crawler	30,000	20	5	
Equipment moving				
truck	1,680/yr			
¾-ton crew cab pickup ..	8,400/yr			
½-ton pickup	7,862/yr			
Chain saws (3)	3,024/yr			
Other labor				
yarding crew........				47.84 (5 men)
foreman				14.11
supervisor (1/3 time)..				4.72

[a]West, includes payroll benefits.

SOURCE: Office of Technology Assessment.

which is not available. Nevertheless, some fuelwood can be had for as little as $10/green ton ($20/dry ton) plus stumpage fees.[26] In unfavorable circumstances, the wood could cost as much as $30/green ton ($60/dry ton for relogging of logging residues in the Northwest).[27] Thus, fuelwood chips may vary in price from about $20 to $60/dry ton which is in substantial agreement with the cost estimates based on harvesting costs.

In each category of wood there will be small businesses or individuals who are willing to work at lower rates, who are figuring only marginal costs, and/or who own the land and assign a zero stumpage fee. In other words, there will always be limited supplies of wood below the average market price.

[26]C. Hewett, *The Availability of Wood for a 50 MW Wood Fired Power Plant in Northern Vermont*, report to Vermont State Energy Office under grant No. 01-6-01659.

[27]Kip Howlett, Georgia Pacific Corp., Atlanta, Ga., private communication, 1979.

Table 9.—Cable Logging System Costs

Annual requirement to pay all expenses and earn 15% aftertax ROI (thousands of dollars)

Equipment	Annual capital cost	Maintenance	Fuel, lube, etc.	Local taxes and insurance	Labor[a]	Miscellaneous equipment[b]	Total
50-ft tower/380-hp chipper investment: $466,000	$200	$43	$44	$ 9	$158	$21	$475
90-ft tower/600-hp chipper $531,000	228	61	47	11	158	21	526

[a]Inlcudes foreman and supervisor.
[b]Pickup trucks, chainsaws, etc.

SOURCE: Office of Technology Assessment.

Table 10.—Factors Affecting Logging Productivity

Road space.
Slope and slope changes—slope steepness and whether logging is uphill or downhill.
Size and shape of landing.
Skidding distances—both loaded and return if different; affected by the tract shape and its relation to the road system.
Skid trail preparation.
Timber character—
    Species—Especially hardwood v. softwood.
    Volume and number of trees per acre to be removed—The size of trees and logs is a very important consideration.
    Quality—More defective timber is likely to result in more breakage, increasing materials handling problems.
    Residual stand, if any, in terms of number of trees and volume per acre. This is prescribed by the silvicultural method.
Cutting policy—appropriate for the stand.
Felling and logging methods—whole tree, shortwood, tree length, etc.
Brush height and density.
Condition and number of windfalls, old stumps, and slash per acre.
Drainage and stream crossings.
Season.
Crew size and aggressiveness.
Wage plan.
Equipment types, functions, and balance—especially the number of places handled per cycle.
Maintenance policy.
Environmental regulations—may prescribe certain practices or preclude certain equipment from areas with sensitive mixes of soils, slopes, and/or drainage thereby reducing production or increasing costs. In the West these regulations have caused a shift in the mix of tractor v. cable logging as well as shifts within each of these general categories.

SOURCE: J. S. Bethel, et al., ''Energy From Wood,'' College of Forest Resources, University of Washington, Seattle, contractor report to OTA, April 1979.

Figure 4.—Supply Curve for Forest Chip Residues for Northern Wisconsin and Upper Michigan

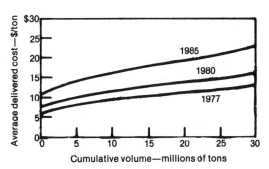

SOURCE: J. A. Mattson, D. P. Bradley, and E. M. Carpenter, "Harvesting Forest Residues for Energy," *Proceedings of the Second Annual Biomass Symposium* (Troy, N.Y.: Rensselaer Polytechnic Institute), June 20-22, 1978.

Factors Affecting Wood Availability

The presence of nearby roads, the concentration of wood on the logging site, and the terrain (steepness of the slope) are the most important physical factors affecting the economics, and thus the availability, of harvested wood. Nevertheless, landownership, alternate uses for the land, taxation, and some subsidiary benefits and constraints also play an important role in wood availability. These other factors are discussed below.

Landownership

One of the more important features distinguishing the various forest regions in the country is landownership. In New England 2 percent of the commercial forestland is federally owned, and public ownership accounts for only 6 percent. In the East as a whole, 14 percent of the commercial forestland is publicly owned, while 7 percent is federally owned. Ownership patterns in the West are reversed, with 68 percent of the commercial forestland being publicly owned (96 percent in Arizona) and 58 percent in Federal ownership.

Although patterns in the West permit logging firms to deal with a limited number of large landowners, other restrictions may be placed on the logging operations. One example is the Federal requirement that logging residues be removed from or otherwise disposed of on national forests in the West, to minimize the risks of forest fires.

Logging firms in the East must deal with a larger number of landowners, and in the Northeast, forestlands are often owned for recreational or investment purposes. It may be difficult to determine who owns the land, to contact the owner, or to interest the owner in using the land for logging. In the South this is less of a problem. Large areas of forestland owned by relatively small landowners are managed by the forest products industry and are available for logging.

Alternate Uses for the Land

The fact that a tract of land is forested and designated commercial does not necessarily mean that it can be logged. The owner may have esthetic objections to logging, may use the land for recreational purposes, or, in the case of an investor, may feel that it would be more difficult to sell the land after logging. In New Hampshire and Vermont, for example, a recent study concluded that only 6 percent of the owners of commercial forestland considered timber production as a reason for owning forestland, and only 1.3 percent listed it as the most important reason.[28] (This 6 percent owns 21 percent of the commercial forestland in the two States). Nevertheless, 10 percent of the private owners (representing 53 percent of the forestland) intended to harvest their timber within 10 years and over one-third of the owners (representing 87 percent of the land) intended to harvest "some day." About half of the landowners (owning 9 percent of the land) indicated that they would not harvest the land because of its scenic value or because their tracts were too small.

Public Opinion

While proper management of a forest can improve the health and vitality of the trees, improper management can have severe environmental consequences. (See "Environmental Impacts".) In any event, an intensively managed forest will look like it is being managed.

_____
[28]N. P. Kingsley and T. W. Birch, "The Forest-Land Owners of New Hampshire and Vermont," USDA Forest Service Resource Bulletin NE-51, 1977.

There will be fewer overmature trees, the trees will be more uniform in appearance and spacing, and the forest floor will have less debris and "extraneous" vegetation. The managed forest will not look like a natural forest, and the difference in appearance can be quite large.

This change in appearance, together with various environmental uncertainties (see "Environmental Impacts"), leads to widely varying opinions about the benefits of forest management. If the citizenry affected by increased management cannot effectively participate in the process of deciding where and how intensively the forests will be managed, and if business and Government officials are not sensitive to the concerns of the citizenry, then the political atmosphere surrounding forest management for energy could become polarized. Public opposition could then seriously restrict the use of forests for energy.

Forest management, however, is not an absolute. There are many ways to manage forestlands, from wood plantations to the occasional gathering of fallen trees and branches. The ability of political leaders to convey this fact to the public, and the ability of Government to aid in striking an equitable balance between environmental and esthetic concerns and the economics of wood harvesting, may prove to be one of the most significant factors affecting an increased availability of wood for energy outside of the forest products industry.

Alternate Uses for Wood

Much of the wood that will be used for energy in the near future is less suitable for materials (e.g., particle board or paper) than the wood currently used for these products. If there is a strong demand for wood products, however, some of this lower quality wood will be drawn into the materials market. Similarly, technical advances in wood chemistry may create an additional demand for wood to be processed into chemicals.

It must be remembered that a strong wood energy market would provide an incentive to increase the number of stand improvements. This will result in an increased supply of what is considered commercial-grade timber. Furthermore, some stands that cannot now be harvested economically for only lumber or pulpwood will become economically attractive for a combined harvest of lumber, pulpwood, and woodchips for energy.

In the very long term, competition for wood may develop between the energy and materials/chemicals markets. For the next 20 years, however, a wood energy market—properly managed—will increase the supply of wood for other uses over what would occur in its absence, and indeed this situation is likely to prevail for at least 50 to 60 years.

Other Factors and Constraints

As noted previously, the Forest Service requires that logging residues on national forests be disposed of to minimize the risks of forest fires. Stumpage fees for logging national forestlands are therefore lower than for comparable private lands in the region, in order to cover the cost of disposing of the residues. In the early 1970's, as a result of a strong demand for paper, some of these residues were collected and chipped for paper pulp. Currently, however, the residues (about 0.2 Quad/yr) are disposed of onsite by burning and other techniques. If a strong energy market existed, much of this could be chipped and used for energy.

It has been common practice in site preparation to use herbicides to kill unwanted plants so that preferred trees could regenerate either naturally or artificially. Increasingly, however, the use of herbicides for this purpose is being restricted and in some cases banned (e.g., 2, 4, 5-T). A strong energy market would provide an additional incentive to harvest the brush and other low-quality wood and thereby minimize the use of these controversial chemicals.

Net Resource Potential

There is no simple way to assess accurately the impacts of the various and sometimes contradictory factors affecting the availability of wood for energy. Many of the important factors, such as public opinion, the way the forests are managed, and the presence of roads, will depend on actions taken in the future. Assuming, however, that 40 percent of the growth potential of the U.S. commercial forestland is eventually accessible, 450 million to 900 million dry ton/yr (7.3 to 14.6 Quads/yr) could be available for harvest.

In terms of energy, the forest products industry currently cuts 5.1 Quads/yr of wood, including logging residues (1.3 Quads/yr) and stand improvement cutting (0.7 Quad/yr). Of this total, 1.7 Quads/yr are converted into products sold by primary or secondary manufacturers, and 1.2 Quads/yr, supplied by wood wastes, satisfies over 45 percent of the industries direct energy needs. This leaves about 2.2 Quads/yr of wood that are currently being cut but not used (see figure 5), and there is at least 40 Quads (total) of unmerchantable standing timber.

Assuming that the demand for traditional forest products doubles by 2000, then 3.4 Quads/yr will be needed for finished wood products, and 3.9 to 11.2 Quads/yr could be used for energy, provided increased forest management occurs. If, however, the forest products industry becomes energy self-sufficient by 2000, it could require as much energy as the lower limit of available wood energy, but three factors will probably alter this simple projection. First, the increased demand for wood products is likely to increase the number of stand improvements. Second, the energy efficiency of the forest products industry will probably increase as a result of higher energy prices and new processes (such as anthraquinone catalyzed paper pulping). Third, if the forest products industry requires most of the available output of 40 percent of the commercial forestlands to supply its needs, then additional roads would be built to access more timberland. Additional wood that is not of high enough quality for lumber, veneer, paper pulp, etc., would therefore become available. In light of these factors, it is likely that significant quantities of wood will become available for energy uses outside of the forest products industry, but this industry could be the major user.

These estimates are admittedly approximate, but a more precise estimate would require a survey of potential logging sites, land capability, road availability, and the costs of harvesting.

The results of such a survey could change these estimates, but 5 to 10 Quads/yr is OTA's best estimate of the energy potential from existing commercial forestland.

Figure 5.—Materials Flow Diagram for Felled Timber During Late 1970's (Quads/yr)

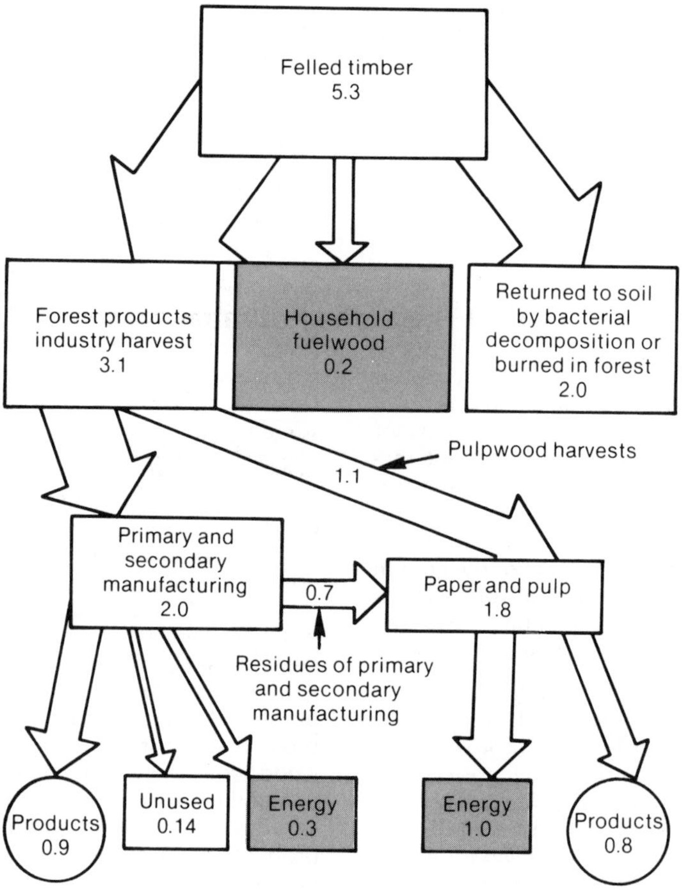

Total left in forest	2.0 Quads/yr	
Total used as energy	1.5 Quads/yr	
Unused residues	0.14 Quad/yr	
Total products	1.7 Quads/yr	

SOURCE: Office of Technology Assessment.

Environmental Impacts

Introduction

A forest may be perceived as:

- a natural ecosystem deserving protection;
- a source of materials—renewable or otherwise;
- a physical buffer to protect adjacent areas from erosion, flood, pollution, etc.;
- a source of esthetic beauty;
- a wildlife preserve;
- a source of recreation—hiking, hunting, etc.;
- a temporary land use;
- a place to retreat from civilization; or
- an obstacle to another desired land use such as mining or agriculture.

This range of perceptions is complicated by the fact that individuals do not perceive all forests to be alike, and few would attach the same perspective—or value—to all forests. Thus, the keenest environmentalist may comfortably accept a managed, single-aged pine forest in the same terms as he accepts a wheatfield, while a lumber company president may view a preserve of giant Sequoias with as much reverence as a Sierra Club conservationist.

These perceptual differences make an evaluation of the environmental effects of a wood-for-energy strategy difficult, because many of the effects may be valued by some groups as positive and by others as negative. In other words, although some potential effects of growing and harvesting operations (e.g., effects such as impaired future forest productivity or extensive soil erosion) are clearly negative or (in the case of restoration of lands damaged by mining) positive, other effects are more ambiguous. Changes in such forest characteristics as wildlife mix, physical appearance, accessibility to hikers, and water storage capabilities may be viewed as detrimental or beneficial depending on one's objectives or esthetic sense. For instance, measures that increase forest productivity by substituting softwood for hardwood production would be considered as strongly beneficial by those who value the forest mainly for its product output,

but may be perceived as detrimental by those who cherish the same forest in its original state. Hence, it is likely that a wood-for-energy strategy that increases the areal extent or intensity of forestry management will promote a wide range of reactions . . . *even if the physical impacts are fully predictable and if forecasts of these physical changes are believed by all parties.*

Environmental evaluation is further complicated both by difficulties in predicting the physical impacts and by the strong possibility that even those predictions that *can* be accurately made will not be accepted as credible by all major interest groups.

The problem of credibility stems largely from the history of logging activities in the United States and the negative impact it has had on public perceptions of logging. The adaptation of the steam engine to logging around 1870 began an era (lasting into the 20th century) when America's forest resource was mined and devastated.[29] The dependence of logging on the railroads and on cumbersome steam engines—capital-intensive equipment that could not easily be moved from site to site—led to the cutting of vast contiguous areas. There was virtually no attention to reforestation. In fact, it was then thought that most of this land would be used for agriculture, and that clearcutting enhanced the value of the land. It also was thought that the timber resource was essentially unlimited and that it was unnecessary to worry about regeneration.

Massive cutting followed by repeated fires led to the destruction of tens of millions of acres of hardwood (in the South and East) and softwood (in the Lake States, Rockies, and part of the Northwest) forest and their replacement by far less valuable tree types or by grassland. This massive destruction led to a considerable public revulsion towards logging, much of which still survives. It also led to a revulsion

---

[29]M. Smith, "Appendix L, Maintaining: Timber Supply in a Sound Environment" in *Report of the Presidents Advisory Panel on Timber and the Environment* (Washington, D.C.: Forest Service, U.S. Department of Agriculture, 1973).

against clearcutting and even-aged management within the forestry profession which lasted for 20 years;[30] although clearcutting (at least the very limited version used today, which involves very much smaller areas than were routinely cut in the past) is now an accepted and even popular practice in the profession, the attitudes formed by attempts at public education about forest values in the 1930's and 1940's linger on. Furthermore, there have been enough reports of unsound forest management and widely publicized environmental fights over such management in the intervening decades to create a sizable constituency that is generally very skeptical about logging practices. As a result, assessments that focus on the potential positive effects of increased forest management may be greeted with skepticism by large segments of the public.

The prediction of environmental changes that might occur in American forest areas if demand for wood energy grows is extremely difficult. The potential for wood energy identified previously is based on a "scenario"—a vision of a possible future—that assumes an increased collection of wood residues that are now left to rot in the forest as well as an assumed intensification of silvicultural management *on suitable land* that would increase growth rates and timber quality, increasing the supply of nonenergy wood products while providing a steady supply of wood fuel. This type of strategy could lessen harvesting pressures on wilderness areas and other vulnerable forestlands. It probably would be perceived by many groups as environmentally beneficial, although it would lead to esthetic and ecosystem changes on those lands where management was intensified. Given the present institutional arrangements, however, there is no guarantee that this assumed "scenario" will unfold as outlined. Instead, a combination of Federal, State, business, and other private interests will respond to a complex market amid a variety of institutional constraints. In order to predict the environmental outcome of such a response, the following factors must be understood:

_____
[30]Ibid.

1. The environmental effects that occur when different kinds of silvicultural operations (including different kinds and intensities of cuts, regeneration practices, roadbuilding methods, basic management practices, etc.) are practiced on different forest types and land conditions.
2. The kinds and amounts of land likely to be harvested and their physical-environmental condition.
3. The types of practices, controls, etc., likely to be adopted by those harvesting this land.

There is an extensive literature describing factor #1. However, the range of forest ecosystems and possible silvicultural practices is far greater than the range of existing research, and there are as well substantial gaps in the knowledge of some important cause-effect relationships such as the effect of whole-tree removal and short rotations on nutrient cycling, or, more generally, the ecosystem response to physical pollutants such as sediments and pesticides.

Identification and characterization of the land base most likely to be affected by increased wood demand (#2) are complicated by a lack of good land resource data, the lack of information on the precise nature of the future wood market, and the complexity of incentives that affect the decisionmaking of small woodland owners.

Predicting the types of practices and environmental controls likely to be adopted (#3) is difficult because State and local regulatory controls generally do not specify or effectively enforce "best management practices." Thus, existing regulations cannot be used as a guide to actual practices. Also, although knowledge about the present environmental performance of the forest industry might provide a starting point for gaining an understanding of what to expect in the future (because most wood-for-energy operations are more intensive extensions of conventional forestry), it is surprisingly difficult to produce a clear picture of how well the forestry industry is performing. With the exception of a few isolated State surveys and a detailed survey of erosion parameters (percentage of bare ground, compaction, etc.) in the Southeast,[31] *there appears to be a severe lack of surveys or credible assessments of actual forestry operations and their environmental impacts.* As a result, a critical part of the basis for an adequate environmental assessment is unavailable.

Because of these limitations, this discussion generally is limited to a description of *potential* impacts, although a few of the impacts described are inevitable. The economic and other incentives that influence the behavior of those engaging in forestry are examined to determine how probable some of these impacts are. The types of controls and practices available to moderate or eliminate the negative impacts also are described.

As discussed above, wood for energy may be obtained from several sources. With the growth of a wood-for-fuel market, the residue of slash from logging may be removed and chipped. Thinning operations may become more widespread because the wood obtained will have considerable value as fuel. Stand conversions—clearing of low-quality trees followed by controlled regeneration—as well as harvesting of low-quality wood on marginal lands may increase, also because of the increased value of the fuelwood gained. New harvesting practices such as whole-tree removal may become more common. Waste wood from milling and other wood-processing operations will certainly be more fully utilized. Finally, wood "crops" may be grown on large energy farms.

Many of these activities are similar to (though usually more intensive than) conventional logging. In addition, other activities associated with using wood as a long-term energy supply—including tree planting, pesticide and fertilizer application, etc.—are similar or even identical to "ordinary" silvicultural activities. This section, therefore, first discusses the general impacts of silviculture and then describes any changes or added effects associated with alternative wood-for-energy systems. In each case, the discussion will attempt to draw a distinction between clearly positive or negative pollution and land degradation and restoration impacts and the more ambiguous ecosystem and esthetic impacts. Because the environmental effects of silviculture are exceedingly varied and complex and because a number of good reviews are available, the discussion highlights only the major and most widespread impacts. It is stressed that few if any of the environmental relationships described in the discussion are applicable to all situations.

Environmental Effects of Conventional Silviculture

The practice of silviculture can have both positive and negative effects on the soils, wildlife, water quality, and other components of both the forest ecosystem and adjacent lands. Table 11 provides a partial list of the potential environmental effects of conventional silviculture. The magnitude of these impacts in any situation, however, depends almost entirely on management practices and on the physical characteristics of the site, i.e., type of trees and other vegetation, age of the forest, soil quality, rainfall, slope, etc. It is also important to remember that most of the negative impacts generally are short term and last only a few years (or less) over each rotational cycle.

Erosion has always been a concern in silviculture, especially in logging operations (and particularly in road construction). Undisturbed forests generally have extremely small erosion rates—often less than 75 lb of soil per acre per year[32]—and in fact tree planting is often used to protect erosion-prone land.* Increased erosion caused by logging, however, varies from negligible (light thinning and favorable condi-

---

[31]G. Dissmeyer and K. Stump, "Predicted Erosion Rates for Forest Management Activities and Conditions Sampled in the Southeast," USDA Forest Service, April 1978.

[32]*Environmental Implications of Trends in Agriculture and Silviculture, Volume 1: Trend Identification and Evaluation* (Washington, D.C.: Environmental Protection Agency, December 1978), EPA-600/3-77-121.

*However, from a historical perspective, all land forms go through natural erosional cycles that produce much higher rates of soil loss. These rates are often driven by natural catastrophic events including wildfire and storms.

Table 11.—Potential Environmental Effects of Logging and Forestry

Water
- increased *flow of sediments* into surface waters from logging erosion (especially from roads and skid trails)
- *clogging of streams* from logging residue
- *leaching of nutrients* into surface and ground waters
- potential *improvement of water quality* and more even flow from forestation of depleted or mined lands
- *herbicide-pesticide pollution* from runoff and aerial application (from a small percentage of forested acreage)
- *warming of streams* from loss of shading when vegetation adjacent to streams is removed

Air
- *fugitive dust*, primarily from roads and skid trails
- *emissions* from harvesting and transport equipment
- effects on atmospheric CO_2 concentrations, especially if forested land is permanently converted to cropland or other (lower biomass) use or vice-versa
- *air pollution* from prescribed burning

Land
- *compaction of soils* from roads and heavy equipment (leading to following two impacts)
- *surface erosion* of forest soils from roads, skid trails, other disturbances
- *loss of some long-term water storage capacity* of forest, increased flooding potential (or increased water availability downstream) until revegetation occurs
- changes in *fire hazard*, especially from debris
- possible *loss of forest* to alternative use or to regenerative failure
- possible reduction in *soil quality/nutrient and organic level* from short rotations and/or residue removal (inadequately understood)
- *positive effects of reforestation*—reduced erosion, increase in water retention, rehabilitation of strip-mined land, drastically improved esthetic quality, etc.
- *slumps and landslides* from loss of root support or improper road design
- temporary degrading of *esthetic quality*

Ecological
- changes in *wildlife* from transient effect of cutting and changes in forest type
- temporary degradation of *aquatic ecosystems*
- *change in forest type or improved forest* from stand conversion

SOURCE: Office of Technology Assessment.

tions) to hundreds of tons per acre per year (poorly managed clearcuts on steep slopes in high rainfall areas).[33]

A recent Environmental Protection Agency (EPA) report suggests the loss of 7 or 8 tons of sediment per acre per year as a mean value for recently harvested forests, although the variation around this mean is very large.[34] To place this rate in perspective, the *continuous* sheet

and rill erosion rate on intensively managed agricultural land averages 6.3 ton/acre-yr.

Most forestland is harvested at most once every several decades and the increased erosion generally lasts only a year or two on the majority of the affected acreage. Increased erosion from poorly constructed roads, however, may last longer.

The processes involved in erosion of forestland are stream cutting, sheet and gully erosion, and mass movement of soil. Erosion danger increases sharply with the steepness of the landscape, and the most common form of this erosion is mass movement. Mass movement "includes abrupt or violent events such as landslides, slumps, flows and debris avalanches, as well as continuous, almost imperceptible creep phenomena."[35] Occurrence of mass movements is most often associated with steep slope conditions where the forest soil is underlaid with impermeable rock.[36] These movements are natural processes associated with the downwearing of these steep slopes, but they can be triggered by man's activities. In contrast, sheet and gully erosion are rare in undisturbed forests, but they can be triggered by soil disturbances caused by careless road construction or logging practices.

The major causes of erosion problems in forestry operations are the construction and use of roads and other activities that may compact or expose soil or concentrate water.[37] The compaction caused by the operation of heavy machinery can reduce the porosity and water-holding capacity of the soil, encouraging erosion and restricting vegetation that eventually would reduce erosion. Roads and skid trails comprise up to 20 percent of the harvest area,[38] and the total area that may be compacted at a site may range up to 29 percent in

---

[33]*Environmental Readiness Document, Wood Commercialization*, draft (Washington, D.C.: Department of Energy, 1979).
[34]*Environmental Implications of Trends in Agriculture and Silviculture, vol. 1,* op. cit.

[35]Earl Stone, "The Impact of Timber Harvest on Soils and Water," *Report of the President's Advisory Panel on Timber and the Environment* (app. M, Washington, D.C.: Forest Service, U.S. Department of Agriculture, April 1973).
[36]*Environmental Implications of Trends in Agriculture and Silviculture, vol. 1.,* op. cit.
[37]Stone, op. cit.
[38]*Draft 208 Preliminary Non-Point Source Assessment Report* (Augusta, Maine: Land Use Regulatory Commission, State of Maine, 1978).

some instances.[39] Although in most areas the thawing and freezing cycle allows compacted soil to recover in 3 to 10 years, recovery takes far longer when, as in parts of the Southeast, this cycle does not occur.[40] Also, when compaction is very severe, recovery may take considerably longer than 10 years; old logging roads are still visible in the Northeast, even with the frost cycle.

The vulnerability of logging roads to erosion is related to topography and soil type as well as to road design. Roads developed on gentle to moderate slopes in stable topography pose few problems with the exception of careless movements of soil during construction. Large areas of forestland served by such roads draw little attention or criticism.[41]

The great majority of difficulties and hazards arise, however, when roads are constructed on steep terrain, cut into erosive soils or unstable slopes, or encroach on stream channels. Steep land conditions present a dilemma for road development, and criteria for location, design, and construction that are satisfactory on even moderate slopes may lead to intolerable levels of disturbance on steep lands. Building a road on a slope involves cutting into the slope to provide a level surface. The soil removed from the cut is used as fill or dumped. The steeper the slope, the more soil that must be disposed of and the more difficult is the job of stabilizing this soil. In the absence of proper attention to soil and geology, road design (especially alinement and drainage), and other factors, surface erosion from road and fill surfaces can continue for years. Road-building on steep slopes may also remove enough support from the higher elevations to cause mass failures; problems created by the road cut may be aggravated by inadequate drainage allowing further cutting away of supporting soil.

Aside from roads, the movement of logs from the harvest site to loading points may present considerable erosion potential. "Skid-

ding" logs may expose the subsoil, or compact the soil. Exposing the surface is a problem when the soil is highly erosive or when water concentrates, but is usually not a major erosion problem. The deeper disturbances of compaction and of cutting into the soil create more significant erosion problems, especially when they occur parallel to the flow of water. Most surveys of logging have concluded that the hauling or skidding of logs "generally does not lead to appreciable soil erosion or impaired stream quality;"[42] however, the same surveys conclude that "exceptions are common," and logging in vulnerable areas, under wet weather conditions, or with inappropriate equipment are thought to be important problems in the industry.

Erosion caused by the actual cutting of the trees generally is considered to be relatively unimportant. Vegetation usually regenerates quickly and reestablishes a protective cover on the land, preventing surface erosion except in areas where other components of the logging operation have damaged the soil. "Many observations and several studies on experimental watersheds demonstrate that sheet and gully erosion simply do not occur as a result of tree cutting alone, even on slopes as steep as 70 percent."[43]

However, land that is vulnerable to mass movements may be damaged by tree cutting. The decay of the old root systems will remove crucial support from a vulnerable slope faster than it can be replaced by the root systems of new growth; within 4 to 5 years after tree cutting (or fire), mass movement potential may increase dramatically. Forests in the Northwest United States and coastal Alaska are the main areas for this type of damage potential.[44]

The method of clearing for forest regeneration may also affect erosion potential. Intensive mechanical preparation of land before tree planting (i.e., use of rakes, blades, and other devices to reduce a forest to bare ground to favor reproduction of pine) can cause very

[39]*Environmental Implications of Trends in Agriculture and Silviculture, vol. 1.,* op. cit.
[40]Ibid.
[41]Stone, op. cit.

[42]Ibid.
[43]Ibid.
[44]Ibid.

serious erosion problems. This practice is occurring on hilly sites in the South that have been depleted by intensive cotton production in the past; it "may foster a dangerous cycle of topsoil and nutrient loss and increased sediment loading in streams."[45] Poorly managed raking may have adverse effects on forest productivity.[46] Area burning can also badly damage forest soils if managed improperly or if used on improper soils. Although suitable for highly porous, moist soils (where much of the surface cover is not consumed), poorly managed burning may consume most of the cover and leave the soil exposed to surface erosion. (However, area burning is considered to have a lesser potential to degrade productivity than raking.[47]) Burning may also represent a significant local source of air pollution. On the other hand, "controlled" burning may reduce future fire hazard by reducing slash buildup and may favor regeneration on the site of fire-resistant trees.

The sediment resulting from the erosion described in this section is "the major cause of *impaired water quality* associated with logging."[48] These sediments are directly responsible for water turbidity, destruction of stream bottom organisms by scouring and suffocation, and the destruction of fish reproductive habitat. Sediments also carry nutrients from the soil. Nutrient pollution is further increased by increased leaching and runoff as increased solar radiation reaches the forest floor and warms it, microbial activity (which transforms nutrients to soluble forms) accelerates and nutrient availability increases (this soil heating effect also has been known to retard regeneration, especially on south-facing slopes, by killing off seedlings). The increased nutrient loading of streams may have a variety of effects, including accelerated eutrophication and oxygen depletion. Fortunately, the increased nutri-

ent loading is usually short-lived, because revegetation of the site slows runoff and leaching, increases nutrient uptake, and, by shading and cooling the soil, slows the decomposition of organic material and consequent nutrient release.

The effects of nutrient enrichment are aggravated by the decomposition of organic matter from slash that is swept into streams, and by any water temperature increases caused by loss of streambank shading* (the temperature increases speed up eutrophication and further reduce oxygen content of the water). Temperature increases may also directly harm some freshwater ecosystems by affecting feeding behavior and disease incidence of cold water fish.

Logging operations affect *water supply* and may decrease a watershed's ability to absorb high-intensity storm waters without *flooding* (although this problem may have been exaggerated somewhat in the past).

The possibility of increased flooding stems from two causes. First, cutting the forest reduces the very substantial removal by transpiration of water from underground storage. During the period before substantial revegetation has taken place, the amount of this long-term "retention storage" capacity available to absorb floodwaters will be lessened and peak stream flows may rise. For example, increases in peak flows of 9 to 21 percent in the East and 30 percent in Oregon following clearcutting have been reported. These increases are usually observed only during or right after the growing season, where continual drawdown of storage would be occurring had the trees not been cut (floods occurring during the winter, as in the Northwest, may be unaffected or less affected because drawdown would not normally be occurring). This decrease in storage capacity apparently is not significant unless at least 20 percent of the canopy is removed.[49] Second,

[45]*Environmental Effects of Trends, vol. 2* (Washington, D.C.: Environmental Protection Agency, December 1978), EPA-600/3-77-121.

[46]Stone, op. cit.

[47]Ibid.

[48]*Silviculture Activities and Non-Point Pollution Abatement: A Cost-Effectiveness Analysis Procedure* (Washington, D.C.: Forest Service, U.S. Department of Agriculture, November 1977), EPA-600/8-77-018.

*The extent of any increases depends on stream volume, degree of removal of understory vegetation, and several other factors. In many cases, no significant effects occur.

[49]*An Assessment of the Forest and Range Land Situation in the United States* (Washington, D.C.: Forest Service, U.S. Department of Agriculture, 1979), review draft.

damage to forest soil increases runoff and inhibits the action of even the temporary "detention storage" potential wherein water is temporarily stored in pores in the upper soil layers and can be delayed from reaching streams for anywhere from several minutes to several days. Although treecutting, even clearcutting, is not likely to affect this temporary storage capacity, compaction of the soil by roadbuilding, log skidding, and operation of heavy machinery may reduce the infiltration of water into the soil if the compaction occurs over a wide area[50] and thus drastically reduce storage. Area burning on coarse-textured soils can create a water-repellant layer that would also decrease this infiltration and thus reduce the soils' capacity for detention storage.[51]

The reduction of transpiration that is caused by timber harvest may be beneficial by increasing stream flow and groundwater supplies in water short areas. Also, carefully structured cuts can be used to trap and maintain snow accumulation, greatly reducing evaporation losses. It is claimed that by using such techniques, water yield from commercial forestland in the West could be increased, supplying millions of additional acre feet at a cost of a few dollars per acre foot.[52]

Large-scale forestry operations often drastically alter local ecosystems, even for the long term. Wetlands in the South are being drained and pine forests are being created with the aid of substantial applications of phosphate fertilizers. In the process, aquatic ecosystems are being replaced by terrestrial ones and some critical wildlife habitats, especially for waterfowl, are being destroyed.[53] In the Pacific Northwest, old stands of Douglas fir are being replaced by single-aged plantings of the same species. Elsewhere, mixed hardwood forests are being replaced by plantations of conifers. In many cases, however, the ecosystems being replaced are themselves the result of past logging and agriculture as well as "unnatural" forest fire suppression that gradually replaced conifer forests with mixed hardwoods.

All types of replanting are accompanied by major changes in habitats available for wildlife. In the short term, any wood-harvesting operation, other than large area clearcutting, usually increases wildlife populations because mature forests normally do not support as great a total population of wildlife as do young growing forests. Many species require both cleared and forested area to survive, and thus, the "edges" created by logging operations are particularly attractive to deer and other species. Other species dependent on subclimax habitats (such as eastern cottontails) will also increase following logging, while species dependent on mature climax forests (e.g., wolverine, pileated woodpecker) will decline.[54]

Although the desirability of the ecosystem changes caused by logging may always be subject to one's point of view, different forestry practices tend to have varying effects that may be judged unambiguously from the standpoint of wildlife diversity and abundance:

> Forest management practices that reduce structural diversity of habitat, such as extensive old growth clearcutting, the removal of snags that provide wildlife food and nesting sites, and conversion to plantation management will generally reduce wildlife abundance and diversity by reducing habitat essential to many species. Conversely, animal diversity and wildlife abundance generally will be increased by opening up dense stands, making small patch cuts, or by conducting other timber management activities that increase structural diversity and provide a wide mix of habitat types.[55]

Current *pesticide and fertilizer use* in U.S. forests is low. In 1972, insecticides were used on only 0.002 percent of commercial forestlands, and fertilizers were used on less than 500,000 acres.[56] Because long-rotation logging and removal of only boles generally do not

[50]Stone, op. cit.
[51]R. M. Rice, et al., "Erosional Consequences of Timber Harvesting: An Appraisal," *Watershed in Transition*, (Urbana, Ill.: American Water Res. Assoc. Proc. Ser. 14, 1972).
[52]*An Assessment of the Forest and Range Land Situation in the United States*, op. cit.
[53]*Vol. II, Environmental Effects of Trends*, op. cit.

[54]*An Assessment of the Forest and Range Land Situation in the United States*, op. cit.
[55]Ibid.
[56]*Vol. 1: Environmental Implications of Trends in Agriculture and Silviculture*, op. cit.

deplete nutrients from forest soils, the most important use of fertilizers is on soils that are naturally deficient in nutrients or that have been depleted by past farming practices. For example, intensive cotton production in the Southeast seriously depleted soils and much of this land was abandoned long ago. Phosphate fertilization has allowed this land to become productive in the growth of softwood forests. Pesticides generally are used in forest management to control weed vegetation during reforestation or to combat serious outbreaks of insect pests. There is considerable controversy over aerial spraying of insecticides to control the gypsy moth and other damaging insects. Also, circumstantial evidence exists that certain herbicides in recent use may have caused outbreaks of birth defects and other damage when inadvertently sprayed over populated areas. Although the existence of these effects has been vigorously denied by the manufacturers, and although pesticide use in forests is a tiny fraction of the use in food production and is likely to remain so,[57] this use is likely to continue to be a source of disquiet accompanying intensive management of forests.

Silvicultural activities, and especially intensive harvesting operations, strongly affect the *esthetic appeal* of forests. The immediate aftermath of intensive logging is universally considered to be visually unattractive, especially where large amounts of slash are left on the site. Therefore, wood harvesting has a strong potential to conflict with other forest uses such as recreation or wilderness.

The significance of any negative effects depends on the nearness of logging sites to activity areas or to scenic vistas, the rapidity of revegetation, and the extensiveness of the operation. Therefore, the Forest Service seeks to route trails away from active harvesting sites, to avoid interrupting vistas, and to plan the extent and shape of the areas to minimize visual impacts.

The negative effect on the esthetic and recreational quality of forests caused by logging may be aggravated by a negative public per-

ception of the environmental effects of clearcutting in particular and logging in general. As noted earlier, this perception has been exaggerated by a number of factors including the grim history (1870-1930) of forest exploitation in the United States, the former revulsion against clearcutting practices within the forestry profession itself during the 1930's and 1940's, and continued attacks against logging by the environmental community. Although a logged-over area may be no uglier, objectively speaking, than a harvested field, the public perception of the two vistas is vastly different.

All reviews of logging and general forestry impacts stress the importance of *regional differences*—as well as extensive site-specific differences—in determining the existence and magnitude of environmental effects. Figure 6 presents a summary of those characteristics of U.S. forest regions that are most relevant to potential silvicultural impacts. Because the descriptions in figure 6 are, of necessity, much oversimplified, they are meant to give some perspective of the general range of environmental conditions and problems in American forestlands and should not be considered as fully representing all of the major conditions and problems in these lands.

Potential Environmental Effects of Harvesting Wood for Energy

This section discusses the activities—harvesting logging residues, whole-tree removal, intensifying and expanding silvicultural management, and harvesting for the residential space-heating market—which are characteristic of an expansion in the use of wood as an energy source.

Harvesting Logging Residues and Whole-Tree Removal

The harvesting of logging residues for an energy feedstock has potential for both positive and negative environmental impacts depending on the nature of the forest ecosystem and the previous manner of handling these residues.

---

[57]Ibid.

Figure 6.—Environmental Characteristics of Forest Regions

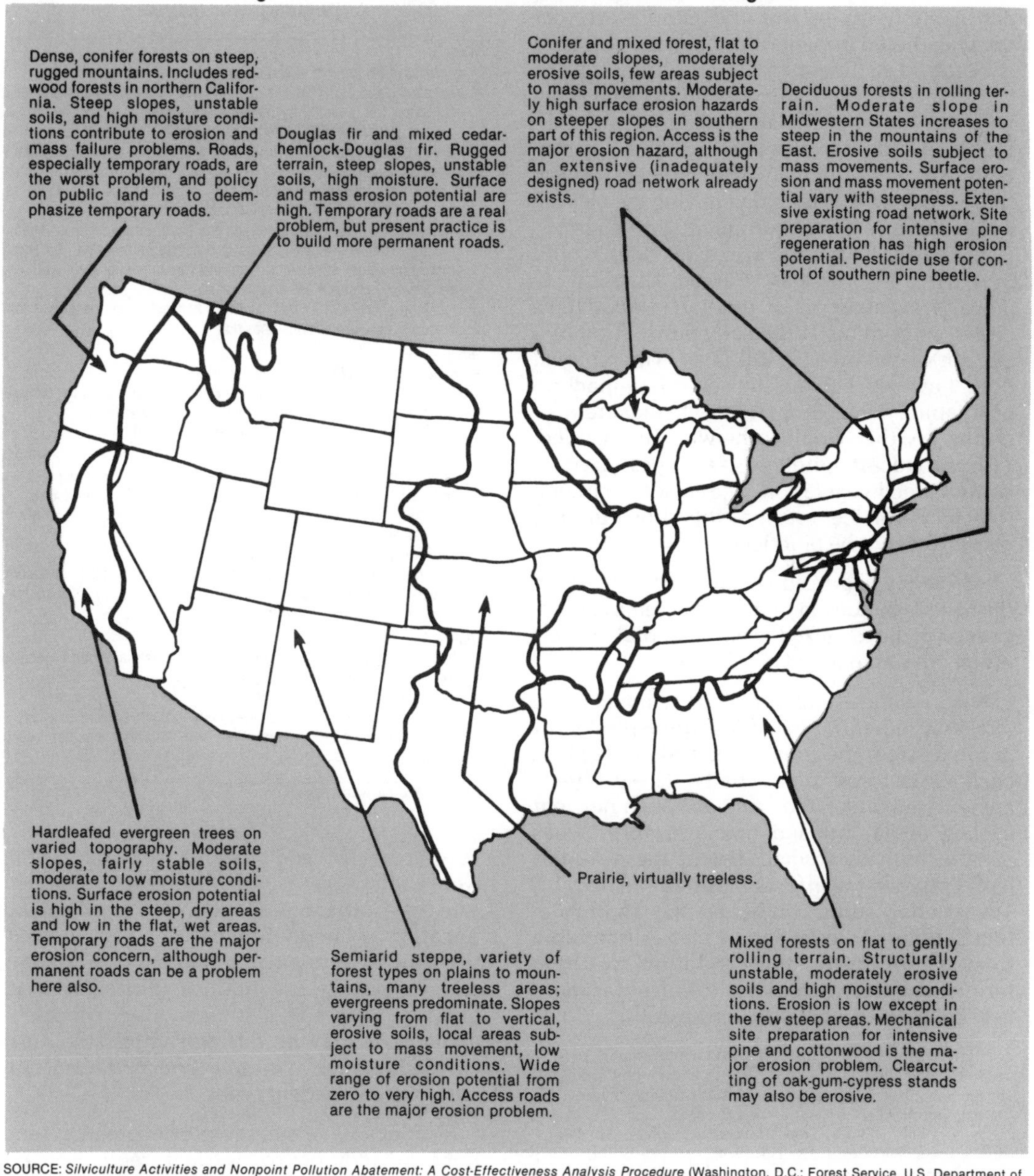

Dense, conifer forests on steep, rugged mountains. Includes redwood forests in northern California. Steep slopes, unstable soils, and high moisture conditions contribute to erosion and mass failure problems. Roads, especially temporary roads, are the worst problem, and policy on public land is to deemphasize temporary roads.

Douglas fir and mixed cedar-hemlock-Douglas fir. Rugged terrain, steep slopes, unstable soils, high moisture. Surface and mass erosion potential are high. Temporary roads are a real problem, but present practice is to build more permanent roads.

Conifer and mixed forest, flat to moderate slopes, moderately erosive soils, few areas subject to mass movements. Moderately high surface erosion hazards on steeper slopes in southern part of this region. Access is the major erosion hazard, although an extensive (inadequately designed) road network already exists.

Deciduous forests in rolling terrain. Moderate slope in Midwestern States increases to steep in the mountains of the East. Erosive soils subject to mass movements. Surface erosion and mass movement potential vary with steepness. Extensive existing road network. Site preparation for intensive pine regeneration has high erosion potential. Pesticide use for control of southern pine beetle.

Hardleafed evergreen trees on varied topography. Moderate slopes, fairly stable soils, moderate to low moisture conditions. Surface erosion potential is high in the steep, dry areas and low in the flat, wet areas. Temporary roads are the major erosion concern, although permanent roads can be a problem here also.

Prairie, virtually treeless.

Semiarid steppe, variety of forest types on plains to mountains, many treeless areas; evergreens predominate. Slopes varying from flat to vertical, erosive soils, local areas subject to mass movement, low moisture conditions. Wide range of erosion potential from zero to very high. Access roads are the major erosion problem.

Mixed forests on flat to gently rolling terrain. Structurally unstable, moderately erosive soils and high moisture conditions. Erosion is low except in the few steep areas. Mechanical site preparation for intensive pine and cottonwood is the major erosion problem. Clearcutting of oak-gum-cypress stands may also be erosive.

SOURCE: *Silviculture Activities and Nonpoint Pollution Abatement: A Cost-Effectiveness Analysis Procedure* (Washington, D.C.: Forest Service, U.S. Department of Agriculture, November 1977).

In forests where wood residues—tops, limbs, and possibly leaves and understory—are routinely gathered into piles for open burning (this is required in forest fire prone areas of the West), residue use for energy production is environmentally beneficial. It eliminates the air pollution caused by this burning and has essentially no additional adverse impacts except those incurred in physically moving the residue out of the forest (and burning it, *with controls,* in a boiler). In forests where residues would otherwise be broadcast burned, physical removal prevents some of the potential adverse effects of burning—especially destruction of a portion of the organic soil layer. The removal does, however, subject the soil to compaction or scraping damage by the mechanical removal process that would otherwise be avoided. Also, broadcast burning is, at times, used to control weed vegetation, and in some circumstances herbicide use may be substituted if burning cannot be practiced.

Where logging residues are normally left in the forest, institution of a residue removal program will have mixed environmental effects which are summarized in table 12.

A worrisome effect of residue removal is the increased potential for long-term depletion of nutrients from the forest soils and consequent declines in forest productivity. These effects are not well understood and although nutrient cycling in natural and managed forests has been extensively studied, few of these studies have included the effects of residue removal.[58] The existing studies indicate that short-rotation Southern forests may be more susceptible to depletion than longer rotation Northern forests, and that marginal sites suffer far more heavily than forests with fertile soils.[59 60 61 62]

[58]C. J. High and S. E. Knight, "Environmental Impact of Harvesting Noncommercial Wood for Energy: Research Problems," Thayer School of Engineering, Dartmouth College paper DSD No. 101, October 1977.
[59]E. H. White, "Whole-Tree Harvesting Depletes Soil Nutrients," Can J. Forest. Res. 4: 530-535, 1974.
[60]J. R. Jorgensen, et al., "The Nutrient Cycle: Key to Continuous Forest Production," J. Forestry 73: 400-403, 1975.
[61]J. R. Boyle, et al., "Whole-Tree Harvesting: Nutrient Budget Evaluation," J. Forestry 71: 760-762.
[62]G. F. Weetman and B. Webber, "The Influence of Wood Harvesting on the Nutrient Status of Two Spruce Stands," Can. J. Forest. Res. 2: 351-69, 1972.

Table 12.—Environmental Impacts of Harvesting Forest Residues

Water
- *decrease in clogging* of streams caused by entry of slash
- increased short-term *flow of sediments* into streams because of loss of erosion control provided by residues, soil damage caused by removal operations; somewhat counteracted by decline in broadcast burning, which at times destroys surface cover and causes erosion potential to increase
- possible changes in long-term flow of sediments where residue removal affects revegetation; this effect is mixed
- *changes in herbicide usage*—on the one hand, chemical destruction of growing residues (valueless trees) will cease; on the other, broadcast burning no longer effective in retarding vegetative competition to new tree growth, herbicide use may increase
- increased short-term *nutrient leaching* because of increased soil temperatures, accelerated decomposition

Air
- *reduction in air pollution* from forest fires
- *reduction in air pollution* from open burning of residues (if the residues normally are broadcast burned or burned after collection)
- *dust* from decreased land cover, harvesting operations

Land
- *potential depletion of nutrients and organic matter from forest soils* and possible long-term loss of productivity (inadequately understood)
- *short-term increase in erosion and loss of topsoil,* possible long-term decrease or increase
- *reduction in forest fire hazard*
- *short-term decreased water retention,* increased runoff (and flooding hazard) until revegetation takes place; aggravated by any soil compaction caused by removal operation

Other
- change in *wildlife habitat*—bad for small animals and birds, good for large animals unless serious erosion results
- changes in *tree species* that can regrow
- *esthetic change,* usually considered beneficial when slash is heavy
- *reduction in bark beetles* and other pathogens that are harbored by residues

SOURCE: Office of Technology Assessment.

Further study and careful soil monitoring would allow the use of fertilizers to compensate for nutrient depletion, but fertilizer application is energy intensive; it may increase the flow of nutrients to neighboring streams, and its correct use may be difficult to administer for smaller stands. Also, successful application may be difficult unless the nutrient depletion is a simple one involving only one or a few nutrient types.

Residues serve a number of ecological functions in addition to nutrient replenishment, and their removal will eliminate or alter these functions. They provide shelter and food to small mammals and birds, provide a temporary food supply for deer and other larger mammals, moderate soil temperature increases that normally occur after logging, provide some protection to the forest floor against erosion, and are a source of organic matter for forest soils. Thus, removal of residues will reduce certain wildlife habitats and may expose the forest floor to some additional erosion above and beyond that caused by conventional logging. Higher soil temperatures resulting from loss of the shade provided by residue cover will accelerate organic decomposition activity and may lead to a period of increased nutrient leaching before revegetation commences. Also, the increased rate of organic decomposition coupled with the removal of a primary source of organic matter may lower the organic content of forest soils. Declines in soil organic matter are expected to be accompanied by declines in nitrogen-fixing capacity, soil microbial activity rates, and cation exchange capacity, all considered to be important determinants of long-term forest health.[63][64] The present scientific understanding of organic matter removal is, however, insufficient to allow a determination of the significance of these possible effects.

The extensive residue left on the forest floor after cutting dense stands can inhibit revegetation, especially in softwood forests. To the extent that residue removal may promote new vegetation, this will counteract the removal's short-term negative erosion and nutrient-leaching effect (as long as removal is not so complete as to eliminate the light mulch necessary to shade the surface and maintain soil moisture).

Residues also provide a habitat for disease and pest organisms such as the bark beetle and, when washed into neighboring streams, may clog their channels and degrade water quality. They add considerably to the incidence and intensity of forest fires, especially in the West. Also, the esthetic impact of residues is generally considered to be negative when they are left at the logging site; when densely forested areas are cut, residues will completely cover the ground with several feet of unsightly slash. Therefore, removal of residue will, in a positive sense, reduce the number and severity of forest fires and pest infestations, improve esthetics, and reduce the potential for stream clogging.

"Whole-tree harvesting" is really a variation of residue removal with the bole and "residue"—branches, leaves, twigs—removed in one integrated operation. It is most likely to occur when the entire tree is to be chipped for fuel or some other use.

The problems of long-term nutrient and organic matter depletion from whole-tree harvesting are basically the same as those of residue removal, and whole-tree logging similarly removes far greater nutrients and organic matter from forest soils than do other conventional methods. Whole-tree removal of Norway spruce, for example, results in a loss of 2 to 4 times more nitrogen, 2 to 5 times more phosphorus, 1.5 to 3.5 times more potassium, and 1.5 to 2.5 times more calcium than conventional logging.[65] In addition, ground disturbance from the actual tree removal is likely to be worse with whole-tree harvesting when the fully branched trees are dragged off the logging site, eradicating understory vegetation in the process. This disturbance, besides promoting erosion, will accelerate organic matter decomposition. As noted previously, however, the effects of these organic matter and nutrient removals on long-term forest productivity are poorly understood.

Intensifying and Expanding Silvicultural Activities

The creation of new energy markets for wood will have a significant effect on the economics of managing forested land, including land not currently considered to be high-grade

[63]E. L. Stone, "Nutrient Removals by Intensive Harvest—Some Research Gaps and Opportunities," *Proceedings: Impact of Harvesting on Forest Nutrient Cycle* (Syracuse, N.Y.: State University of New York, College of Environmental Science and Forestry, 1979).

[64]E. H. White and A. E. Harvey, "Modification of Intensive Management Practices to Protect Forest Nutrient Cycles," *Proceedings: Impact of Harvesting on Forest Nutrient Cycle* (Syracuse, N.Y.: State University of New York, College of Environmental Science and Forestry, 1979).

[65]E. Malkonen, "The Effects of Fuller Biomass Harvesting on Soil Fertility," *Symposium on the Harvesting of a Larger Part of the Forest Biomass* (Hyvinkaa, Finland: Economic Commission for Europe, Food and Agriculture Organization, 1976).

forest. New lands will be harvested and silvicultural practices will intensify.

One effect will be the expansion of logging onto lands that are not now in the wood marketplace. The operational costs of logging some of these lands cannot, at present, be recouped through increased property values, the sale of the harvested wood, or the value of future growth of a regenerated forest. Additional lands that currently are economically attractive targets for logging activities (stand conversion, clearing for nonforest use, etc.) are withheld by their owners for a variety of reasons (their higher valuation of the land's recreational potential, fear of environmental damage, etc.). As an energy market for wood develops, however, harvesting part or all of the wood resource on these lands will become increasingly attractive.

The logging of some forests that would otherwise be untouched (or, perhaps more realistically, that would only be logged at some later time) may be viewed as beneficial by some groups. Most reviews depict American forests as being characterized by "overmature stands of old-growth timber, especially in the West, and . . . many stands, mainly in the East and South, that were repeatedly mined of good trees in earlier, more reckless times."[66] Conversion of such stands is often characterized as a step towards a healthier forest, because tree growth generally is enhanced and more "desirable" tree species are introduced. Where whole-tree harvesting or residue removal is practiced, the forest may become more accessible to hikers and may be more esthetically appealing. The extent to which all this is considered a benefit depends heavily on one's perspective, however, and optimizing commercial value is not necessarily synonymous with optimizing other values such as ecosystem maintenance or wildlife diversity.

As discussed later, expansion of silvicultural management onto suitable lands, combined with an increase in the intensity of management on existing commercially managed lands,

may provide important environmental benefits in the form of decreasing logging pressures on lands that combine high-quality timber with competing values that would be compromised by logging. Unfortunately, a decrease in logging pressures on one segment of America's forests may be coupled with an undesirable increased pressure on another segment.

A particular fear associated with the rise in demand for "low quality" wood is that marginal, environmentally vulnerable lands with stands of such wood may become targets for logging. Much of this land that may be vulnerable to logging for energy, although "poor" from the standpoint of commercial productivity, is valuable for esthetic, recreational, watershed protection, and other alternative forest uses. These forest values may be lost or compromised by permanent clearing or by harvesting on sites where regeneration may be a problem. For example, forests in areas with marginal rainfall—e.g., in the Southwest— may be particularly vulnerable to regeneration failures and thus may be endangered by a growth in wood demand. On lands with poor soils and steep slopes, clearcutting and other intensive forms of harvesting create a high potential for nutrient depletion, mass movement, and other problems as described earlier. Because, as discussed later, the Federal Government maintains supervisory control over forest operations on federally owned lands, this potential problem is likely to be concentrated on private lands. The overall danger is somewhat mitigated, therefore, by the Federal Government's ownership of a significant percentage of the most vulnerable land.

It is difficult to predict whether wood-for-energy operations will tend to gravitate to the poorer quality and more vulnerable lands. The several factors that will determine the tendency of wood-for-energy harvesting to gravitate to vulnerable lands include:

1. *The direct cost of wood harvesting.* —Development of more versatile harvesting equipment can lower the cost of operating on steep slopes and promote harvesting on vulnerable lands.

[66]Smith, op. cit.

2. *The stringency and enforcement of environmental standards.*—The stronger the controls, the more likely it is that loggers will avoid the more vulnerable stands.
3. *The price of woodchips for energy.*—At a high enough price, the "value-added" to the land by clearing will become less important, and poorer quality lands will become more attractive targets for harvesting.
4. *The price of agricultural land and "high value" forestland.*—At high prices, wood harvesting for energy would tend to gravitate to higher quality, less erosion-prone/depletion-prone lands because clearing for agriculture or stand conversion will be more profitable.
5. The *distribution* of different soil/slope/rainfall conditions in forestland potentially available for cutting.
6. The *attitude of private landowners,* who currently own much of the land available for clearing but who often are reluctant to allow harvesting.
7. *The cost of transporting wood.*—Because the higher this cost, the more likely it is that local shortages could force harvesting onto vulnerable lands.

Except for (1) and (5), these factors may be extremely volatile and will themselves depend on the availability of alternate fuels, the state of the economy, etc. Except for forestland in the Southeast, the data necessary to define (5) are not available.

The Department of Energy, in its draft "Wood Commercialization Environmental Readiness Document,"[67] asserts that the sites with "nutrient deficiencies and delicate nutrient balances, and subsequently low productivity . . . are the non-commercial forests that often are considered available for whole-tree harvest for energy." And a recent EPA report asserts that "areas previously left unlogged . . . are most often increasingly steep with difficult terrain."[68] Both of these statements imply that an areal expansion of logging to satis-

fy energy demands could be expected to lead to exploitation of lands particularly vulnerable to environmental damage.

These references may have overlooked several factors, however:

1. As noted previously, there is considerable forest acreage of high quality—low slopes, rich and nonerosive soils, adequate rainfall—with low-quality timber growing on it. This is especially true in the East.
2. The cost of harvesting timber on flatter—and thus less erosive—slopes is considerably less than on steep-sloped lands. These flatter lands presumably would be the first choice for harvesting.
3. The higher quality, less vulnerable sites offer the landowner the economic incentive of an added return from regrowth of high-quality timber or else alternative land uses such as farming.
4. Increases in land prices for rural acreage with high recreational and esthetic value have increased the economic incentive to guard against environmental damage that would compromise these values.

On balance, it would appear that market pressures would tend to favor the harvesting of the less environmentally vulnerable lands. However, variations of land availability from region to region, landowner decisions based on other than land suitability grounds, and other factors are likely to lead to some level of inappropriate harvesting—especially if the current state of regulatory "laissez-faire" continues (see discussion on "The Institutional Climate for Environmental Control").

A second effect of new energy markets for wood will be an *intensification* of forest management—especially of thinning—because part or all of its cost will be recouped through use or sale of the collected wood. Residue removal or whole-tree harvesting, discussed previously, are likely to be another facet of this management intensification.

The process of removing trees that are dead or diseased, stunted, poorly shaped, or of "undesirable" species is considered by foresters to

---

[67]*Environmental Readiness Document, Wood Commercialization,* op. cit.
[68]*Vol. II, Environmental Effects of Trends,* op. cit.

be beneficial to the forest. Thinning allows increased growth in the remaining trees, esthetically and physically "opens up" the forest, and may allow some additional growth of understory vegetation if the thinning is extensive enough. If heavy machinery is used, however, resulting soil compaction can cause adverse impacts, and care must be taken during the thinning operation to avoid damaging the trees that remain.

A critical argument in favor of thinning and other logging operations is that these activities result in increased wildlife populations and diversity. The detinition of "diversity" is critical to this argument. There is a substantial difference between maximizing diversity in a *single forest stand* and maximizing it *in the forest system* composed of many forest stands in a region. The first definition may be well served by more intensive management because such management provides more "edges" and understory vegetation for browse. On the other hand, many species will suffer from such management. A great many species depend for their food and shelter on "unhealthy"—dead, dying, rotten—trees that would be removed in a managed forest, and other species cannot tolerate the level of disturbance that would be caused by thinning operations. Maintaining diversity in a forest *system* must include protecting these species by deliberately leaving unmanaged substantial portions of the forest or a percentage of the individual stands within the entire system. In regions where officially designated wilderness areas or other protective measures are adequate, intensive management on the remaining stands may be considered (even by environmental groups) as benign or beneficial if good management practices are carefully followed. In other regions, especially in the East, intensive management may conceivably work to the detriment of species diversity although it may increase the total wildlife population. Even in these regions, however, there is a possibility that large numbers of property owners may choose to leave their lands unmanaged because of personal preferences. This would serve to protect diversity.

The potential for added growth of high-quality timber from stand conversions of low-quality forest and the increased use of thinning on commercial forestlands may have, as its most important effect, a decrease in the pressures to log forests that have both high-value timber and strong nontimber values—recreation, esthetic, watershed protection, etc.—and that may be quite vulnerable to environmental damage. Analysts such as Marion Clawson of Resources for the Future have long argued that the management of American forestland is extremely inefficient, that by concentrating intensive management practices on the most productive lands we could increase harvest yields while withdrawing from silviculture less productive or more environmentally vulnerable lands.[69] [70] An expansion of wood use for energy and the consequent creation of a strong market for "low quality" wood may have this beneficial effect.

OTA estimates that placing 200 million acres of commercial forestland into intensive management (full stocking, thinnings every 10 years, 30- to 40-year rotations) could allow wood energy use to reach 10 Quads annually while the availability of wood for nonenergy products might double its 1979 value. Alternatively, the same result might be achieved by using less intensive management on a larger acreage. The nature of any actual benefits, however, are dependent on the following considerations:

- Major effects on the availability of high-quality timber probably would not occur for a number of years. Some additional high-quality wood might be available immediately from stand conversions and harvest of noncommercial timber, and some in about 20 years from timber growth in stands that required only thinning for stand improvement. The quantities would not peak, however, before about 30 to 40 years as stands that had been cleared and replanted began to reach harvesting age. By this time, most of the old-growth stands accessible to logging already may have been harvested, al-

[69]M. Clawson, "The National Forests," *Science,* vol. 20, February 1976.
[70]M. Clawson, "Forests in the Long Sweep of American History," *Science,* vol. 204, June 15, 1979.

though significant benefits from reducing logging pressures on other valuable or fragile lands would still be available.

• Although the increased availability of high-quality timber might negate arguments that these valuable or fragile stands *must* be cut to provide sufficient wood to meet demand, there is no guarantee that the wood made available from intensified management will be less expensive than that obtainable from these stands, and economic pressure to harvest them might continue.

Although the long-range economic goals of intensive management provide an incentive against poor environmental practices, careless logging and regeneration practices will still occur on a portion of the managed sites. Poor management may be practiced on a smaller proportion of sites than would have been the case without an expansion of wood for energy, but the effects of such management may be aggravated with such an expansion because:

• more acreage will be logged each year,
• most affected sites will have fewer years to recover before they are logged again, and
• the removal of maximum biomass and subsequent soil depletion may reduce the sites' ability to recover.

Thus, the impacts associated with conventional logging—including erosion and soil degradation, damage to water quality, esthetic damage, and other impacts—are likely to occur with even greater severity on a portion of those lands devoted to wood production for energy. Unfortunately, because of the lack of data on logging practices and the very mixed nature of the incentives for good management, it is impossible to make a good quantitative prediction of the size of this portion.

A basic—and difficult to resolve—issue concerning the wisdom of moving to a very high level of intensive management of U.S. forestland is the possibility that the long-term viability of these forests may be harmed. The possibility of soil depletion is only one aspect of this. The cycles of natural succession occurring in an unmanaged forest give that forest substantial resilience, because the diversity of vegetation and wildlife of the more mature states of the forest cycle as well as the diversity created by the heterogenous mix of stages tend "to buffer the system against drastic change as by diluting the effects of pests on single species."[71] Ecologists often have argued that man pays a significant price in moving too far from this natural state:

> The whole history of agriculture, and later, forestry, is basically a continuous effort to create simplifed ecosystems in which specialized crops are kept free of other species which interfere with the harvest through competition . . . diversified systems have built-in insurances against major failures, while the simplified systems need constant care.[72]

> In relation to human needs, the human strategy can be viewed as a reversal of the successional sequence, creating and maintaining early successional types of ecosystem where gross production exceeds community respiration. Such . . . ecosystems, despite their high yield to mankind, carry with them the disadvantages of all immature ecosystems, in particular they lack the ability to perform essentially protective functions in terms of nutrient cycling, soil conservation and population regulation. The functioning of the system is thus dependent upon continued human intervention.[73]

There are, of course, counterarguments to the thesis that this simplification of ecosystems places these systems under significant risk. One argument is that much of silviculture duplicates natural events, and purposely so; for example, clearcutting, sometimes followed by broadcast burning, is said to duplicate the effects of severe storms or catastrophic fires.[74] Another is that professional silviculturalists can compensate for any tendency towards a

[71]Smith, op. cit.

[72]A. H. Hoffman, "Comprehensive Planning and Management of the Countryside: A Step Towards Perpetuation of an Ecological Balance," *Global Perspectives on Ecology,* Thomas C. Emmel, ed. (Mayfield Publishing Co., 1977).

[73]R. Manners, "The Environmental Impacts of Modern Agricultural Technologies," *Perspectives on Environment,* I. R. Manners and M. W. Mikesell, eds. (Association of American Geographers, 1974), publication No. 13.

[74]Smith, op. cit.

decline in resiliency. In its extreme, this argument is particularly unacceptable to those who are skeptical of placing too great a faith in science:

> We ought to believe that we can excel over nature; and if we do, we should not be restricted to blind imitation of her methods . . . we have the chance to sift nature's truths, and recombine them into a new order in which not only survival, but enhanced productivity are the ruling criteria . . . (we) must look to near-domestication of our forests . . . we must move forestry close to agriculture.[75]

The strongest argument that can be made, however, is that past forestry experience has demonstrated that temperate forests can absorb an unusual amount of stress without suffering long-term damage. For example, large acreages in Europe as well as the United States that today are densely forested were intensely exploited as agricultural land in the past. In many instances, foresters can point to intensive management practices in European forests that have continued to provide high productivity of lumber for a hundred or more years. In counterpoint to these arguments, some environmentalists are worried about the future of Europe's forests and point to increasingly high external costs in terms of polluted water and increasing incidences of disease epidemics.[76] Also, insufficient data is available to indicate whether or not small but significant drops in long-term productivity may have occurred because of such past practices.

A similar argument rages about high-yield agriculture: yield levels in the Western countries have climbed steadily over the past century, with temporary setbacks that have thus far been dealt with by further adjusting the system, but environmentalists as well as many agronomists are worried about increasing numbers of pesticide-resistant insects and rising environmental costs.

Pursuit of the evidence on both sides of this argument may be worthwhile, but it is beyond

the resources of this assessment. Also, the high level of emotional commitment that is attached to the alternative views of how far nature can be safely manipulated makes it unlikely that such a gathering of evidence will change many minds. However, it is at least clear that a substantial increase in intensive management must be accompanied by a thorough research program stressing examination of such critical factors as nutrient cycling, the role of soil organic material vis-a-vis resistance to tree disease, and other factors affecting system resiliency. The possibility that forest viability might be at excessive risk if hundreds of millions of acres in the United States were placed in intensive management should not be automatically rejected, even though some degree of success in such management apparently has been achieved elsewhere.

Harvesting for the Residential Market

The rapidly expanding demand for wood fuel for residential use currently is satisfied largely by harvesting of wood by homeowners and by local entrepreneurs. The high price of wood for residential use is an incentive for larger scale loggers to enter the market, and a trend in this direction probably should be expected in the future. The identity of the supplier may be an important component in determining the environmental effects of satisfying a high residential demand for wood fuel.

An expansion of the residential wood market represents an opportunity for improved forest management because of the value it places on lower quality wood, which in turn should stimulate an increase in thinning activities. The potential benefits are the same as those described for the increase in intensive management: an increase in productivity and timber value on the affected lands. This opportunity exists on woodlands ranging from small private woodlots to federally- and State-managed forests. The latter could use homeowners as a "free" work force to harvest selected trees, a practice that is already in operation in many areas.

Unfortunately, a rising demand for wood will bring with it a potential for significant

---

[75]G. Staebler, "The Forest and the Railroad," brochure published by Weyerhaeuser Co., December 1975.
[76]Goldsmith, "The Future of Tree Diseases," The Ecologist, No. 4/5, July-August 1979.

negative effects on woodlands. High prices for wood fuels are likely to stimulate an increased incidence of illegal cutting of wood. "Timber rustling" apparently is frequently encountered in stands of very high-quality timber such as redwood and walnut. More substantial cutting involving multiple acres at a time must be expected as wood demand grows and prices increase; remote areas, or areas where property boundaries are not well marked should be particularly vulnerable. (Illegal mining of coal may be an analogous and somewhat prophetic example. Although it takes considerable time and effort to expose and mine a coal seam, coal poaching is not at all unusual in Appalachia, and some examples involving millions of dollars worth of coal have been reported recently. Poaching timber is going to be a lot easier than poaching coal.) In areas where wood stoves are oversold or where forest products companies occasionally enter the (lower quality) wood market, temporary fuelwood shortages or price escalation may further stimulate illegal cutting, especially among poorer homeowners or those who cannot shift to an alternative fuel for space heating.

The same forces that stimulate illegal cutting, especially where coupled with ignorance of forest management, are likely to result in a variety of poor practices: improper harvesting techniques leading to damage to adjacent trees or to forest soils, incorrect tree selection, overcutting, etc.

The balance between beneficial and adverse effects of a rising demand for wood as a residential fuel is uncertain. Positive measures such as an increased availability of trained foresters to provide assistance to small woodlot owners, better dissemination of information on woodlot management, and the organization of efficent and competitive retail suppliers would help to limit adverse impacts. On the other hand, the combination of a sharply increased demand for wood coupled with a resource base that is accessible and vulnerable to illegal or poorly managed cutting appears to be virtually a guaranteed source of trouble.

Tree Plantations

The concept of an energy farm or plantation where trees are grown and harvested on short rotations like agricultural crops is a logical extension of current intensive single-aged management of forests. In fact, the growing of Christmas trees on plantations is a more intensively managed activity than an energy farm is likely to be, because the level of "management"—including pesticide and fertilizer use—will tend to increase with the unit value of the crop. In addition, a Christmas tree farmer cannot tolerate relatively minor levels of pest or drought damage because his crop value is strongly dependent on appearance, and thus he must apply pesticides or irrigation water during episodes that the energy "farmer" may be able to ignore.

The land requirements, growing needs and harvesting techniques associated with energy farms appear to be very similar to those of a large agricultural enterprise growing perennial food crops. Because of this resemblance, the environmental impacts are not treated in this section. The chapter on agricultural biomass production should provide sufficient information about these impacts.

Controlling Negative Impacts

A common theme running through reviews of silvicultural practices by the forestry establishment—the wood products industry, schools of forestry, and the Forest Service—is that these practices may have negative environmental consequences but that the consequences are readily controlled, that significant environmental damages today are the exception rather than the rule, and that in those cases where damages occur they are almost always short lived, i.e., the forest quickly recovers and normal forest dynamics are restored.

The President's Advisory Panel on Timber and the Environment reported that:[77]

[77]Fred A. Seaton, et al., *Report of the President's Advisory Panel on Timber and the Environment* (Washington, D.C.: President's Advisory Panel, April 1973).

A careful review . . . revealed that most of . . . (the environmental) damage caused by logging can be avoided or minimized. Many of the fears that have been expressed are unfounded, misleading, or exaggerated, often due to extrapolation from an isolated case to forest lands in general.

Properly executed timber harvesting and other silvicultural procedures need not result in important long-term losses of soil nutrients, deterioration of the soil, nor cause other physical environmental damage. Damage that has occurred resulted primarily from erosion associated with logging road construction and use, skidding of logs downhill or across streams, or harvesting on steep slopes where removal of vegetative cover caused slides. *With updated methods, such difficulties will become rare exceptions. Such damage as has occurred will be corrected through natural processes as the forest grows back.* (Emphasis added.)

The problem with statements such as these is that they do not acknowledge the current paucity of information on actual logging practices and effects. As noted in the introduction to this section, there are few credible assessments of forestry operations on a statewide or regional basis. The few that have been attempted are limited in scope; for instance, a survey of practices in Maine in support of the 208 program (sec. 208, Public Law 92-500/ Federal Clean Water Act) is limited to recording the occurrence of gullying and the use or nonuse of simple erosion controls.[78]

The limited information that is available seems to indicate that the generally optimistic tone of most reviews of forestry impacts should be viewed with caution. An interesting conclusion of the Maine study was that "the area wide magnitude of the (erosion and sedimentation) problem is somewhere between the positions espoused by the industry representatives on the one hand, and groups and agencies concerned with maintaining environmental quality on the other hand.[79] The survey found that simple—and supposedly standard—erosion

control techniques such as using water bars and artificially seeding erodible areas "are (done) so infrequently that the role of these convenient erosion control devices in preventing postlogging degradation of water quality is minimal at present."[80]

Given the lack of knowledge of current forest practices and the hints of environmental problems provided by the limited data, Congress should consider both the availability of control measures and the institutional climate for putting these measures into practice *before* attempting to stimulate the increased use of wood for energy.

Control Capability

The technical capability exists to control or reduce the negative effects of logging and, more generally, of all silvicultural activities. Table 13 presents a partial list of the control methods available to the forester. Some of the more critical are:

- *Site selection/identification and possibly avoidance of problem areas.* — Because many of the environmental problems of logging are strongly site-dependent, identification of problem areas followed by revision or abandonment of logging plans is a critical environmental control strategy. Avoidance of steeply sloped sites with unstable soils is important for minimizing erosion. This often coincides with economic incentives, because the more efficient heavy equipment cannot operate on steep slopes. Geologic surveying of the site can often detect vulnerable soil/rock/ slope formations, although this capability is not fully developed. Temporary avoidance of some areas, for example, during rainy conditions, can avoid major problems of soil compaction and destruction of soil structure. Other site conditions that must be treated with special care or avoided include nutrient-deficient and thin soils, and sites in immediate proximity to lakes and streams. In the latter case, a buffer strip of smaller trees and shrubs

---

[78]"A Survey of Erosion and Sedimentation Problems Associated With Logging in Maine," Land Use Regulation Commission, State of Maine, for the Maine Department of Environmental Protection, May 1979.

[79]Ibid.

[80]Ibid.

Table 13.—Control Methods

Mitigative[a]	Preventive
Surface protection: • Access: seeding, mulching, riprap, or mat on cut-and-fill slopes • Timber harvest: maintenance of vegetative cover; distribution of slash • Cultural treatments· seeding; planting; fertilization **Flow diversion and energy:** • Access: berms above cut slopes; benches on cut slopes; checkdams in ditches; drop structure at culvert ends; water bars on road surface; flow diversion from potential mass failures or at mid-slope • Timber harvest: buffer strips; water bars on skid trails • Cultural practices: plowing, furrowing, bedding **Access design modification**	**System design and maintenance:** • Access: minimize cuts and fills, roadway widths and slopes; control road density • Timber harvest: minimize soil compaction from equipment operation; use site-compatible log removal system; control harvested volume within a watershed; limit harvest on unstable slopes; shape openings for minimum esthetic impact, avoid cutting next to recreational activity areas • Cultural treatments: minimize reentry disturbances; fire control **Timing:** • Access: closure of temporary roads; limited access; closure during adverse conditions • Timber harvest: limit operation during adverse climatic conditions; site preparations during favorable conditions • Cultural treatments: intensity and number of thinnings

[a]Controls can be described as "preventive" or "mitigative" according to the mode of applications. *Preventive controls* apply to the preimplementation phase of an operation. These controls involve stopping or changing the activity *before* the soil-disturbing activity has a chance to occur. *Mitigative controls* include vegetative or chemical measures or physical structures which alter the response of the soil disturbing activity *after* it has occurred.

SOURCE: *Silviculture Activities and Nonpoint Pollution Abatement: A Cost-Effectiveness Analysis Procedure* (Washington, D.C.: Forest Service, USDA, November 1977).

along the shoreline may be sufficient to provide shading and some sediment protection to the body of water.
• *Selection of harvesting system.*—Control of erosion, esthetic, and other impacts can be achieved by matching the harvesting system to the site conditions. For example, the type of forest regenerated at the site can be controlled by the harvest system, because different degrees of disturbance favor different tree species. Clearcutting and residue removal favor species that need maximum disturbance to grow (e.g., Douglas fir, jack and lodgepole pine, paper birch, red alder, and cottonwood[81]) and shelterwood cutting (which leaves residual trees in sufficient numbers to shade new seedlings) favors species (such as true firs, spruces, and maples) that require light shade to thrive. The harvesting system may also be used to avoid some of the negative effects to which the site is particularly vulnerable. Clearcutting, for example, would be indicated for old, decrepit stands in which residual trees would be likely to blow down in the first severe storm following harvest. Shelterwood cutting would be appropriate for stands important to scenic views. A light selection cut may be the

[81]Smith, op. cit.

only harvesting allowed on soils subject to mass movement.
• *Erosion/sediment control measures.*—Although a certain amount of erosion from soil compaction and mineral soil exposure is inevitable in logging operations, it can be reduced by using lighter equipment to avoid compaction, by using overhead or even aerial (balloon or helicopter) log collection methods (although these methods are economically feasible only for very high-quality timber), by properly designing roads and minimizing their overall length, by mulching the site, and by a variety of other methods. Furthermore, the erosion that cannot be controlled can be prevented from damaging water quality by using buffer strips, sediment traps, and other means.

The Institutional Climate for Environmental Control

Despite the generally resilient nature of the forests and considerable scientific knowledge of forest ecology and regeneration, forest environments may be threatened in the future because certain market forces or institutional constraints discourage adequate environmental protection. These problems include a lack of expertise in the logging community, a volatile market that hinders adequate planning in certain segments of the industry, and a lack of sufficient incentives to practice environmentally sound management.

1. *Lack of expertise.* — Although the majority of negative impacts may occur because of failure to follow well-recognized guidelines, others occur because of failures of judgment; forest environments are extremely complex and often require expert judgment about site conditions to select correct harvesting strategies. Some important impacts can be avoided only if the logger can recognize subtle clues to the existence of vulnerable conditions. For example, many unstable soil conditions may be recognizable only to a soils expert. This type of expertise usually is not available to the small operator, except possibly where local and State governments offer preoperation inspections and guidance (e.g., in Oregon). This poses a special problem if the residential market for wood expands considerably, because small operators may be expected to satisfy much of this new demand.

2. *Insufficient time for proper planning.* — In current mill operations in Maine, many "mill managers commonly call on short notice for a certain volume of a given type of product from the firms' logging division . . . A common result is that a considerable amount of the haul road construction is done on short notice . . . (without) . . . proper planning and correctly installed and maintained drainage structures."[82] It is not clear that problems of this nature will be as severe for wood harvesting for energy, because demand for the wood as a feedstock may be more uniform and predictable than the demand for traditional forest products (it also is not certain that the Maine experience is widely applicable). Nonetheless, most operations will combine lumber and energy feedstock operations—removing the high-quality wood, and then clearing to harvest the remainder of the biomass for energy users. To the extent that the timing of these operations depends on the demand for the (higher value) lumber, this problem may remain.

---

[82]"A Survey of Erosion and Sedimentation Problems Associated With Logging in Maine," op. cit.

3. *Lack of incentive.* — These are four reasons why a logger would pay strict attention to minimizing environmental damages:
- personal environmental or esthetic idealism,
- economic incentive,
- regulatory controls, or
- public relations

Idealism — and the role of education in fostering it — should not be ignored in predicting impacts and attempting to mitigate them. The strengthening of existing programs to educate potential wood harvesters about the adverse environmental effects of careless harvesting may be useful in tapping the vein of environmental idealism in the United States. Idealism is clearly insufficient to assure environmental protection, however, and more selfish incentives are needed.

The long time period needed to recoup the benefits of protective measures and the tendency of many of the benefits to accrue to adjacent landowners or the general public reduce the *economic incentive* of environmental protection. The shorter rotation periods that may be used for obtaining wood for energy may enhance the economic incentive, especially for owners of large tracts of land (because *they* are the "adjacent landowners"). Also, some "best management" measures do yield immediate returns to loggers, for example, measures that minimize road length or that prevent roadbeds from washing away.

Finally, to the extent that poor management of logging does long-term physical and esthetic damage to the forest, the value of forested land as a recreational and esthetic asset offers a strong incentive to the landowner to insist on sound practices. This incentive will be particularly strong in areas that have seen recent increases in market value because of their environmental value. This incentive will be effective, however, only where the landowner maintains close supervisory control over the logger.

Regulatory control of wood harvesting operations in the United States is very uneven. Although the Forest Service can exert considerable control of logging operations on Federal lands, logging on private lands is largely uncontrolled or very loosely controlled.

The 1976 National Forest Management Act includes requirements that federally owned timber "be harvested only where soil or . . . water conditions will not be irreversibly damaged, that harvests be on a sustained yield basis, that silvicultural prescriptions be written to ensure that stands of trees will generally not be harvested until they are mature (although thinning and other stand improvement work is permitted), that clearcutting meet certain standards, and that land management plans be written with public participation."[83] The Multiple Use Act of 1960, by defining environmentally oriented uses (such as wildlife protection) as legislated uses of the national forests, requires management practices in these forests to consider environmental protection as a direct requirement. In response to these mandates, the Forest Service enforces strict standards for harvesting lumber on Federal lands.

The degree of control exerted on non-Federal forests—especially privately owned forests—is noticeably weaker. Water quality impacts from wood harvesting theoretically should be regulated through the development of nonpoint source control plans under section 208 of the Federal Water Pollution Control Act. As discussed in volume I, however, implementation of section 208 generally has been disappointingly slow, and the eventual effectiveness of the 208 plans is highly uncertain. Also, few States have comprehensive forest practices legislation or the manpower to enforce such legislation. A major problem facing States wishing to control forest practices is the complexity and site-specific nature of the environmental impacts, forcing the difficult choice of using either a substantial force of highly trained foresters enforcing loosely written performance guidelines or else a more (economically) manageable agency enforcing rigid,—and perhaps impractical—rules. This

problem is discussed with insight in Brown 1976:[84]

> The difficulty is that rules specific to the wide variety of situations encountered would often be difficult to write and cumbersome to enforce for a great many problem areas, particularly within the context of our present state of technology. Field personnel recognize the dilemma of rules so vaguely written that they provide no control versus rules so specific that they prohibit flexibility and prevent forest practice officers and operators from adjusting methods to meet complex or highly varying situations. Given the option, most field people prefer to have flexibility at the risk of losing some control.

Finally, many State forestry agencies have concentrated their attention on forest fire prevention and control and not on forest management. Hence, the experience, interest, and expertise of present State forestry personnel may not provide a good base on which to build a strong management-oriented program.

The public's increasing awareness of environmental problems and willingness to act may serve as a strong incentive for the larger forest products companies to consider the *public relations implications* of their decisions. Companies like Weyerhaeuser spend large sums of money explaining their activities in sophisticated advertisements; presumably, this awareness of the importance of public approval affects their decisionmaking and operations.

Potential Environmental Effects— Summary

The use of wood as an energy feedstock holds considerable potential for reducing the adverse impacts associated with fossil fuel use. It also offers the potential for some important environmental benefits to forests, including:

* decreased logging pressures on some environmentally valuable forests;

---

[83]*Environmental Readiness Document, Wood Commercialization,* op. cit.

[84]Brown, et al., *Meeting Water Quality Objectives Through the Oregon Forest Practices Act,* (Oregon State Department of Forestry, 1976).

- improved management of forests that have been mismanaged in the past, with consequent improvements in productivity, esthetics, and other values; and
- reduced incidence of forest fires.

There is considerable uncertainty, however, about the extent to which a significant increase in the use of wood for energy will actually result in these benefits and avoid the negative impacts that could also accompany such an increase. There *are* important economic incentives for good management, including increased production of high-value timber and avoidance of losses in land values. There are a number of factors, on the other hand, that must be interpreted as warning signals:

1. Environmental regulation of forestry operations, especially on private lands, generally is weak or nonexistent.
2. Some of the existing economic incentives may induce cutting of vulnerable lands or neglect of best management practices.
3. Important gaps in the knowledge of the effects of intensive silvicultural activity— for example, of the nutrient and organic matter changes in the soil caused by whole-tree logging—may deter environmentally sound choices from being made.
4. The complexity and site-specificity of the harvesting choices that must be made may complicate adoption of environmentally sound harvesting plans, especially by small operators.

If careful environmental management is not practiced, the result might be:

- increased erosion of forest soils and consequent degradation of water quality,
- significant losses in esthetic and recreational values in forested areas,
- possible long-term drop in forest productivity,
- decline in forested area, and
- reduction of forest ecosystem diversity and loss of valued ecosystems and their wildlife.

Because the quality of forest management and the capacity for environmental regulation currently span the entire range from very low (or nonexistent) to high, the expected result of a "business as usual" approach to wood-for-energy environmental management would undoubtedly be a complex mix of the above impacts and benefits—with the marketplace determining the balance between positive and negative effects. Government action—including improved programs for local management assistance, increased research on the effects of intensive management, and increased incentives (economic or regulatory) for good management—may be capable of shifting this balance more towards the positive.

R&D Needs

The primary R&D needs in the area of wood supplies from forests fall into the categories of harvesting technology, growth potential, environmental impacts, and surveys. Traditional harvesting technologies are geared toward removing large pieces of wood in a way that is appropriate for lumber or paper pulp production. The wood that can be harvested for fuel, however, is considerably more varied, involving brush, rough and rotten timber, and the smaller pieces associated with logging residues. Although the whole-tree chip method seems to work well on relatively flat land, there is a need to develop low-cost techniques and equipment for harvesting smaller pieces of wood and brush on more varied terrains and at greater distances from roads.

Most research into forest growth potential is aimed at producing large straight trees suitable for the traditional forest products industry. Although some of this is research applicable to the production of wood for energy, the conditions and techniques for enhancing commercial timber growth are not the same as those for enhancing total biomass growth. As an example, thinning of tree stands reduces the total leaf surface and, with it, the amount of sunlight that is being captured by plants. This reduces the total biomass growth on the stand, although it tends to increase the growth of commercial timber. If a strong wood energy market develops, the ideal forest composition could involve a mix of tree types, sizes, and qualities. Various strategies for achieving integrated and economical energy-commercial timber operations need to be investigated. Tree hybrids, for example, should be developed with both commercial timber production and biomass production as dual goals.

There are a number of uncertainties regarding the environmental impacts of increased logging for energy. The nutrient balance in forests, as noted, needs to be better understood in order to better define the types and quantities of wood that can be removed without depleting the soil's nutrients. The effects of high biomass removal on soil carbon content and any subsequent long-term impacts on productivity or on forest viability require considerable research. The relationship between the diversity of tree and understory species in a forest and the forest's resilience to environmental stresses must be better understood before highly intensive management is allowed to expand to a majority of the commercial forest acreage. Alternative harvesting techniques such as strip cutting (or the cutting of strips of trees through the forest rather than clearcutting a large area) should be pursued in order to provide a repertoire of techniques that can be used where soil erosion may be a problem, such as in steeper slope terrains. Harvesting techniques that decrease the degree of soil compaction should also be developed. Furthermore, the entire forest ecosystem needs to be better understood if the environmental impacts of various types of forest activities are to be appropriately managed.

The national forest survey is primarily intended as a survey of commercial timber. The assumptions as to what is commercial should be separated from the survey of the biomass inventory and growth potential, in order to have an accurate assessment of the quantities available for all uses. The survey should incorporate noncommercial forestlands which are classified that way because of low growth potential.

A thorough assessment of the energy potential of the forests should also include a qualitative assessment of the conditions of the stocking on forestlands and the silvicultural activities (e.g., stand improvements) that could be carried out to increase the yield. The survey data should include environmental conditions such as soil types, rainfall, and other parameters. Finally, the size of tract is an important factor affecting the availability of the wood. Consequently, the farm and miscellaneous forest landowner classifications in forest surveys should be subdivided according to tract size and ownership.

Chapter 3
AGRICULTURE

Chapter 3.—AGRICULTURE

AGRICULTURE

Introduction

Agriculture was originally developed to provide a reliable source of food. Later feed was included in farm production and animals provided large parts of the population's energy needs. Although animals are rarely used for energy on U.S. farms today, agriculture has expanded to include the production of nonfood commodities, including cotton, tobacco, paint solvents, specialty chemicals, and various industrial oils. In 1977, these nonfood products accounted for over 13 percent of total farm production.

Many of the food and feed crops as well as farming byproducts can also be used to produce fuels or be combusted directly. In this chapter, the technical aspects of conventional agriculture are considered, leading to estimates of its potential for supplying energy.

Plant Growth, Crop Yields, and Crop Production

Harvested yields of many crops have increased dramatically over the past 30 years as a result of the development of genetically improved crop strains, as well as increased use of fertilizers and irrigation. Also, increased application of chemicals for control of insects, diseases, and weeds; further mechanization so that operations can be timely; improved tillage and harvesting operations; and other forms of improved management have also helped to raise yields.

Photosynthesis is the basic process providing energy for plant growth. Solar energy is absorbed by the green chlorophyll in the leaf and used to combine carbon dioxide (CO_2) from the air with water from the soil into stored chemical energy in the form of glucose. Glucose is used in the formation of compounds like adenosine triphosphate which provides energy for the synthesis of the various materials needed in the plant such as cellulose and lignins for cell walls and the structural parts of the plant and various amino acids (protein components). Glucose is respired to provide energy for production of other compounds, plant growth, and absorption of nutrients from the soil. As the plant matures, carbohydrates are stored in the seed to provide energy for the growth of new plants.

Sixteen nutrients are essential for plant growth and two or three more may increase yields but are not essential for the plant to complete its growth cycle. Of the 16, nitrogen, phosphorus, and potassium are the 3 main nutrients needed in large quantities to supplement the soil supply in order to obtain high crop yields. Calcium and magnesium are applied where needed as finely ground limestone. Sulfur is added as elemental sulfur or as sulfates when needed. Carbon, hydrogen, and oxygen come from the air and water and the remaining seven are used in extremely small amounts and are absorbed from the soil. All of these nutrients play essential roles in the growth processes within the plant.

Theoretical Maximum Yield

The theoretical maximum photosynthetic efficiency can be estimated as follows:

Ten percent of the light striking a leaf is reflected. Only 43 percent of the light that penetrates the leaf is of a proper energy to stimulate the chlorophyll. The basic chemical reactions (10 photon process) which use stimulated ("excited") chlorophyll to convert CO_2 and water to glucose have an overall efficiency of 22.6 percent. The net result of these factors is, in theory, a maximum photosynthetic efficiency of about 9 percent. A summary of the various cases of photosynthetic efficiency is presented in table 14.

Table 14.—Photosynthetic Efficiency Summary

	Average PSE[a] during growth cycle (percent)
Maximum theoretical	8.7
Highest laboratory short-term PSE[a]	~ 9
Laboratory single leaves, high CO_2 or low O_2, C-3 plants, <7% full sunlight	6.3
Same as above, for C-4 plants	4.4
Corn canopy, single day, no respiration	5.0
Record U.S. corn (345 bu of grain/acre, 120-day crop)	3.0
Record sugar cane (Texas)	3.0
Record Napier grass (El Salvador)	2.5
Record U.S. State average for corn (128 bu/acre, Illinois, 1979)	1.1
Record U.S. average for corn (108 bu/acre, 1979	0.9

[a]PSE-photosynthetic efficiency.

SOURCE: Office of Technology Assessment.

An efficiency approaching the theoretical maximum appears to have been achieved for a short time under laboratory conditions using an alga or single-celled water plant.[1] These results are controversial, however, and in practice there are always several other factors that limit the efficiency of photosynthesis, and the transformation of glucose into plant material. The most important of these factors, many of which are interdependent, are listed in table 15. For example, light saturation can be influenced by the CO_2 concentration, which is affected by other things. The key factors are light saturation, soil productivity (its ability to hold water, supply oxygen, release nutrients, and allow easy root development), weather (amount and timing of rainfall, absence of severe storms, length of growing season, temperature and insolation during the growing season), and plant type (leaf canopy structure, longevity of the photosynthetic system, sensitivity to various environmental stresses, etc.).

---

[1]V. C. Goedheer and J. W. Kleinen Hammans, *Nature,* vol. 256, p. 333, 1975.

Table 15.—Factors Limiting Plant Growth

- Water availability.
- Light saturation—a tendency for the photosynthetic efficiency to drop as the incident light intensity increases above values as low as about 10% of peak solar radiation intensity.
- Ambient temperature, especially wide fluctuations from ideal.
- Mismatch between plant growth cycle and annual weather cycle.
- Length of photoperiod (hours of significant illumination per day).
- Plant respiration.
- Leaf area index—completeness of coverage of illuminated area by leaves or other photosensitive surfaces.
- Availability of primary nutrients—especially nitrogen, phosphorus, and potassium.
- Availability of trace chemicals necessary for growth.
- Physical characteristics of growth medium.
- Acidity of growth medium.
- Aging of photosynthetically active parts of plants.
- Wind speed.
- Exposure to heavy rain or hail or icing conditions.
- Plant diseases and plant pests.
- Changes in light, absorption by leaves due to accumulations of water film, dirt, or other absorbers or reflectors on surfaces of leaves or any glazing cover.
- Nonuniformity of maturity of plants in crop.
- Toxic chemicals in growth medium, air, or water, such as pollutants released by human activity.
- Availability of CO_2.
- Adjustment to rapid fluctuations in insolations or other environmental variables—i.e., "inertia" of plant response to changing conditions."

SOURCE: Office of Technology Assessment.

In an untended system, the environmental factors are left to chance. Consequently, in any given year, some areas of the country will experience a favorable combination of factors, resulting in more plant growth, while in other areas environmental factors will be unfavorable, resulting in less plant growth. The exact places where the growth is favorable or unfavorable will also change from year to year, as will the exact growth at the most favorable area in each year. In the absence of long-range environmental changes, however, such as weather changes or soil deterioration, the average growth over a very large area and for many years will remain relatively constant.

Some of the environmental factors can be controlled, while others cannot within the present state of knowledge. Managing a plant and soil system consists of artificially maintaining some of the environmental factors, such as nutrients or water, at a more favorable level than would occur naturally. The many remaining factors, however, are still left to chance. Furthermore, there is a limit to how much plant growth (or other characteristics such as grain yield) can be influenced by changes in given environmental factors. Once some of the factors have been optimized for plant growth (or, e.g., grain yield) the plant's performance will not be improved by further changes of these factors. Too much water or fertilizer, for example, could actually inhibit growth rather than increase it.

Because plants vary in their sensitivity to growth-limiting factors yields can be improved by selecting or breeding for plants that are relatively insensitive to environmental factors that cannot be controlled and/or that respond well to factors that can.

A dramatic example of the success of breeding and management is corn. The history of U.S. average corn yields from 1948-78 is shown in figure 7. While the national average yields have not been analyzed in detail, Duvick has analyzed the changes in yields from hybrids grown in various Midwestern locations.[2] He

Figure 7.—U.S. Average Corn Yield

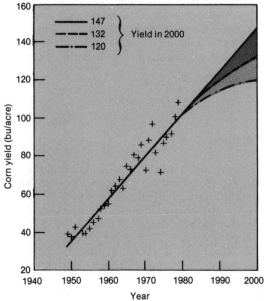

SOURCE: *Agricultural Statistics* (Washington, D.C.: U.S. Department of Agriculture, 1978).

[2]D. N. Duvick, *Maydica XXII*, p. 187, 1977.

concluded that 60 percent of the increase on these plots was due to genetic improvements while 40 percent was attributed to improved management. The management tends to reduce the environmental stresses, while hybrids were developed that are less sensitive to adverse environmental factors and more responsive to the factors that can be controlled (e.g., fertilizers, weed control, insect control, etc.).

Historical Yield Trends

Past yield trends can be used as a guide for projection of future yields. A period of at least 15 years should be considered because of weather fluctuations since the desired value is the yield trend if weather remained constant. A period of dry years from 1973 to 1976 tended to exert some influence on data variability. During the 1948-78 period, corn yields increased an average of 2 bu/acre-yr giving a 1978 yield just over 100 bu/acre. Similarly soybean yields showed an increase of about 0.4 bu/acre-yr and a 1978 yield of 29.2 bu/acre. National wheat yields are somewhat more variable since wheat is grown primarily in areas that are more affected by drought than is corn. Nonetheless, the yield trend indicates an increase of 0.5 bu/acre-yr and a 1978 average yield of 31.6 bu/acre. The U.S. Department of Agriculture (USDA) has calculated a summary of all crop yields per acre using a relative value of 100 for the 1967 yield. The trend for increase over this period was 1.4 units per year, but the uncertainty in this number is large (see figure 8).

Yield increases in the future as in the past will come from a combination of improved crop varieties and improved cultural practices. Since current fertilization practices have reached near optimum rates, increases in yield due to increases in fertilization rates will be less than for the past 20 years. As yield potentials of varieties are increased, however, increased rates of fertilization will be needed to keep pace with the increased yields. Since yield increases result from a combination of practices, it is difficult to attribute yield increases to any one practice.

Figure 8.—U.S. Average Crop Output

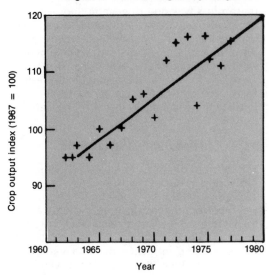

SOURCE: *Agricultural Statistics* (Washington, D.C.: U.S. Department of Agriculture, 1978).

Record Yields

Projection of yields and determination of where yield increases will diminish may be judged on the basis of yields that have been obtained. For example, the average U.S. corn yield has reached over 100 bu/acre, but the average yield in Illinois in 1979 was 128 bu/acre and in Iowa in 1979 it was 127 bu/acre.* If county averages within a State are examined, average yields are found to approach 140 bu/acre. If individual farms of 500 acres of corn are considered, yields of 175 bu/acre have occurred. And on selected areas of 2 or 3 acres yields of 345 bu/acre have been noted.

Future Yields

Over the past 30 years corn yields have increased at an average rate of about 2 bu/acre-yr. One would not expect this rate of increase to rise, and it may decline. Therefore, in projecting corn yields in the year 2000, 140 bu/acre would be optimistic. A less optimistic projection, based on annual average yields increasing at one-half the rate that they have in the recent past, is 120 bu/acre in 2000. A study by the National Defense University in 1978 gave a projected corn yield for 2000 of 132 bu/acre.

Future increases in the yields of other crops are also expected, but each crop together with the cropland on which it will be grown must be considered separately. Dramatic increases, such as a doubling in crop yields by 2000, however, are not expected for conventional crops.

---

*The weather in 1979 was ideal for corn growing.

Land Availability

Cropland is land used for the production of adapted crops for harvest, alone or in a rotation with grasses and legumes, and includes row crops (e.g., corn), small grain crops (e.g., wheat), hay crops, nursery crops, orchard crops, and other similar specialty crops. Cropland is generally categorized into the agricultural production regions shown in figure 9.

Of the U.S. total land area of 2.3 billion acres, 413 million or 467 million acres are currently classified as cropland depending on whether one uses the Soil Conservation Service (SCS) or the other USDA classification system.

The second is a broader classification that includes some land not currently cropped that is rotated into cropland but may now be in pasture or other use. The percentage of the total land area of each State that is cropland is shown in figure 10 and the cropland used in 1977 is shown in table 16. Both of these are based on the more restrictive SCS classification of cropland.

Cropland, however, is **not** a static category. The location of cropland may shift even though the quantity of cropland remains rela-

Figure 9.—Farm Production Regions of the United States

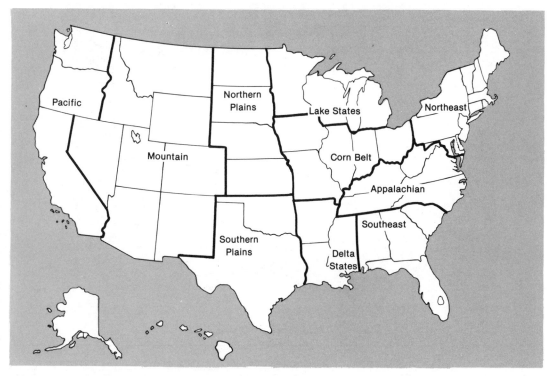

SOURCE: Soil Conservation Service, U.S. Department of Agriculture.

tively constant. Over time there are both additions and deletions to the cropland inventory.

The quality of land and cropping systems may shift as well. One such change has been the increase in irrigation, in areas like the Texas high plains, from 1.2 million acres in 1948 to 6.4 million acres in 1976. This represents both a trend in improving the productivity of existing cropland and a trend towards opening new, marginal land that is only productive and economic with irrigation. To some extent, the United States has been replacing rainfed arable land that is lost to agriculture with irrigated land in dry areas. This trend, however, is likely to change due to increasing energy costs and depletion of some Western ground water.

Over time, the content of a land inventory can be influenced by the way that a given sta-

tistic is enumerated. For example, up to 1964 the agricultural census was personally enumerated and in 1969 it was done by mail. According to the broader USDA classification, cropland pasture increased by over 30 million acres between these surveys, and the suspicion is that the farmer applied a less strict definition to cropland pasture which resulted in the inclusion of 30 million acres of pastureland and grassland into the cropland pasture category even though the actual usage had not changed.

One strong influence on the land inventory has been the Government's agricultural programs. The land retirement programs of the 1960's reduced planted cropland and had the net effect of moving less productive land out of crop production temporarily or even permanently in the case of very low-quality land. As

Figure 10.—Cropland as a Percentage of Total Land Area by Farm Production Region

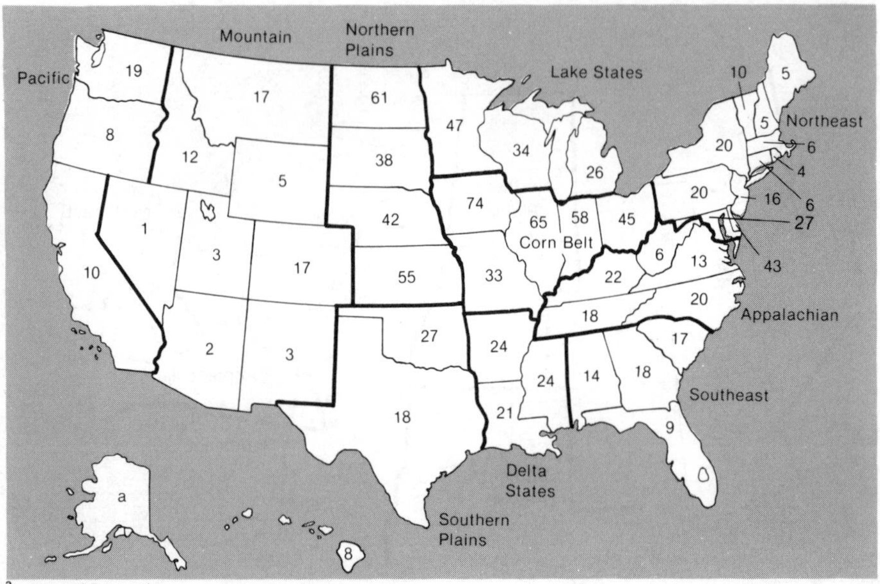

Mountain · Northern Plains · Pacific · Lake States · Northeast · Corn Belt · Appalachian · Southeast · Delta States · Southern Plains

19 17 61 10 5
8 12 5 47 34 26 20 5 6 4
1 42 74 65 58 45 20 16 6 27
10 3 17 55 33 22 6 13 43
2 3 27 24 18 20 17
18 24 14 18
21 9
a
8

aLess than 0.5 percent.
SOURCE: Soil Conservation Service, U.S. Department of Agriculture.

Table 16.—Cropland Use in 1977 (thousand acres)

Region	Row crops Irrigated	Nonirrigated	Close-grown crops Irrigated	Nonirrigated	Rotation hay Irrigated	Nonirrigated	Occasionally improved hayland Irrigated	Nonirrigated
Northeast	254	6,771	14	1,033	9	3,339	5	3,286
Appalachian.	349	14,445	33	543	7	1,248	10	3,151
Southeast	1,681	12,108	23	342	24	74	8	604
Delta States	2,294	15,358	1,489	285	170	299	0	397
Corn Belt	1,035	70,291	31	7,228	8	5,987	4	4,116
Lake States	799	20,930	80	9,140	36	7,753	8	2,802
Northern Plains	8,641	18,062	920	40,007	690	5,387	325	3,885
Southern Plains	5,935	13,908	2,354	15,784	33	802	286	725
Mountain.	4,117	962	3,316	14,021	2,269	380	3,821	1,276
Pacific.	4,477	132	2,757	5,556	1,124	502	1,092	295
Totala.	29,750	173,493	11,025	93,865	4,398	25,818	5,559	20,839

Region	Native hay Irrigated	Nonirrigated	Summer fallow	Orchards, etc. Irrigated	Nonirrigated	Other	All cropland Irrigated	Nonirrigated
Northeast	0	789	24	84	508	493	372	16,534
Appalachian.	0	86	112	0	149	690	406	20,339
Southeast	0	0	54	699	661	1,324	2,449	15,053
Delta States	0	84	179	6	136	489	3,979	17,207
Corn Belt	0	117	749	22	72	876	1,115	88,739
Lake States	0	719	466	42	291	1,073	972	43,167
Northern Plains	33	2,493	13,825	0	15	268	10,790	83,733
Southern Plains	0	405	1,061	33	153	755	9,011	33,223
Mountain.	1,460	320	9,449	53	0	641	17,208	26,111
Pacific.	261	208	4,082	2,242	183	277	12,261	10,927
Totala.	1,754	5,221	28,319	3,295	2,225	6,806	57,647	355,520

aAlso includes Caribbean and Hawaii.

SOURCE: Soil Conservation Service, U.S. Department of Agriculture.

programs were terminated in the early 1970's, some of these acres came back into crop production.

USDA's SCS surveyed non-Federal lands in 1977 and identified the land that potentially could become cropland.[3] The survey classified potential croplands according to whether the land was judged to have a high, medium, low, or zero potential for conversion. Figure 11 summarizes the quantities of land that have a

Figure 11.—Present Use of Land With High and Medium Potential for Conversion to Cropland by Farm Production Region

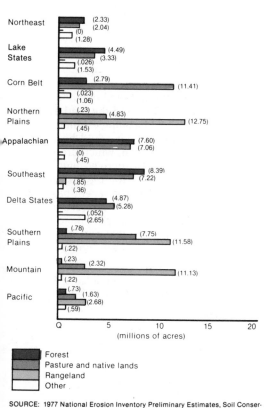

SOURCE: 1977 National Erosion Inventory Preliminary Estimates, Soil Conservation Service, U.S. Department of Agriculture, April 1979.

---

[3]*1977 Soil Conservation Service National Erosion Inventory Estimate* (Washington, D.C.: Soil Conservation Service, U.S. Department of Agriculture, December 1978).

high or medium potential for conversion to cropland. Of the total potential cropland in 1977, 40 million acres have a high probability to be converted and another 95 million acres are classified as having medium probability.*

The breakdown of the potential cropland into high and medium potential for conversion is an attempt to define a crude cost curve for the availability of new cropland. It was judged by SCS that the land with a high potential will enter agriculture as a matter of course, if price relationships are somewhat more favorable to the farmer than the 1976 prices on which the survey was based. The medium potential, however, is a category involving lands with a wide variety of problems but which cannot be categorically excluded from conversion if farmland prices increase sufficiently.

The SCS survey, however, does not include a quantitative measure of the price increases necessary to bring potential cropland under cultivation. A conservative approach would be to assume that only land with high potential can be included in the cropland base without excessive inflation in food prices above that which would occur normally due to increased demand for food. A more optimistic approach would be to include those types of medium-potential land that probably will be considered high potential in the future, as increased demand for food raises cropland prices. This is the approach that was taken.

Two major factors, mentioned above, that affect crop productivity are water availability and soil quality. Therefore, land was included from the medium-potential category that has greater than 28 inches of annual rainfall and potentially has good productivity (capability classes 1 and 2 of the eight agricultural land capability classes). These land types are the most likely to be brought into cultivation if demand exceeds the high-potential category.

With these assumptions, the potential cropland is shown in table 17. This together with existing cropland provides the cropland base, to-

---

*As of November 1979, these numbers were 36 million and 91 million acres, respectively.

Table 17.—Potential Cropland (thousand acres)

Region	Source			
	Forest	Pasture	Rangeland	Other
High potential with over 28 inches rainfall				
Lake States.	868	1,206	22	443
Delta	1,482	1,781	0	129
Corn Belt	664	4,451	23	368
Northeast	268	562	0	371
Southeast.	2,111	2,926	133	151
Appalachian	1,981	2,974	0	183
	7,374	13,900	178	1,645
Total of forest, pasture, and rangeland: 21,452				
Medium potential, class 1 & 2 land only with over 28 inches rainfall				
Lake States.	1,463	723	0	315
Delta	986	1,423	0	0
Corn Belt	974	2,590	0	356
Northeast	656	433	0	306
Southeast.	2,082	1,256	0	17
Appalachian	2,012	1,157	0	119
	8,137	7,582	0	1,113
Total of forest, pasture, and rangeland: 15,755				
High potential with less than 28 inches rainfall				
Arid regions	312	5,503	9,549	—
Total of forest, pasture, and rangeland: 15,364				

SOURCE: Otto C. Doering III, ''Cropland Availability for Biomass Production,'' contractor report to OTA, August 1979.

taling 520 million acres. (This is based on the broader cropland classification as used in USDA's *Agricultural Statistics* and all Economics, Statistics, and Cooperatives Service publications.) Although it is impossible to predict exactly how the cropland base will develop in the future, one plausible scenario is shown in table 18 based on continuation of past trends to 1984 and on USDA's National Interregional Agricultural Projections System (NIRAP) for 1990 and 2000.[4]

[4]L. Quance, A. Smith, K. Liu, and L. Yao-Chi, ''Adjustment Potentials in U.S. Agriculture,'' Vol. 1—National Interregional Projections System (Washington, D.C.: Economics, Statistics, and Cooperatives Service, U.S. Department of Agriculture, May 1979).

By examining the detailed demand for various crops from NIRAP and the land available, Doering has derived baseline estimates for the quantity of cropland that could be available for bioenergy production, which are shown in table 19.[5] Doering, also derived high and low food demand scenarios for 1984 based on extrapolation of trends in the recent past and increased this demand range proportionately to the increase in baseline crop demand from the NIRAP projections for 1990 and 2000. Finally these demand ranges were combined with

[5]Otto C. Doering III, ''Cropland Availability for Biomass Production,'' contractor report to OTA, August 1979.

Table 18.—Cropland Balance Sheet (million acres)

Year	1977[a]	1979	1984	1990	2000
Cropland (except cropland pasture and hayland)	393	395	404	407	439
Cropland pasture and hayland	74	72	65	80	60
Noncropland pasture	27	27	22	10	2
Other potential cropland	26	24	22	13	4
Total	520	518	514	510	505
Total land lost to other uses to date	0	2	6	10	15
Total	520	520	520	520	520

[a]With SCS acreage counting system, the 1977 acreages (in millions of acres) would be as follows: cropland except cropland pasture and hayland, 343; cropland pastaure and hayland, 63; noncropland pasture which is potential cropland or is periodically rotated into cropland, 88; and other potential cropland, 26. The major differences are that the cropland categories total 406 million acres with the SCS classification rather than 467 million acres and the noncropland categories are increased accordingly. In both classification schemes there are additional noncropland pature categories which are neither potential cropland nor periodically rotated into cropland, primarily because the terrain is too rocky or rough to allow mechanized harvests.

SOURCE: Deduced from Otto C. Doering III, ''Cropland Availability for Biomass Production,'' contractor report to OTA, August 1979.

NIRAP high and low productivity (yield/acre) estimates to determine plausible ranges of demand for cropland for food and feed production and thus the ranges of land available for bioenergy production. These estimates are shown in table 19.

Table 19.—Cropland Available for Biomass Production

	Million acres
1984	
From cropland pasture	10
From high potential	10
From medium potential	10
	30
Range of uncertainty	30-70
1990	
From cropland pasture	25
From high potential	5
From medium potential	5
	35
Range of uncertainty	9-69
2000	
From cropland pasture	10
From high potential	NA
From medium potential	NA
	10
Range of uncertainty	0-65

NA = none available.

SOURCE: Otto C. Doering III, ''Cropland Availability for Biomass Production,'' contractor report to OTA, August 1979.

It should be emphasized that these are not predictions, but rather plausible estimates given the current state of knowledge. The ranges are less than ±10 percent of the cropland base, so it is unlikely that more accurate estimates can be made. Furthermore, unexpected increases in crop productivity, in world food demand, or in demand for cropland for nonagricultural uses could increase or decrease the quantity of cropland available for bioenergy production beyond the ranges shown. Also, since this only refers to cropland capable of producing more or less conventional crops, the development of unconventional crops could open new land categories not considered here.

In addition to the physical availability of cropland, one must consider the cost of bringing it into production. Four major factors influence this cost. First the land is currently being used for some purpose that the owner considers to be more valuable than crop produc-

tion. Second an investment is sometimes necessary to convert the land to crop production, such as installation of drainage tiles or removing trees occupying the site. These costs can vary from virtually nothing to as much as $600/acre.[6] Third, the land that can be brought into production is generally less productive, on the average, than cropland currently in production. Finally, this land also typically suffers from problems of drought or flooding that make crop yields extremely sensitive to weather (particularly the rainfall pattern). Consequently, farming this land involves a larger cost and risk than with average cropland; and, from the national perspective, using it will increase the year-to-year fluctuations in food supplies and prices.

As a result of these added costs and risks, farm commodity prices will have to rise before it will be profitable to bring new land into crop production. Eventually this raises the cost of all farmland, the cost of farming, and food prices. The exact price rise needed to increase the cropland in production by a given amount is unknown, but some things can be deduced from this analysis. During the next few years, bioenergy production from cropland is not likely to be constrained by the availability of cropland. However, the quantity of land that can be devoted to energy production without reducing food production is likely to decrease in the future. Furthermore, since the marginal cost of bringing new cropland into production increases as the quantity of cropland in production expands, the added cost in terms of higher food prices needed to keep a given amount of cropland in energy production is likely to increase with time. In other words, it is likely to be increasingly expensive to produce energy crops, even if the energy output remains constant.

The above comments are particularly applicable to grains and sugar crops. Considerable quantities of land, however, already are devoted to forage grass production and the yields can be raised through increased fertilization (see below). Furthermore, grass yields tend to be less sensitive to poor soil quality than grains

---

[6]Ibid.

and sugar crops. Consequently, the economic barriers to increased grass production are considerably lower than for increased grain or sugar production, and one would expect the indirect costs of grass production to be less than those of grains and sugar crops.

Nevertheless, in the long term, there may be little cropland suitable for food/feed production that can be devoted to energy, and any energy crops would have to be grown on land totally unsuited to food and feed production.

Types of Crops

There are over 300,000 plant species in the world, but less than 100 are grown as crops in the United States. Of the various crops, three basic types are currently of interest for immediate energy production: starch, sugar, and forage.

The major starch crops are corn (for grain) and wheat, accounting for 21 percent each of the total acreage of harvested crops, or about 70 million acres each. The annual production and disposition of corn and wheat are shown in tables 20 and 21. In addition, oats, barley, grain sorghum, and rice are other important starch crops.

The main sugar crops currently grown in the United States are sugarcane and sugar beets. About 760,000 acres are devoted to sugarcane (in Florida, Louisiana, Texas, and Hawaii) and sugar beets were grown on 1.2 million acres in 1977. Also, a smaller acreage is devoted to sweet sorghum production, primarily for sorghum syrup. The sugar yields averaged about 3.7 ton/acre for sugarcane (some growing seasons were 18 to 24 months) and 2.6 ton/acre for sugar beets. The very limited commercial acreage of sweet sorghum has yielded about 1.9 ton/acre of sugar, however, the acreage is too small to accurately reflect the yields that would occur from large-scale production of this crop.

Forage crops are grown for feed and bedding. Including alfalfa, the area under forage crop production is about 60 million acres.[7] Forage crops include orchard grass, brome grass, tall fescue, alfalfa, clover, and reed canary-

Table 20.—Annual Production and Disposition of Corn for Grain in the United States, 1966-75 (million bushels)

Year	Production	Domestic consumption	Exports	Stocks
1966	4,167	3,697	487	826
1967	4,860	3,885	633	1,169
1968	4,450	3,966	536	1,118
1969	4,687	4,189	612	1,005
1970	4,152	3,977	517	667
1971	5,641	4,387	796	1,126
1972	5,573	4,733	1,258	706
1973	5,647	4,631	1,243	483
1974	4,664	3,641	1,149	359
1975	5,797	4,049	1,711	398

SOURCE: Agricultural Statistics 1977 (Washington, D.C.: U.S. Department of Agriculture, 1977).

Table 21.—Annual Production and Disposition of Wheat in the United States, 1966-75 (million bushels)

Year	Production	Domestic consumption	Exports	Stocks
1966	1,967	683	771	513
1967	2,202	626	765	630
1968	2,188	740	544	904
1969	2,350	764	603	983
1970	2,336	772	741	823
1971	2,442	848	610	983
1972	2,530	798	1,135	597
1973	2,305	748	1,217	340
1974	2,140	686	1,019	435
1975	2,572	735	1,173	664

SOURCE: Agricultural Statistics 1977 (Washington, D.C.: U.S. Department of Agriculture, 1977).

grass. Yields average about 1.5 to 2.5 ton/acre but could be increased to 4 to 5 ton/acre by relatively straightforward changes in management practices.

Most crops can be grown in several different areas of the country. However, each crop has unique characteristics that enable it to do well under certain combinations of soil type, growing season, rainfall, etc. Since these parameters vary widely throughout the United States, it is unlikely that any one crop could prove to

---

[7]Agricultural Statistics (Washington, D.C.: U.S. Department of Agriculture, 1978).

be the correct energy crop for a given product. Rather, the available cropland can best be utilized for energy by growing various different crops suited to the various soil types, climates, and other conditions. Nevertheless, there are some striking differences when national average yields are compared. (See "Energy Potential From Conventional Crops" below.)

The crops mentioned here do not exhaust the possibilities, even for starch, sugar, and cellulosic products. Other crops may be superior to these under certain circumstances. But these crops do serve to illustrate U.S. agriculture's energy potential, costs, and impacts.

Current Agricultural Practices and Energy and Economic Costs

As mentioned above, the purpose of managing a plant system is to provide an artificially favorable environment for plant growth. Since increasing the intensity of management costs money, there is always a tradeoff between the increased cost and the expected increase in yields. As price relationships change, the intensity of management will also change. A summary of some current agricultural practices and their costs and energy usage is given below.

Aside from weather and soil type, the planting date, planting density, weed, disease, and insect control, and tillage practices can all affect crop yield. Different plants have different sensitivities to these various factors. Practices also have to be suited to the climate and soil type that is being farmed. Consequently, the direct costs of farming will vary depending on the crop and region. There can even be significant differences for the same crop within a given region (e.g., erosion control measures, irrigation, etc.).

A "typical" farming operation for annual crops such as corn and soybeans, however, might be as follows: After harvest of the crop in the fall the residues are chopped or the soil is disked to reduce the size of the residues and to level the soil. Phosphate and potassium fertilizer are broadcast and the residues and fertilizer are plowed under. In the spring, surface tillage to level the soil is done soon after the soil becomes suitable for tillage. Nitrogen fertilizer—anhydrous (dry) ammonia, etc.—if needed, is applied to the soil. Five to ten days later the soil is surface tilled with a cultivator

or disk and the crop planted. During planting some additional fertilizer may be added, an insecticide may be applied, and herbicides may be broadcast on the soil surface. The crop may be cultivated for weed control once or twice within the first month of growth. No additional operations occur until the crop is harvested with a harvestor that separates the grain and leaves the residue on the field. If the grain has a moisture content above that needed for storage without spoilage, it is dried. The grain may be fed on the farm, stored and sold later, or sold directly to a grain company at harvest.

Minimum or reduced tillage operations are used to reduce soil erosion. With their use the soil may be chisel-plowed rather than moldboard-plowed so that much of the residue remains in the surface. Herbicides may be used to give complete weed control so that no further cultivation is needed.

Forage crop management is considerably simpler. Since forage crops are usually perennials, crop planting is done only once every 4 to 5 years, or longer. Aside from planting, the only management is the application of fertilizers and the harvesting of the forage crop.

The estimated costs for producing corn (a row crop) and wheat (a close-grown crop) are shown in table 22. These costs are fairly representative of what can be expected for the production costs per acre for annual crop production, with intensive agriculture. Costs will vary from place to place, but where costs other than land costs are higher and/or yields are lower, the land will be worth less and land costs will be lower.

Table 22.—Estimated per Acre Production Costs in Indiana, 1979

Production cost item	Corn	Wheat
Yield per acre....................	110 bu	50 bu
Direct cost per acre		
Fertilizer and lime[a]	$ 32.00	$ 22.50
Seed and chemicals	20.00	10.00
Machine operating and drying	25.50	11.25
Interest on operating capital.........	9.00	7.00
Total direct costs................	$ 86.50	$ 50.75
Indirect costs per acre		
Machinery and equipment	$ 43.00	$ 18.00
Labor and management.............	31.00	20.00
Grain storage (bin only)............	11.00	—
Land cost[b].....................	92.00	92.00
Total indirect costs	$177.00	$130.00
Total costs per acre	$263.50	$180.75
Cost per bushel	2.40	3.62

[a]Nitrogen prices at $0.12/lb for corn and $0.20 for wheat. Phosphorus pentoxide priced at $0.18/lb; potassium monoxide priced at $0.09 for all crops. A corn-soybean rotation is assumed. Thus soybeans produce a nitrogen credit for corn production and no insectide is used.
[b]Land costs approximate current cash rental rates.

SOURCE: Barber, et al., ''The Potential of Producing Energy From Agriculture,'' Purdue University, contractor report to OTA, May 1979.

Table 24.—Energy Inputs and Outputs for Corn in U.S. Corn Belt

	Energy units	
	Nonirrigated[a] (10⁵ Btu)	Sprinkler[b] (10⁵ Btu)
Output		
Grain	543.7	666.4
Residue	543.7	666.4
Total output	1087.4	1332.8
Input		
Irrigation pumping	—	60.0
Fertilizer...................	47.0	57.6
Drying fuel	19.4	23.8
Equipment fuel	10.0	10.5
Pesticides.................	6.0	6.0
Total input	82.4	157.9

[a]Grain yield: 139 bu/acre; residue yield: 7,770 lb/acre.
[b]Pump irrigated 15 inches water, 100 ft depth. Grain yield and residue yield are 170 bu/acre and 9,520 lb/acre, respectively.

SOURCE: Barber, et al., ''The Potential of Producing Energy From Agriculture,'' Purdue University, contractor report to OTA, May 1979.

The energy used for farming varies considerably. Typical energy inputs per acre for various crops are shown in table 23. These energies are for cultivation without pumped irrigation. A comparison of the energy inputs for irrigated and nonirrigated corn is shown in table 24. Overall, the energy per ton of grain can vary at least from 1.2 million Btu/ton for oats in Iowa to 6.5 million Btu/ton for grain sorghum in Texas. For corn the variation is at least from 2.6 million Btu/ton of grain (Illinois average) to 4.6 million Btu/ton (Nebraska). The U.S. average for corn is 3.1 million Btu/ton of corn grain.

These differences reflect not only differences in cultivation practices and yields, but also the presence or absence of pumped irriga-

Table 23.—Energy Inputs for Various Crops (10⁵ Btu per acre)

	Corn					
	Conventional tillage	Minimum tillage	No tillage	Soybeans	Wheat	Alfalfa
Nitrogen[a]..............	43.75	43.75	43.75	—	20.00	1.25
Phosphorus pentoxide + potassium monoxide[b].....	3.20	3.20	3.20	2.70	3.00	6.56
Drying[c]...............	19.35	19.35	19.35	—	—	—
Diesel[d]						
Ground preparation	7.36[e]	5.13[f]	2.21[g]	5.67[h]	3.15[i]	—
Planting..............	1.34	1.34	1.34	1.34	1.34	—
Cultivation	1.34	1.34	—	1.34	—	21.07
Harvest	2.15	2.15	2.15	1.69	1.54	—
Herbicides[j]	4.20	4.65	6.00	4.80	—	—
Insecticide[e]	1.80	1.80	1.80	—	—	5.60
Total	84.49	82.71	79.80	17.54	29.03	34.48

[a]25,000 Btu/lb nitrogen.
[b]3,000 Btu/lb phosphorus pentoxide and 2,000 Btu/lb potassium monoxide.
[c]93,500 Btu/gal LP-gas, 3,414 Btu/kW-hr electricity.
[d]132,250 Btu/gal diesel fuel.
[e]Spread fertilizer, plow, disk, apply anhydrous ammonia, disk.
[f]Spread fertilizer, chisel plow, apply anhydrous ammonia, field cultivate.
[g]Spread fertilizer, spray, apply anhydrous ammonia.
[h]Spread fertilizer, plow, disk, disk.
[i]Disk, disk, spread fertilizer in spring.
[j]120,000 Btu/lb active ingredient.

SOURCE: Barber, et al., ''The Potential of Producing Energy From Agriculture,'' Purdue University, contractor report to OTA, May 1979.

tion and the fuel used for irrigation. Examples of energy-intensive crops range from corn grown in Nebraska which is irrigated with ground water brought to the surface by electric pumps and is dried with liquefied petroleum, to grain sorghum which has relatively low yields compared to energy inputs throughout the United States. Other crops, such as rice, can be even more energy intensive (7.8 million Btu/ton, U.S. average).

For most corn cultivation, over half of the energy input comes from fertilizer, principally nitrogen. However, without nitrogen fertilizers, average corn yields would drop from about 100 bu/acre to less than 30 bu/acre. In the example in table 23, the energy used would increase from 3.0 million to 4.9 million Btu/ton if nitrogen fertilizers were not used, assuming the above yield change.

The other big energy input for some areas— irrigation—can have the opposite effect. In the example given in table 24, the use of irrigation raises the energy input from 2.2 million to 3.4 million Btu/ton. And in some areas (e.g., west Texas and southern Arizona), the energy

required for pumped irrigation is more than twice that shown in table 24.[8] In all, 85 percent of the 58 million acres of irrigated cropland are in the West (Northern Plains, Southern Plains, Mountain, and Pacific farm production regions) and 94 percent of the 0.26 Quad/yr used for pumped irrigation in the United States in 1974 was in the West.[9] On the average, the energy needed to pump the equivalent of 22 inches of rainfall in the West is 6 million Btu/acre. Consequently, this is a reasonable average figure for the energy input due to irrigation.

Another possible type of energy crop is forage grass. Currently, little or no fertilizer is used to cultivate forage grass; and yields are about 2 ton/acre. However, if fertilizers were used and the crops harvested more times per year, additional biomass could be obtained. Table 25 shows the costs of producing grass

---

[8]D. Dvoskin, K. Nicol, and E. O. Heady, "Irrigation Energy Requirements in the 17 Western States," *Agriculture and Energy*, W. Locheretz, ed. (Academic Press, 1977).

[9]G. Sloggett, "Energy Used for Pumping Irrigation Water in the United States, 1974," *Agriculture and Energy*, W. Locheretz, ed. (Academic Press, 1977).

Table 25.—Estimated Costs of Producing Grass Herbage at Three Yield Levels

	Yield level (ton/acre)		
	2	3	4
Growing costs ($/acre)			
Fertilizer[a]	—	19.45[b] - 24.42[c]	41.59[d] - 50.70[e]
Seed and seeding	2.30	2.30	2.30
Interest and miscellaneous[f]	0.22	2.07 - 2.54	4.17 - 5.04
Total	2.52	23.92 - 29.26	48.06 - 58.04
Harvest costs ($/acre)			
Machine operating	8.00	12.00	16.00
Interest and miscellaneous	0.76	1.14	1.52
Machine investment	34.06	34.06	34.06
Hay storage[g]	0 - 8.72	0 - 13.08	0 - 17.44
Labor @ $4/hr[h]	3.68 - 11.04	5.52 - 16.56	7.36 - 22.08
Total	46.50 - 62.58	52.72 - 76.84	58.94 - 91.10
Total non-land costs			
$/acre	49.02 - 65.10	76.54 - 106.10	107.00 - 149.14
$/ton[i]	27.23 - 32.55	28.35 - 35.37	29.72 - 37.29

[a]Includes cost of application.
[b]60 lb nitrogen, 20 lb phosphorous pentoxide, 50 lb potassium monoxide/acre.
[c]60 lb nitrogen, 30 lb phosphorous pentoxide, 90 lb potassium monoxide/acre.
[d]150 lb nitrogen, 30 lb phosphorous pentoxide, 80 lb potassium monoxide/acre.
[e]150 lb nitrogen, 50 lb phosphorous pentoxide, 150 lb potassium monoxide/acre.
[f]9-percent interest = 0.5% miscellaneous costs.
[g]Range from no cost if large hay package stored outside to new barn costs for rectangular bales.
[h]High labor values for rectangular bale handled by hand, low labor bales for large hay packages.
[i]Assumes 10% additional storage loss if hay stored outside (average storage period).

SOURCE: Barber, et al., "The Potential of Producing Energy From Agriculture," Purdue University, contractor report to OTA, May 1979.

herbage at various levels of fertilization and grass production. The additional production is estimated to cost $28 to $37/dry ton.[10] Note particularly that no land charges are included in these cost calculations, because the use of the land has not changed. Only the output has been increased. Nevertheless, some farmer profit in addition to the labor charge may be needed to induce farmers to increase production. Furthermore, obtaining the full potential

---

[10]S. Barber, et al., "The Potential of Producing Energy From Agriculture," Purdue University, contractor report to OTA, May 1979.

from this resource would require a 50- to 100-percent increase in fertilizer use in agriculture.

With no fertilization the energy used to produce the grass is about 0.1 million Btu/ton of grass at the present estimated level of 2 ton/acre. At 3 and 4 ton/acre, the additional energy use is about 1.9 million Btu for the third and 2.4 million Btu for the fourth ton. About 0.1 million to 0.2 million Btu/ton should be added to these energy inputs for a 15-mile transport of the grass.[11]

---

[11]Ibid.

Energy Potential From Conventional Crops

Aside from crop residues, the two major near- to mid-term sources of bioenergy from conventional crops are grains and sugar crops for liquid fuels production and increased forage grass production. On the land capable of supporting grain and sugar crop production, grasses could also be grown; and a comparison of these choices is considered first.

The mechanism through which food and fuel production compete is the increase in farm commodity prices. Since farm commodity prices must rise in order to make it profitable for farmers to increase the quantity of land under intensive production, it is important to examine the net quantities of premium fuels that can be displaced, through liquid fuels production, by each of the alternatives when new cropland is brought into production. (For details of the energy balances, see ch. 11.)

The calculations for sugar crops and grasses are relatively straightforward, since these feedstocks have very little protein in them and, consequently, the byproduct probably has little value as an animal feed (see "Byproducts" in ch. 8). The distillation of grains, in contrast, produces a protein concentrate byproduct that can displace significant quantities of soybean meal and thus soybean production. Additional grains could then be grown on the land formerly devoted to soybean production. Estimates of the effect of this substitution are calculated below.

First, let X represent the number of acres of average soybean production that can be displaced by growing 1 acre of average corn production, converting the corn to ethanol, and feeding the byproduct to livestock. Assuming that the corn yield on marginal cropland (i.e., the new cropland that can be brought into production) is y times as great as on average cropland, then 1 acre of marginal cropland grown with corn for ethanol production results in a byproduct that can displace yX acres of soybean production. Planting this yX acres with corn for ethanol and using the distillery byproduct for animal feed displace an additional yX^2 acres of soybeans, etc. In all, the total acreage of average soybean production displaced by this marginal acre of corn is:

$$yX + yX^2 + yX^3 \ldots = \frac{yX}{1 - X} \qquad (1)$$

If Nm and Na are the net premium fuels displacement per acre of marginal and average corn production, then the total net premium fuels displacement attributable to bringing 1 marginal acre into corn production is:

$$N = Nm + Na \left(\frac{yX}{1 - X} \right) \qquad (2)$$

The ideal crop switching technique would be where $X = 1$, i.e., where one can simply switch to another crop which produces all of the products of the first crop and liquid fuels as well, without expanding the acreage under intensive cultivation. Several imaginative suggestions for crop switching have achieved this ideal but none are proven.[12] The closest to this ideal that has been demonstrated is the corn-soybean switch, in which X = 0.77, based on national average yields of the respective crops.*[13] Nevertheless, even this switch is limited by the quantity of land suitable for corn production and the fact that the corn distillery byproduct is not a perfect substitute for soybeans in all of its uses. As a fuel ethanol industry is first developing, however, these limitations are probably of minimal importance.

Assuming, then, that the distillery byproduct is fully utilized and that marginal cropland produces 75 percent of the yield of average cropland, OTA has calculated the net premium fuels displacement per acre of marginal (new) cropland brought into production for various liquid fuels options. These include ethanol from various grains and sugarcane and both methanol and ethanol from grass. The energy inputs were assumed to be national average energy inputs for the various grains and sugarcane and 1 million Btu/dry ton for grass** and the alcohols are assumed to be used as octane-boosting additives to gasoline. The results are shown in figure 12. Although the exact numbers cannot be taken too literally because of the various assumptions required to derive them, the relative values are fairly insensitive to the assumptions chosen, provided the alcohols are used as octane-boosting additives.* Also, utilization of crop residues does not substantially change the results.

Among the grain and sugar crops, corn appears to be the best choice, as long as the dis-

tillery byproduct is fully utilized to displace soybeans. In this calculation, 2.5 acres of average soybean land plus 1 acre of marginal land, all grown in corn for ethanol production, can produce an equivalent amount of animal feed protein concentrate as 2.5 average acres grown with soybeans, and provide the ethanol as well. However, as the utilization (i.e., X in equation 2) drops, then grass quickly becomes a superior alternative. If, for example, 1 lb of distillery byproduct displaces 0.5 lb of soybean meal instead of the maximum of 0.67 lb (see ch. 8), then grass and corn would be roughly equivalent. Similarly if grass yields increase to 8 dry ton/acre-yr, then grass would be as good or better than corn regardless of the byproduct utilization. (It should be noted, however, that these calculations do not take the economics of producing ethanol from grass or the difficulties of using methanol as an octane-boosting additive into consideration.)

Sugarcane appears to be roughly equivalent to grass, but sugarcane can be grown on only a limited amount of U.S. cropland and the ethanol produced from it would be considerably more expensive than corn-derived ethanol. Other sugar crops have considerably lower yields than sugarcane.

The exact point where the byproduct utilization will drop is unknown. Some analyses have put it at 2 billion to 3 billion gal/yr of ethanol when distillers' grain is the distillery byproduct.[14] Producing corn gluten meal could, however, increase this to as much as 7 billion gal/yr, based on the total domestic use of soybean meal for animal feed.[15] As mentioned above, however, the byproduct is not equivalent to the soybean products, so it is unlikely that one can reach this level with full byproduct utilization to displace other crops. For the purposes of these estimates, it is assumed that 2 billion to 4 billion gal/yr of ethanol from corn

---

[12]R. Carlson, B. Commoner, D. Freedman, and R. Scott, "Interim Report on Possible Energy Production Alternatives in Crop-Livestock Agriculture," Center for the Biology of Natural Systems, Washington University, St. Louis, Mo., Jan. 4, 1979.

*The byproduct of 1 bu of corn can displace the meal from about 0.25 bu of soybean. See "Byproducts" under "Fermentation."

[13]*Improving Soils With Organic Wastes,* op. cit.

**One-half that derived above for increased grass production, because here it is assumed that the entire grass production goes to energy.

*If the alcohols are used as standalone fuels, the relative values are similar, but the net displacement is about half that shown in figure 12.

---

[14]R. C. Meekhof, W. E. Tyner, and F. D. Holland, "Agricultural Policy and Gasohol," Purdue University, May 1979. This reference reports a 3-billion-gal limit based on a 2-lb substitution of distillers' grain for 1 lb of soybean meal. Other studies, however, have put the feed ratio at 1.5:1, which would reduce the limit to 2.25 billion gal/yr.

[15]*Improving Soils With Organic Wastes,* op. cit.

Figure 12.—Net Displacement of Premium Fuel (oil and natural gas) per Acre of New Cropland Brought Into Production

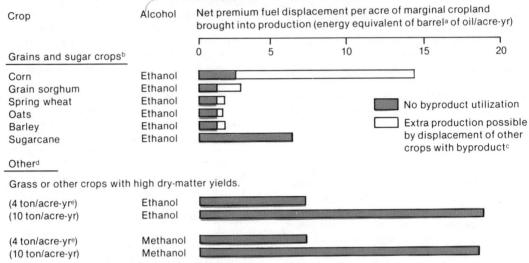

Crop   Alcohol   Net premium fuel displacement per acre of marginal cropland brought into production (energy equivalent of barrel[a] of oil/acre-yr)

Grains and sugar crops[b]

Corn	Ethanol
Grain sorghum	Ethanol
Spring wheat	Ethanol
Oats	Ethanol
Barley	Ethanol
Sugarcane	Ethanol

No byproduct utilization

Extra production possible by displacement of other crops with byproduct[c]

Other[d]

Grass or other crops with high dry-matter yields.

(4 ton/acre-yr[e])	Ethanol
(10 ton/acre-yr)	Ethanol
(4 ton/acre-yr[e])	Methanol
(10 ton/acre-yr)	Methanol

[a]Based on 5.9 million Btu/bbl. alcohol used as octane-boosting additive to gasoline.
[b]Assumes national average energy inputs per acre cultivated and yields (on the marginal cropland) of 75% of the national average yields between 1974-77. Yields on average cropland are assumed to be the average of 1974-77 national averages. This methodology is internally consistent; raising the average cropland yield to 1979 yields would not significantly change the relative results. If usable crop residues are converted to ethanol, the lower value (no distillery byproduct utilization) would be increased by about 1.2 bbl/acre-yr or less for the grains and 2.6 bbl/acre-yr or less for sugarcane.
[c]Economic and physical opportunities for full byproduct utilization diminish with greater quantities of byproduct production.
[d]Uncertainty of ±30% for methanol and more for ethanol from grass, since the ethanol processes are not well defined at present. Assumes 1 million Btu/dry ton of grass needed for cultivation, harvest, and transport of the grass, and conversion process yields (after all process steam requirements are satisfied with waste heat or part of the feedstock) of 84 gal/dry ton of grass for ethanol and 100 gal/dry ton of grass of methanol.
[e]Four ton/acre-yr can be achieved with current grass varieties grown on marginal cropland.

SOURCE: Office of Technology Assessment; yields from USDA. *Agricultural Statistics*, 1978.

Data Used in Figure 12

				Net premium fuels displacement[a] (bbl of oil equivalent/acre)			
		National average farming energy (10³ Btu/gal of ethanol)		Average land		Land that is 75 percent as productive as average land	
Crop	Average 1974-77 national average yields (gal of ethanol/acre)	Average land	Land that is 75 percent as productive as average	With byproduct credit	Without byproduct credit	With byproduct credit	Without byproduct credit
Corn	220	33.3	44.4	4.4	4.0	3.0	2.7
Grain sorghum.....	130	54.5	72.7	2.1	1.9	1.3	1.1
Spring wheat	73	23.8	31.7	1.6	1.5	1.1	1.0
Oats	74	24.2	32.3	1.6	1.5	1.1	1.0
Barley	79	29.4	39.2	4.6	1.5	1.1	1.0
Sugarcane	504	30.3	40.4	NA	9.7	NA	6.4
Grass	400[b]	NA	10	NA	NA	NA	7.3

NA = none available.
[a]Assumes gross displacement of 140,000 Btu/gal of ethanol, byproduct credit of 10,500 Btu/gal, and 5.9 million Btu/bbl of oil. For methanol, 117,000 Btu/gal gross displacement.
[b]Assumes 4 ton/acre on marginal land and 100 gal methanol per ton.

Crop	Displacement of soybean production[c] (average acres of soybeans displaced per average acre of grain = x)	$\frac{0.75x}{1-x}$ (total acres of soybeans displaced by 1 marginal acre of grain and additional cultivation of grain on former soybean land)	Net premium fuels displacement from 1 acre of marginal land plus $\frac{0.75x}{1-x}$ acres displaced soybean land (bbl oil equivalent/acre of marginal land)
Corn..........	0.77	2.5	13.9
Grain sorghum .	0.46	0.64	2.7
Spring wheat...	0.26	0.26	1.5
Oats..........	0.26	0.26	1.5
Barley	0.28	0.29	1.6
Sugarcane.....	0	0	6.4
Grass.........	0	0	7.3

[c]Assumes average soybean yield of 27.1 bu/acre, a displacement of 1 lb of soybean meal per 1.5 lb of distillers' grain, and 48 lb of soybean meal per bushel of soybeans.

SOURCE: S. Barber, et al., "The Potential of Producing Energy From Agriculture," Purdue University, contractor report to OTA; and *Agricultural Statistics, 1978* (Washington, D.C.: U.S. Department of Agriculture, 1978).

(about 0.2 to 0.4 Quad/yr) can be produced while utilizing the byproducts fully. This would require about 2 million to 5 million additional acres in intensive crop production and expansion of corn production by over three times this acreage. It is not certain that cropland will be available for energy production by 2000; but if it is, it is assumed that any further production above this level will use grass as the energy crop.

In the near to mid-term, increased production of forage grass can be obtained on about 100 million acres of hayland, cropland pasture, and noncropland pasture. Assuming a 1- to 2-ton/acre-yr increase in yields, this would result in 100 million to 200 million tons of grass or about 1.3 to 2.7 Quads/yr. Deducting the energy needed to cultivate and transport the grass reduces the output to about 1.1 to 2.2 Quads/yr.

By 2000 anywhere from zero to 65 million acres could be used for energy crops. Assuming that grasses with average yields of 6 dry ton/acre-yr on this cropland have been developed, then the energy potential would be zero to 5 Quads/yr.

Although adding this to the ethanol yield from corn involves a small amount of double counting, the uncertainty in the actual cropland availability and future grass yields is too great to warrant a detailed separation. Consequently, OTA estimates that 1 to 3 Quads/yr can be obtained in the near to mid-term and zero to 5 Quads/yr in the long term from the production of conventional crops for energy.

The above mix of corn and grass was chosen as the one that appears to be the least infla-tionary to food prices in the long term per unit of liquid fuel produced. However, if 65 million acres are available for energy production in 2000, one could conceivably produce over 15 billion gal of ethanol from corn* or about 1.3 Quads/yr of liquid fuel. Grass production, on the other hand, would yield about 2.5 Quads/yr** of liquid fuel from this same cropland and with the same or lower inflationary impact.

Judging when the emphasis should shift from corn to grass is likely to be difficult. As a fallible rule of thumb, however, any significant increase in corn prices relative to the other grains would be an economic signal to distillers and/or animal feeders to use grains other than corn, which would make grass a superior option for energy production. Similarly a significant drop in the price of distillery byproduct, relative to the alternatives, would be an economic signal that the distillery byproducts are not being fully utilized and, again, grass would be superior. Consequently, if there is a significant rise in corn prices or drop in distillery byproduct prices, relative to the alternatives, then these could be indications that the cropland could better be utilized by producing grass.

---

*Seven billion gal with complete substitution of soybean meal and requiring about 10 million of the 65 million acres. The remaining 55 million acres, with yields of 65 bu/acre, could produce an additional 9 billion gal/yr of ethanol.
**5 Quads/yr of grass could yield slightly less than 2.5 Quads/yr of methanol.

Energy Potential From Crop Residues

Crop residues are the plant material left in the field after a crop harvest. Their major function is to protect the soil against wind and water erosion by providing a protective cover, and they have a modest fertilizer value[16] and a soil-conditioning value through maintenance of soil organic matter. (See also "Environmental Impacts.")

Barber, et al., have calculated the total quantities of residues by multiplying the crop yields reported by USDA by residue factors, i.e., the ratio of residue to the yield of tradi-

---

[16]Residues are about 0.7 percent nitrogen, 0.2 percent phosphorus, and 4 percent potash. See Barber, et al., op. cit.

tional crop for the various types of crops.[17] The results of these calculations are shown in table 26. The total quantity of residues generated is about 400 million ton/yr or about 5 Quads/yr.

Table 26.—Total Crop Residues in the United States for 10 Major Crops (based on 1975-77 average production)

	Acres k acres	Total residue k tons
Corn.	69,530	171,084
Wheat	68,789	99,890
Soybeans	53,616	67,556
Sorghum	14,714	21,123
Oats	12,831	20,677
Barley	8,772	13,341
Rice	2,515	8,584
Cotton	10,990	3,578
Sugarcane	660	4,700
Rye	715	708
U.S. Total	243,132	411,240

SOURCE: Barber, et al., "The Potential of Producing Energy From Agriculture," Purdue University, contractor report to OTA, May 1979.

During fall plowing many farmers turn under the residues, rendering them useless as a protection against erosion. These residues could be collected and used for energy without worsening the erosion on these lands. However, current agriculture policy is to encourage farming practices that limit soil erosion to the soil-loss tolerance levels, or the levels of erosion that are believed not to impair the long-term productivity of the land (see "Environmental Impacts"). Consequently, a more detailed consideration of crop residues is appropriate.

Using data supplied by Dr. W. Larson,[18] the total crop residues were calculated for each of the major land resource areas or subregions of States. Using standard equations for soil erosion,* the quantities of residues that could be removed without exceeding standard soil-loss tolerance values were calculated. These were then modified to take into consideration the quantities that can be physically collected with current harvesting equipment (about 60 percent in field trials at Purdue University). In

[17]Ibid.
[18]W. E. Larson, "Plant Residues—How Can They Be Used Best," paper No. 10585, Science Journal Series, SEA-AR/USDA, 1979.
*Universal soil-loss equation and wind erosion equation.

addition, a 15-percent storage loss was assumed. The results of these calculations are shown in table 27 as the usable crop residues, which are about 20 percent of the total crop residues.

Table 27.—Total Usable Crop Residue by Crop

Crops	Amount (k tons)	Harvestable acres (k acres)	Average yield (ton/acre)
Corn	37,098	39,122	0.95
Small grains	33,623	36,324	0.93
Sorghum	1,452	4,100	0.35
Rice	5,457	2,516	2.17
Sugarcane	590	331	1.78
Total	78,220	82,393	0.95

SOURCE: Barber, et al., "The Potential of Producing Energy From Agriculture," Purdue University, contractor report to OTA, May 1979.

Harvesting crop residues would typically consist of moving the residues into windrows, or long thin piles of residues. The windrows would then be collected with baling machinery and the bales dumped at the roadside. The windrows would be collected and transported to a place where they would be stored or used.

Crop residues typically contain 40- to 60-percent moisture 2 days after the grain harvests. In favorable weather conditions, the residues dry to about 20-percent moisture in 18 days.[19] With these moisture contents, bacteria will gradually consume the residues. If the residue bales are compacted too tightly, the heat generated from the bacterial action can cause the material to spontaneously combust. However, with relatively loose bales, the bacterial heating will dry the material to a moisture content at which the bacteria do not consume the material. Some loss, however, is inevitable (15-percent loss has been assumed in table 27).

The extra fertilizers necessary to replace the nutrients in the residues removed cost about $7.70/ton of residue removed.

In addition, one of the main problems with harvesting residues is that it delays the fall ground preparation. If winter rains come too soon, there may not be sufficient time to collect the residues and prepare the ground for the spring planting. The spring planting is then

[19]Barber, et al., op. cit.

delayed and yields for the following year may suffer. Using computer simulations of farming operations and the actual weather conditions in central Indiana between 1968 and 1974, it was found that residue harvests reduced corn yields by an average of 1.6 bu/acre-yr.[20] If this cost is attributed to the residues, then it raises the residue costs by $2.70/ton. This factor is less of a problem with most other grains, however, since they are less sensitive to the exact planting time.

Adding these various costs and assuming a markup of 20 percent above costs gives the State average costs for various residues (table 28). Care should be exercised when using this table, however, since the costs within a State can vary considerably. In favorable cases the

[20]Ibid.

Table 28.—State Average Estimated Usable Group Residue Quantities and Costs

State	Total usable crop residues (million tons/yr)	Delivered cost[a] ($/ton) (estimated uncertainty: 20%))
Corn		
Illionois	8.0	32.16
Indiana	4.6	32.42
Iowa	6.9	32.77
Minnesota	4.2	38.67
Nebraska.	1.8	41.68
Ohio	2.6	35.18
Small grains		
California	1.8	28.29
Illinois	1.0	31.53
Minnesota	6.1	30.54
South Dakota	1.8	33.05
Washington	3.0	31.01
Wisconsin	2.0	26.93
Sorghum grain		
Colorado	0.12	35.60
Kansas	0.72	57.62
Missouri	0.28	36.87
Rice		
Arkansas.	1.9	36.32
California.	1.1	34.82
Texas	1.2	36.08
Sugarcane		
Florida.	0.53	30.93

[a]Including 15-mile transport, labor at $5/hr, $0.80/gal diesel fuel, yield penalty of $2.70/ton of residues for corn, additional fertilizers for $7.70/ton of residues, and profit of 20 percent of costs.

SOURCE: Barber, et al., ''The Potential of Producing Energy From Agriculture,'' Purdue University, contractor report to OTA, May 1979.

costs might be as low as $20/dry ton and, in unfavorable cases, as high as $60/dry ton or more.

Crop residues contain about 13 million Btu/ton. The energy costs for harvesting and transporting the residues are about 0.9 million Btu/ton for a 15-mile transport and 1.8 million Btu/ton for a 50-mile transport. (With integrated residue and crop harvests the energy costs would be slightly less, but this may not be a practical alternative because it delays the harvest.) In addition, the energy content of the additional fertilizer needed to replace the nutrients lost in the residues is about 0.6 million Btu/ton. Thus, the total energy use associated with collecting and transporting the residues is about 1.5 million to 2.5 million Btu/ton of residue.

National average crop yields can fluctuate by ± 5 percent or more from year to year and the usable crop residues will fluctuate by about ± 10 percent, since an absolute quantity of residue should be left regardless of the crop yield. On a local basis, usable crop residues can vary considerably. Within a county located in a humid region of the country, the fluctuation may be ± 20 percent and for crops that are not irrigated in dry regions, the year-to-year variations can reach ± 100 percent. The areas with the largest fluctuations, however, also have the lowest quantities of usable residues.

In summary, the total crop residue production in the United States is about 5 Quads/yr, of which about 3 Quads/yr can be collected with current harvesting equipment. The quantity that can be collected while maintaining current soil erosion standards is about 1 Quad/yr. Considerations of a reliable supply, however, would reduce this to roughly 0.7 Quad/yr* of reliable feedstock, if soil erosion standards are strictly adhered to. By 2000, increases in crop production could raise this by 20 percent to 0.8 to 1.2 Quads/yr.

*Calculated by assuming that the total quantity of residue can fluctuate by ± 20 percent at the local level; i.e., by subtracting 20 percent of the total residues from the usable residues on a State-by-State basis.

Environmental Impacts of Agricultural Biomass Production

Introduction

American agriculture, with its astonishing productivity and reliability, bestows critically important benefits on the economy and general well being of the United States. Unfortunately, it also has serious negative environmental impacts. Any substantial increase in land cultivation or intensification of present crop production to produce energy crops—biomass—will cause an extension and intensification of many of the impacts of the present system.

There are substantial uncertainties in the understanding of the consequences of relying on agricultural feedstocks for energy production. These uncertainties stem from a lack of complete understanding of present impacts, the potential for changes in crop production methods in the future, and uncertainty as to the pace of development. This section attempts to place the potential impacts of large-scale biomass production from agriculture into perspective by briefly describing what is known of the impacts of food crop production (the energy feedstock production system should resemble the food production system), describing how the pace of development may intensify impacts, and finally identifying those differences between food and energy feedstock production that are most critical in determining impacts.

The Environmental Impacts of American Agriculture

Agriculture is a major source of pollution and causes serious environmental impacts. Table 29 lists the major environmental impacts associated with present forms of large-scale mechanized agricultural production. Most of the impacts apply to the majority of farming situations (although with varying magnitude), but some impacts are negligible or nonexistent in certain situations. Also, most of the impacts are more or less controllable, but for a variety of reasons (a high perceived cost or negative

Table 29.—Environmental Impacts of Agriculture

Water
- Water use (irrigated only) that can conflict with other uses or cause ground water mining.
- Leaching of salts and nutrients into surface and ground waters, (and runoff into surface waters) which can cause pollution of drinking water supplies for animals and humans, excessive algae growth in streams and ponds, damage to aquatic habitats, and odors.
- Flow of sediments into surface waters, causing increased turbidity, obstruction of streams, filling of reservoirs, destruction of aquatic habitat, increase of flood potential.
- Flow of pesticides into surface and ground waters, potential buildup in food chain causing both aquatic and terrestial effects such as thinning of egg shells of birds.
- Thermal pollution of streams caused by land clearing on stream banks, loss of shade, and thus greater solar heating.

Air
- Dust from decreased cover on land, operation of heavy farm machinery.
- Pesticides from aerial spraying or as a component of dust.
- Changed pollen count, human health effects.
- Exhaust emissions from farm machinery.

Land
- Erosion and loss of topsoil from decreased cover, plowing, increased water flow because of lower retention; degrading of productivity.
- Displacement of alternative land uses—wilderness, wildlife, esthetics, etc.
- Change in water retention capabilities of land, increased flooding potential.
- Buildup of pesticide residues in soil, potential damage to soil microbial populations.
- Increase in soil salinity (especially from irrigated agriculture), degrading of soil productivity.
- Depletion of nutrients and organic matter from soil.

Other
- Promotion of plant diseases by monoculture cropping practices.
- Occupational health and safety problems associated with operation of heavy machinery, close contact with pesticide residues and involvement in spraying operations.

SOURCE: Office of Technology Assessment.

effect on crop yields are almost certainly the most important) many control techniques are rarely used.

Water pollution and land degradation due to erosion are American agriculture's primary problems, and the two impacts are intimately linked. The action of wind and water strips farmland of its productive topsoil cover, and much of this soil ends up in the Nation's waterways. Thus, estimates of soil erosion are critical to understanding the effects of agriculture on both soil productivity and on water ecosystems.

SCS has recently revised downward its estimates of cropland erosion. Its *1977 National Erosion Inventory* estimates average annual sheet and rill erosion from all cropland to be 4.77 ton/acre-yr (or a total of about 2 billion ton/yr).[21] Previously, it had estimated cropland sheet and rill erosion at about 9 ton/acre-yr,[22] and other sources had estimated total erosion (including wind erosion) from croplands to be as high as 12 ton/acre-yr.[23] SCS attributes the decrease to the greatly improved data base recently made available by the *1977 Inventory*. Also, the original 9-ton/acre-yr estimate apparently referred only to land in row crops, close-grown crops, and summer fallow, whereas the more recent estimate includes lands that are in less intensive (and less erosive) uses such as rotation hay and pasture, or native hay.

Data from the *1977 Inventory* has only recently begun to be released to the general public, and it seems likely to generate controversy — especially because its estimate of average erosion is under the 5 ton/acre-yr that SCS generally considers to be a tolerable level (i.e., a level that will not affect long-term productivity) for much U.S. cropland. However, the lower estimate is not especially comforting for a number of reasons:

- National (sheet and rill) erosion rates for cropland in intensive production are estimated by SCS to be 6.26 ton/acre-yr.
- The national estimate tends to hide several important food-producing areas with uncomfortably high erosion rates (e.g., Missouri averages 11.38 ton/acre-yr; Iowa averages 9.91 ton/acre-yr).
- The estimates do not include wind erosion and alternative forms of water erosion. Cropland wind erosion in 10 western States averages 5.29 ton/acre-yr. Thus, although Texas cropland has a sheet and rill erosion rate of only 3.47 ton/acre-yr, its total erosion rate is greater than 18 ton/acre-yr because of wind erosion.

- Although SCS generally considers 5 ton/acre-yr as an (average) annual erosion at which long-term productivity on good soils will not suffer, it is not certain that soil is actually replaced this fast. Authoritative estimates of soil replacement rates do not exist, but average rates of as low as 1.5 ton/acre-yr have been claimed.[24] However, the SCS rates do represent the general consensus of the agronomy community.
- Even the new lower erosion rate implies that about a billion or more tons of sediment from croplands are entering the Nation's waterways each year.[25]
- Erosion rates from croplands are many times higher than those of natural ecosystems. Forests typically erode at a rate of less than one-tenth of a ton/acre-yr, and grassland at less than half a ton/acre/yr.[26]

As a result of the mismatch between erosion and soil replacement, the United States has lost a considerable portion of its topsoil and, some have claimed, its production potential. Pimentel estimates that U.S. cropland has lost about one-third of its topsoil and 10 to 15 percent of its production potential over the last 200 years.[27] Bennett estimates that, during the period prior to 1935, 100 million acres of cropland were lost to erosion and an additional 100 million acres were stripped of more than half of their topsoil.[28] At best, however, these values represent extremely rough estimates, and the new SCS erosion inventory may cause their downward revision.

It appears likely that the process of land degradation will continue for the immediate future. Although USDA has spent nearly $15 billion in its soil conservation programs since their inception in 1935,[29] only 36 percent of the

[21]*1977 SCS National Erosion Inventory Estimate*, op. cit.
[22]Ibid.
[23]D. Pimentel, et al., "Land Degradation: Effects on Food and Energy Resources," *Science*, vol. 194, Oct. 8, 1976.

[24]Ibid.
[25]*Environmental Implications of Trends in Agriculture and Silviculture — Volume I: Trend Identification and Evolution* (Washington, D.C.: Environmental Protection Agency, October 1977), EPA-600/3-77-121.
[26]Ibid.
[27]D. Pimentel, op. cit.
[28]H. H. Bennett, *Soil Conservation* (New York: McGraw-Hill, 1939).
[29]*To Protect Tomorrow's Food Supply, Soil Conservation Needs Priority Attention* (Washington, D.C.: General Accounting Office, Feb. 4, 1977), CED-77-30.

472 million acres of cropland in 1967 were judged to have adequate conservation treatment[30] and the programs have been criticized as inadequate by the General Accounting Office.[31]

A reason for the inability of USDA conservation programs to satisfy their critics may be the difficulty of demonstrating to the farmer (in all but the more severe cases) the benefits of additional conservation measures. Because an inch of topsoil weighs about 150 ton/acre, a net loss of 5 ton/acre-yr would result in a loss of 1 inch of soil every 30 years. During that time, farming procedures would be gradually changing, obscuring the effects of any soil loss. For example, during the past 30 years, more intensive use of fertilizers, pesticides, and other inputs, better information on future weather and other critical factors, and improved crop varieties more than compensated for erosion-caused losses on most lands. Also, the actual effect on productivity may not be large in some circumstances because the effect of soil loss is very sensitive to soil conditions: while loss of soil from a very shallow soil over rock in Kentucky may cause the land to be withdrawn from production, on some deep loess soils of Iowa, the loss of several inches of topsoil may have little effect on productivity. Few if any agricultural scientists would argue that net soil loss can continue indefinitely without major losses in productivity. However, on many lands the damages of erosion may never become visible to the farmer; rather they will be perceived by his children or grandchildren. Moreover, short-term economic constraints may compel a farmer to discount the future benefits of conservation by much more than he would personally prefer.

Aside from the long-term consequences in land degradation, soil erosion represents a severe water pollution problem. Not only is soil itself a serious pollutant, it also acts as a carrier of other pollutants: phosphorus, pesticides, heavy metals, and bacteria.[32] The soil lost to agricultural erosion represents more than half of the sediment entering the Nation's surface waters.[33] [34] Sediment causes turbidity, fills reservoirs and lakes, obstructs irrigation canals, and destroys aquatic habitats. Yearly material damages have been estimated at over $360 million,[35] not including damage to aquatic habitats and other noneconomic costs. Adding the flooding damage caused by the decrease in storage capacity of reservoirs and streams would increase annual costs to over $1 billion.[36]

The effects on aquatic ecosystems of the enormous flow of sediments into the Nation's waterways have never been satisfactorily estimated. Research on the impacts of "nonpoint" sources of water pollution—agriculture, construction, etc.—has not been given a high priority within the Environmental Protection Agency (EPA) or USDA, and the result is a scarcity of information from which to draw conclusions about either present impacts or future impacts associated with the devotion of millions of additional cropland acres to biomass production.

The other major water pollution problems of agriculture involve the toxic effects of pesticides and inorganic salts and the nutrient influx into the Nation's waterways associated with American agriculture's increasing use of fertilizers.

Pesticide use in American agriculture has grown from 466 million lb in 1971[37] to 900 million lb in 1977.[38] By 1985, American farmers are expected to be using as much as 1.5 billion lb.[39] Much of this increase can be traced to the growth in minimum tillage practices[40] which substitute increased herbicide use for tillage to control weeds. These practices include leaving crop residues on the soil surface, and these residues harbor plant pests and pathogens and generally increase pesticide requirements (al-

[30]"Potential Cropland Study," Statistical Bulletin No. 578, Soil Conservation Service, U.S. Department of Agriculture, 1977.
[31]To Protect Tomorrow's Food Supply, op. cit.
[32]Environmental Implications of Trends, op. cit.

[33]Ibid.
[34]Pimentel, op. cit.
[35]1977 SCS National Erosion Inventory Estimate, op. cit.
[36]Ibid.
[37]Environmental Implications of Trends, op. cit.
[38]1977 SCS National Erosion Inventory Estimate, op. cit.
[39]Ibid.
[40]Environmental Implications of Trends, op. cit.

though they offer substantial benefits in erosion control). Recent growth in the practice of single- and double-cropping may also account for some of the increase. Although less than 5 percent of the pesticides enter the surface and ground water systems,[41] pesticide use has been associated with fish kills and other damage to aquatic systems as well as reproductive failures in birds and acute sickness and death in animals. Under conditions of high exposure—in accidental spills, improper handling by applicators, etc.—pesticides have been associated with the sickness and death of humans. Recent research has implicated some widely used pesticides as possible carcinogenic agents when ingested or inhaled, and EPA has removed certain of these—including Aldrin, Dieldrin, and Mirex—from the marketplace under the Federal Insecticide, Fungicide, and Rodenticide Act (FIFRA). Amendments to FIFRA have considerably tightened the requirements for testing and registering pesticides. However, the tremendous variety of pesticide compounds ("1,800 biologically active compounds sold domestically in over 32,000 different formulations"[42]) and the difficulty of detecting damages in human populations and in the environment will greatly complicate successful enforcement of the Act. At present, the long-term impacts of pesticides on the environment and on man are poorly understood.

The problems of pesticide use in agriculture are becoming particularly visible because of a recent rash of instances where pesticides thought to be safe have been accused of causing severe injuries—including birth defects, miscarriages, and other acute physical disorders—and death to exposed populations. The controversy surrounding the use of the herbicide 2, 4, 5-T in Oregon and its suspension by EPA is a widely publicized—but by no means unique—example of rising national concern. Resolution of the conflicting claims about the safety (or lack of it) of these pesticides is well beyond the scope of this report. Based on current interest, however, it is likely that a major public concern associated with any large in-

crease in crop cultivation will be the concurrent increase in pesticide use on the new lands. There is a distinct possibility that rising public concern over pesticide usage could put a severe constraint both on the continuing increase in this usage and on the expansion of crop production for energy feedstocks.

Salinity increases caused by irrigated agriculture present another substantial impact. Irrigated land produces one-fourth of the total value of U.S. crops, mostly in the 17 western States.[43] Increased salinity in streams in these areas is caused by the salts added to irrigation water from upstream farms and by the concentrating effect of the high evaporation rates in arid climates (evaporation leaves the salts behind). The same mechanisms can lead to increasing salt concentrations in the soils of downstream farms unless sufficient water can be obtained to periodically flush excess salts out of the soil profile. Damages associated with increased salinity of soils and irrigation water include reduced crop yields, inability to grow salt-sensitive crops, increased industrial treatment costs, and adverse effects on wildlife, domestic animals, and aquatic ecosystems. Trends in irrigated agriculture are leading to improvements in irrigation efficiency and decreased salt loadings in streams, but these trends could be overwhelmed by substantial increases in irrigated acreage either to grow crops for energy or to compensate for competition between food and biomass production in other areas.

Fertilizer use is of extreme importance in calculating the environmental impacts of agriculture. Large amounts of energy—one-third of the energy consumed by the agricultural sector and its suppliers—are needed to produce fertilizer. The Haber-Bosche process for the synthesis of anhydrous ammonia fertilizer requires around 21 ft³ of natural gas to produce 1 lb of nitrogen in fertilizer (and more for other forms of nitrogen);[44] current U.S. nitrogen fertilizer production is 10 million metric

[41]Ibid.
[42]*1977 SCS National Erosion Inventory Estimate,* op. cit.

[43]Ibid.
[44]C. H. Davis and G. M. Blouin, "Energy Consumption in the U.S. Chemical Fertilizer System From the Ground to the Ground," *Agriculture and Energy,* W. Lockeretz, ed. (New York: Academic Press, 1977), pp. 315-371.

tonnes per year consuming 3 percent of total U.S. natural gas production. If current trends of increased rates of fertilizer applications continue and food demands increase by 3 percent per year, natural gas requirements for fertilizer production will triple by 2000.[45]

The application of large quantities of chemical fertilizers also represents a water pollution problem because much of the nutrient value ends up in the Nation's waterways. Wittwer estimates that only 50 percent of the nitrogen and less than 35 percent of the phosphorus and potassium applied as fertilizer are actually recovered by crops;[46] other estimates for nitrogen range from 46 to 85 percent.[47] Although a portion of that which is lost is due to volatilization (and consequent loss to the atmosphere), much is lost to surface and ground waters via runoff, leaching, and erosion processes. The amount of nitrogen and phosphorus (potassium is not considered to have significant environmental impacts[48]) entering the waterways from agricultural lands in the early 1970's has been estimated at 1,500 million to 15,000 million lb/yr and 120 million to 1,200 million lb/yr, respectively.[49]

This nutrient pollution from fertilizers may be toxic to humans and wildlife in high concentrations; nitrate poisoning of wells from contaminated ground water is not unusual in some agricultural areas. The more common impact, however, is to speed up eutrophication of streams and the problems of oxygen demand and algae growth associated with eutrophication.

The remaining major water-associated impact of agriculture is water use. The appropriation water rights system in the West offers little incentive to use water efficiently.* The

combination of artificially low prices for water and the requirement of the appropriation doctrine that the holder of a water right must maintain that right through use ("use or lose") has led to the cultivation of water-intensive crops in arid climates. This has led to water shortages in many Western basins and to aggravation of salinity problems in several major rivers.

Several water use trends will affect agricultural production capabilities in the near future. First, large-scale energy development— especially electrical generating stations and, possibly, synthetic fuel plants—will consume substantial quantities of water and, in some cases, compete directly with agricultural interests for the limited supply. Second, expanded acreage for food production will occur, including projects on Indian land that may have priority rights to the limited water supply. On the other hand, improvements in irrigation efficiency will have some conserving effect on water consumption even though this is not a primary goal of efficiency increase (the primary goal is to reduce water withdrawals and return flows and to improve water quality rather than to reduce consumptive use). For example, SCS estimates that irrigated acreage in the critical Upper Colorado River Basin could increase from 1,370,000 acres in 1975 to 1,442,000 acres in 2000 while water consumption declines by 93,000 acre-ft with a concerted program to improve irrigation practices.[50] Further decreases in water consumption are possible by "crop switching"—shifting to less water-intensive crops where markets are available—and removing marginal, low-productivity land from cultivation.[51] Also, substantial potential for water conservation exists in energy production.

Much of the agricultural land in the United States was obtained by forest clearing or plowing native grasslands, and the consequent replacement of natural ecosystems with intensively managed monocultures must be consid-

[45]S. H. Wittwer, "The Shape of Things to Come," Biology of Crop Productivity, P. Carlson, ed. (New York: Academic Press, 1978).

[46]Ibid.

[47]Environmental Implications of Trends, op. cit.

[48]Ibid.

[49]Ibid.

*For an excellent review of Western water law see E. Radosevitch, "Interface of Water Quantity and Quality Laws in the West, in Proceedings of the National Conference Irrigation Return Flow Quantity Management, J. P. Law and G. V. Skogerboe, eds. (Fort Collins, Colo.: Colorado State University, 1977).

[50]Conservation Needs Inventory (Washington, D.C.: Soil Conservation Service, U.S. Department of Agriculture, 1976).

[51]S. E. Plotkin, H. Gold, and I. L. White, "Water and Energy in the Western Coal Lands," Water Resources Bulletin, vol. 15, No. 1, February 1979.

ered a major environmental impact of agricultural production. (This process is not a one-way street. A combination of changing crop patterns, alternative producing areas, increasing average productivity, and, especially in the South, depletion of soils has led during this century to the abandonment of considerable farmland acreage and, in many cases, reversion to second-growth forest. Principle areas involved in this transformation include the Piedmont areas of the Southeast, the hillier areas of the Northeast, and the upper lake States. However, farm abandonment no longer appears to be a significant force.[52]) Aside from the loss of esthetic and recreational values, this replacement represents a substantial decline in wildlife diversity, loss of watershed protection, and the loss of the alternative wood (or other) resource. At present, this loss involves a bit over 400 million acres of designated cropland[53] and will probably increase unless crop production efficiency can keep pace with the rising demand for food. Also, because millions of acres of cropland are lost each year to roadbuilding and urban development, merely the maintenance of the status quo demands continued clearing of unmanaged and lightly managed lands for crop production.

1973-74: A Case Study in Increased Cropland Use[54]

In 1973, USDA told American farmers that they would be free to plant as many acres of wheat, corn, and feed grains as they wished during the 1973-74 season. In response, 8.9 million additional acres were planted and harvested during that season:

- 3.6 million acres from grassland,
- 0.4 million acres from woodland, and
- 4.9 million acres from idle cropland and set-aside land.

The results of this new agricultural production may provide a basis on which to predict

the potential impact of a surge in production caused by incentives to grow crops for biomass energy production.

Of the 8.9 million acres, SCS estimated that 5.1 million acres had inadequate conservation treatment and water management, and 4 million acres had inadequate erosion control. These problems in land selection and environmental planning were soon translated into severe erosion losses. Although poor weather conditions (fall and winter drought in the southern high plains, spring floods in the northern Great Plains, torrential spring rains followed by drought in the Corn Belt) aggravated these losses, most observers appear willing to place a major blame on the farmers' land selection and inattention to erosion control practices.

Soil losses on the additional acreage during the 1973-74 season averaged over 6 ton/acre over and above expected losses without production. Those lands designated as suffering from inadequate conservation treatment lost an average of more than 12 ton/acre above expected losses. First-year erosion losses are expected to be lighter than subsequent years because the root structures of the original cover crops are not totally destroyed by tilling and provide some protection to the soil until they decompose; thus, erosion rates would be expected to rise still further unless conservation practices were begun.

The hardest hit of the agricultural regions were the Corn Belt (390,000 acres, 15 to 100 ton/acre additional soil loss on the new land), western Great Plains—North Dakota, Montana, Wyoming, eastern Colorado (325,000 acres, 5 to 40 ton/acre), eastern Great Plains—Nebraska, Kansas, South Dakota (260,000 acres, 5 to 55 ton/acre), Great Lakes (195,000 acres, 5 to 55 ton/acre), and the southern Coastal Plains of Florida, Georgia, Louisiana, Alabama, and Mississippi (210,000 acres, 5 to 70 ton/acre). In addition, a number of other producing regions experienced high soil losses on the additional acreage.

High as these soil losses were, however, they are not unusual when compared to losses suf-

---

[52]M. Clawson, "Forests in the Long Sweep of American History," *Science,* vol. 204, June 15, 1979.

[53]*Potential Cropland Study,* op. cit.

[54]Adapted from K. E. Grant, *Erosion 1973-74: The Record and the Challenge.*

fered by land in continuous production. As noted previously, many areas that are critically important to U.S. grain production routinely lose soil at rates well above the 5-ton/acre-yr maximum recommended by SCS. Assuming that much of the converted land was taken out of relatively nonerosive uses (the 4 million acres of grassland and woodland, nearly half the total, would have suffered virtually no erosive losses if left undisturbed), the erosion experienced on the additional acreage was only slightly worse than the average erosion rates on all U.S. cropland. On the other hand, the lands designated as inadequately protected did have much higher erosion than average. The conclusion appears to be that a rapid increase in land under production will not necessarily cause proportionately more erosion than our current experience would lead us to expect, but that conservation planning and treatment will be required to keep erosion rates from escalating beyond current rates.

Potential Impacts of Production of Biomass for Energy Feedstocks

Most proposals for using the agricultural system to produce energy feedstocks do not contemplate growing and harvesting systems that appear to be radically different from current large-scale mechanized food-growing systems found in the Corn Belt and other centers of American agriculture. Proposals centering around gasohol, for example, assume that at least the near-term feedstock (after food wastes and spoiled grains are used up) will be corn and other conventional starch or sugar crops. Even the more radical systems—for example, tree plantations—can be viewed as variations of common agricultural systems.

The key to identifying the impacts of implementing the various approaches to energy feedstock production is to identify those differences from today's systems that are most critical to causing differences in the impacts. These differences in impacts primarily depend on differences in:

- quality and previous use of the land,

- production practices, and
- type of crop grown.

Land Quality and Previous Use

The land available for growing biomass crops consists of cropland that is presently not in intensive use—for instance, land used to grow native hay—and land currently in range, forest, or other use that can be converted to cropland. Table 30 presents SCS estimates of cropland not currently being utilized to its maximum production potential, and land available for conversion to cropland in 1977. (The acreage "not in intensive use" includes land where the current use meshes with the farmers' desired mix of livestock and crops and thus is unlikely to be converted to more intensive use; thus, the table may overestimate the acreage available for switching to biomass production.) Table 31 presents SCS estimates of the rates of erosion on these lands, by land use and capability class.

The data shows that there is a very substantial amount of land available for biomass production that could be cultivated with few environmental problems. For example, table 30 shows well over 3 million acres of the highest quality (class I) land with high and medium conversion potential. Over 10 million acres of high-quality class II (for brief definitions, see table 30) land requiring some drainage correction is available. However, there currently is no guarantee that land for biomass production will be selected for its environmental characteristics. Erosion potential, which is of critical environmental importance, is only one of several characteristics used by farmers to decide whether to put land into production. Characteristics such as contiguity of land, current ownership, and the cost of conversion may be the deciding factors.

According to table 30, farmers currently have biased their choice of land for row crop cultivation somewhat in favor of the less erosive lands. Over 11 percent of land in row crops is prime class I land with both high productivity and minimal erosion. In contrast, other land uses typically have about 4 or 5 per-

Table 30.—1977 Cropland and Potential Cropland Erosion Potential (in thousand acres, % of total acreage)

Class	Present cropland in intensive use		Present cropland not in intensive use (rotation hay and pasture, occasionally improved/native hayland)	Potential cropland	
	Row crops	Close-grown crops		High potential	Medium potential
I. Excellent capability, few restrictions	23,034	4,471	2,389	2,186	1,412
	(11.3)	(3.4)	(4.2)	(5.5)	(1.5)
II. Some limitations, require moderate conservation practices					
Erosive	45,954	23,463	11,718	10,543	13,921
	(22.6)	(22.4)	(20.8)	(26.3)	(14.8)
Other problems	58,657	22,762	9,855	8,278	10,750
	(28.9)	(21.7)	(17.5)	(20.7)	(11.3)
III. Severe limitations, reduced crop choice and/or special conservation practices required					
Erosive	28,054	26,997	12,561	7,893	25,142
	(13.8)	(25.7)	(22.3)	(19.7)	(26.7)
Other	27,676	10,811	6,557	4,797	12,703
	(13.6)	(10.3)	(11.6)	(12.0)	(13.4)
IV. Severe limitations, more restricted than above					
Erosive	9,159	9,324	5,701	1,896	11,531
	(4.5)	(8.9)	(10.1)	(4.7)	(12.3)
Other	5,436	2,933	3,154	1,601	7,210
	(2.7)	(2.8)	(5.6)	(4.0)	(7.6)
V-VIII. Generally not suited	5,728	345	4,479	2,888	12,248
	(2.6)	(0.3)	(7.9)	(7.2)	(13.0)
Total	203,243	104,890	56,414	40,082	94,917
Percent of land that is erosive	40.9	57.0	53.1	50.7	53.4

SOURCE: *1977 Soil Conservation Service National Erosion Inventory Estimate* (Washington, D.C.: Soil Conservation Service, U.S. Department of Agriculture, December 1978).

Table 31.—Mean National Erosion Rates by Capability Class and Subclass (rates are in ton/acre-yr)

Class/subclass	Row crop	Close grown	Nonintensive	Potential	
				High	Medium
Class I. Excellent capability, few restrictions	3.46	1.75	0.66	0.31	0.35
Class II. Some limitations, require moderate conservation practices/erosive	6.51	3.67	0.96	0.67	0.71
Class II/other	3.46	2.55	0.43	0.30	0.31
Class III. Severe limitations, reduced crop choice and/or special conservation practices required/erosive	12.39	6.62	1.51	1.08	1.28
Class III/other	3.41	2.51	0.51	0.21	0.28
Class IV. Severe limitations, more restricted than III/erosive	17.88	12.20	2.93	2.01	2.28
Class IV/other	4.52	1.85	0.45	0.46	0.43
Classes V-VIII. Generally not suited/erosive	46.82	19.61	5.42	2.38	4.15
Classes V-VIII/other	14.26	3.27	0.80	1.51	0.38

SOURCE: *1977 Soil Conservation Service National Erosion Inventory Estimate* (Washington, D.C.: Soil Conservation Service, U.S. Department of Agriculture, December 1978).

cent of their land classified as class I. In land quality classes I through IV, 43 percent of the row-cropped acreage is erosive, whereas over 50 percent of every other land use category is erosive. Because row crop cultivation is generally the most vulnerable to erosion, this bias towards use of less erosive land is not surprising.

Close-grown crop cultivation is considerably less erosive than row cropping. Apparently in response to this, farmers have placed close-grown crops on lands that are more vulnerable to erosion; 60 percent of close-grown cropland acreage is erosion-prone.

It is important to look beyond these overall percentages and examine the percentage of land in each land use capability class. The erosivity of lands categorized as E (erosive) by SCS appears to be a strong function of the capability class. For example, the average 1977 annual sheet and rill erosion rates on erosive croplands in intensive use were estimated to be (from table 31):

Class I .... 3.18 ton/acre-yr
Class IIE................. 5.55
Class IIIE 9.56
Class IVE 15.02
Class V-VIIIE 34.70

Thus, the erosion danger appears to increase markedly as land capability declines. If the erosive portions of the land with future biomass potential (present cropland not in intensive use and land with switching potential) were skewed towards the lower quality classes, then an examination of the overall erosive potentials would underestimate the erosion danger presented by massive shifts to intensive cultivation. An examination of table 30 indicates that the erosive portions of the present cropland not in intensive use and the high-potential land are somewhat skewed towards the lower quality lands when compared with present cropland, but the differences do not appear to be substantial. For example, whereas 53 percent of erodible land in intensive use is class IIIE or below, 60 percent of erodible land with high biomass potential is in this erosivity range.

The surprising implication of the statistics presented in table 30 is that **the land available for agricultural biomass production is not radically different in its erosion qualities from land currently being utilized for intensive agricultural production.** Although clearly some selection has been made in utilizing the best lands and keeping idle the worst, this selection process appears to have been skewed by other physical attributes and economic and social factors that are as important or more important than erosion potential. It appears that erosion prob-

lems will be significant in adding new lands to intensive agricultural production, but it does not appear on a national basis that these problems will be very much worse than those that could be predicted by extrapolating from current erosion rates.

It is possible to estimate quantitatively the general erosion danger from an expansion of intensive production by utilizing the data in tables 30 and 31 and by making the following simplifying assumptions:

• The 1977 erosion rate for land under intensive production, for each land capability subclass, is representative of the erosion that would occur if additional land in that subclass were to be put into intensive production.

• Given a desire to place additional land into intensive production, farmers will select land mainly from cropland not now in intensive production and "high potential" land, and their selection will be random (this is probably a "worst case" assumption but may not be seriously in error judging from the discussion above).

• A mix of row and close-grown biomass crops will be grown, with the mix being about the same as the 1977 food crop mix.

Under these conditions, *the average erosion rate on the new land put into intensive production will be about 7.5 ton/acre-yr.* For comparison, the 1977 erosion rate on intensively cultivated lands was 6.26 ton/acre-yr. In other words, erosion from additional acres devoted to growing biomass crops may be about 20 percent worse than similar acreages of food crops in production today (this assumes no Government action to improve land selection). Given the substantial uncertainties in this estimate, the 20-percent differential is well within the range of possible error. It is, however, consistent with what is known about agricultural land selection and the quality of available (but undeveloped) farmland.

Because land quality is affected by net rather than gross erosion—i.e., by the difference between erosion and soil replacement—the effect on the land of relatively small changes in erosion rates may be greater than would be apparent at first glance. For example, if the average topsoil replacement rate is 5 ton/acre-yr,* the 7.5 ton/acre-yr biomass erosion rate yields a 2.5 ton/acre-yr net soil loss, versus 1.26 ton/acre-yr net loss from food production. Thus, while a large-scale expansion of acreage for biomass production may have effects on waterways that are similar in magnitude to the effects of present intensive agriculture, this acreage may lose its topsoil layer at twice the rate of current agricultural land. However, it should be noted that the rate of loss is (on the average) fairly low.

Aside from new biomass cropland's capability to resist erosion and its productive potential, an important factor determining the environmental impact of the conversion to intensive production is the nature of the previous land use. For example, the conversion of land in rotation hay and pasture to intensive crop production would clearly be valued differently from a conversion from forest. Because different groups value alternative land uses differently, it is difficult to place more or less weight on the conversion of one land use relative to another. It seems likely, however, that most environmentally oriented groups would prefer to see the conversion of lands that are manmade monocultures (e.g., improved haylands) before more natural and diverse ecosystems were converted.

The cost of conversion will play an important role in determining which lands will be chosen. At the present time, conversion of pastureland and hayland is likely to be less expensive than conversion of forest, and land conversions may be expected to be skewed away from forests. Least expensive of all to convert are lands currently in set-aside, and these are likely to be the first to be taken. The cost of forest conversion may, however, be lowered significantly if the demand for wood-for-energy rises with the demand for energy crops (because the value of the now-worthless cull wood and slash can be traded off against clearing and site preparation costs). Thus, there is

no guarantee that forests—which make up about one-quarter of the high- and medium-potential cropland[55]—will not be cleared in significant quantities if large-scale conversion to biomass crop production occurs.

Production Practices

A variety of practices are available to control the erosion and other impacts of farming. These range from crop rotation to conservation tillage to scouting for pest infestations. Table 32 provides a partial list of these prac-

Table 32.—Agricultural Production Practices That Reduce Environmental Impacts

Runoff and erosion control
Contour farming or contour stripcropping
Terraces and grass waterways
Minimum tillage and no-till
Cover crops
Reducing fall plowing

Reducing chemical pollution
Scouting (monitoring for pest problems)
Disease- and insect-resistant crops
Crop rotation
Integrated pest management
Soil analysis for detecting nutrient deficiencies
Nitrogen-fixing crops
Improved fertilizer and pesticide placement, timing, and amount
Improving irrigation efficiency-trickle irrigation, etc.
Incorporating surface applications into soil

SOURCE: Office of Technology Assessment.

tices. Their future use will play a critical role in determining the environmental impacts of biomass energy production.

The availability of these controls should not be confused with the probability that impacts will not occur. In fact, it is unwise to assume that the use of many of the practices listed in table 32 will be widespread. There are a number of reasons for this.

First, the costs of the controls may considerably exceed the farmer's perceived benefits. The effects of erosion on water quality are largely "external" effects; although the farmer may benefit from the control efforts of others, he is unlikely to benefit from any water quality improvements caused by his own efforts. This

---

*This is almost certainly very optimistic, but SCS guidelines define 5 ton/acre-yr as an acceptable rate for many lands.

[55]*National Erosion Inventory Estimate,* op. cit.

problem of "external" benefits is endemic to American agricultural practices. It is, in fact, merely one aspect of the "tragedy of the commons" that hinders voluntary environmental control in virtually all of man's activities. Also, any success in delaying or preventing productivity declines from erosion effects may be masked by improvements in other production practices and in any case would be very long term in nature. The farmer must balance these benefits against very high erosion control costs. SCS has examined the effects on farm production costs of requiring reductions in current erosion rates on croplands. For example, requiring a 10-percent reduction in each of 105 producing areas would raise corn production costs by $0.07/bu in 1985. Requiring all acreage to conform to a maximum allowable erosion rate of 10 ton/acre-yr (twice the "no productivity loss" rate) would cost $0.31/bu or a 16-percent increase over the projected 1985 cost without controls. Further constraints could raise costs astronomically (a 5-ton/acre-yr constraint leads to a $23.70/bu production cost) because heroic efforts must be made on some acreage in order to meet the constraints.[56] Although these estimates are sharply dependent on a number of critical assumptions (e.g., the role of Federal soil conservation assistance is ignored), they demonstrate the large potential cost (and price) increases that erosion control requirements could cause.

Second, there are substantive scientific disagreements about the actual environmental benefits achieved by these controls. Some of the controls may reduce one environmental impact at the expense of increasing others. A primary example of this is the effect of some erosion controls—reductions in fall plowing and conservation tillage—on pesticide use. These controls leave crop residues on the surface, and the residues in turn act to break the force of raindrops on the soil and drastically decrease erosion and runoff. Because the residues harbor plant pathogens and insect pests, pesticide requirements will go up sharply. Also, increased applications of herbicides are used for weed control to compensate for the

reduced tillage. The net effect on the environment is not entirely clear because a large source of pesticide entry into surface waters— adsorption on soil particles and transport in runoff—is considerably reduced by the controls, but EPA has identified increased pesticide use with conservation tillage as a significant problem.[57] Tables 33 and 34 identify in greater detail the environmental tradeoffs involved in erosion controls.

Third, some of these controls may appear to be incompatible with the present agricultural system and may not be accepted by farmers. For example, the use of nitrogen-fixing crops, cover crops, and crop rotations conflict with today's large-scale, highly mechanized, chemical-oriented farming although they were widely practiced in the past. Although some scientists argue that the economic advantages of present methods will evaporate (or have already evaporated) in the face of rising prices for energy and energy-intensive agricultural chemicals, and that the long-term environmental viability of the methods is questionable, the relative advantages and disadvantages of the present system and its alternatives are a subject of intense controversy in the agricultural community—with defense of the present system having the upper hand at present. It appears virtually certain that in the absense of Government intervention the provision of feedstocks for energy production will rely primarily on a mechanized, chemical-oriented philosophy modified only by any economic pressures arising from increases in energy prices. Any substantive changes from this philosophy would represent essentially a revolution from established practice and would be unlikely because the present system has clearly succeeded in providing a reliable supply of food at (comparatively) moderate prices.

Crop Types

The environmental impacts of growing and harvesting agricultural crops for energy will vary strongly with the type of crop grown, since different crops have different fertilizer and pesticide requirements, water needs, soil

[56]B. English, Iowa State University, personal communication, June 15, 1979.

[57]Environmental Implications of Trends, op. cit.

Table 33.—Environmental Pollution Effects of Agricultural Conservation Practices

Extensiveness	Resource use	Pollutant changes in media: surface water — sediment	Nutrients	Pesticides	Pollutant changes in media: ground water— nutrients—pesticides	Pollutant changes in media: soil	Pollutant changes in media: air
Contour farming/contour stripcropping							
Acreage of crops farmed on the contour or strip-cropped decreased 25% between 1964 and 1969 and continued to decrease slightly to 1976. Contour farming is more widely used in nonirrigated crop production than in irrigated crop production.	Fertilizer and herbicide use remain constant. Insecticide use will remain constant or very slight increases.	Sediment loss can be reduced substantially on moderate slopes, but much less on steep slopes. Reductions up to 50% are possible, but average reductions will be about 35%. Contour stripcropping can reduce sediment losses more than contour alone. (Note: research shows substantial loss can occur with contour watersheds with some soil types, with long slopes and/or with steep slopes.)	Nutrients associated with sediment will be reduced, but reductions may be proportional to the amount of sediment lost.	Pesticide reductions will be less than that for nutrients since a greater amount of pesticide is lost through surface water than bound to sediment.	Loss of nutrients and pesticides through ground water will remain constant or decrease slightly. However the amount of N leached is small compared to amount that can be lost in runoff and loss of pesticides to ground water is minor with proper application rates.	Erosion losses can be reduced up to 50% with average reductions of 12% (see conclusions on sediment).	Pesticide losses through volatilization will decrease if they are incorporated into the soil by mechanical means.
Terraces and grass waterways							
Terraces and grass waterways are not important in irrigated production, but are important for nonirrigated crops. However, only 6% of all acres in 1969 had terraces. The acres with terraces in 1976 could have increased or decreased slightly.	Fertilizer, herbicide, and insecticide use is not expected to increase (fertilizer could increase if production per cropped acre is expected to increase to compensate for land taken out of production). However, terrace practices will not require more fertilizers. Costs and maintenance increase for terraces.	Substantial reductions in sediment and runoff can usually be expected.	Reductions in nitrates and phosphates are expected with decreased soil loss and surface runoff. Reductions could be substantial with some soils and cropping systems.	Reduction of pesticide residues in surface water could be substantial with terrace systems, since both surface runoff and soil loss are reduced.	N in ground water may be reduced, based on limited research data. Leaching of pesticides is not likely to result in significant loss with normal applications rates.	Substantial reductions in erosion can result.	No change.

SOURCE: U.S. Environmental Protection Agency, *Environmental Implications of Trends in Agriculture and Silviculture, Volume II: Environmental Effects of Trends,* EPA-600/3-78-102, December 1978.

Table 33.—Environmental Pollution Effects of Agricultural Conservation Practices—continued

Extensiveness	Resource use	Pollutant changes in media: surface water sediment	Nutrients	Pesticides	Pollutant changes in media: ground water—nutrients—pesticides	Pollutant changes in media: soil	Pollutant changes in media: soil
Conservation tillage: no-till Approximately 2.6% of all cropped land was no-till in 1977. While this practice is expected to increase to limited use in 2010, current projections (up to 55% of crops under no-till in 2010) seem high. Extensiveness may only be 10 to 20% in 2010.	Fertilizer and herbicide use increases by 15%, insecticide use by 11%. An estimated 5 million acres of land could be shifted to crop production with no-till and reduced-till methods. Labor costs are reduced. More water will be conserved with no-till, as much as 2 inches per year.	Sediment reductions of 50 to 90% will result.	While large soil loss reductions will tend to reduce nutrient losses, fertilizer use will increase by 15%. There will probably still tend to be reductions in total nutrient loss, but reduction will not be proportional to reductions in soil loss. N content of soil may also increase from weathering of crop residues.	Effect of no-till on pesticide losses is not well documented. Loss to surface water is greater when the compound is surface applied and not incorporated in the soil, and 11% more insecticides and 15% more herbicides will be used for no-till. While reductions of pesticides in surface water could occur, current research does not prove this. Increased use and surface application, even with reduced soil loss with no-till, could even cause slight increases in pesticide losses.	Nitrates in ground water will show no change to slight increases. Pesticide loss to ground water will not be significantly changed with no-till practices.	Erosion losses will be decreased 50 to 90%. Crop residues will increase which may result in increased N loss to the soil or available for runoff. Additionally, residues may provide a hiding place for pests and increase the incidence of pests.	With some pesticides, increased volatilization will occur with surface applications. The vapor pressure, molecular weight, and other properties of a pesticide will determine the extent of vaporization.
Conservation tillage: reduced tillage In 1977, an estimated 58.8 million acres (19% of total cropped acres) will be reduced tilled. An additional 40 million acres will be classified as less till includes chisel plowing, disking once instead of twice, and planting in rough ground. In 2010, a total of 40% of all cropland may be classified as reduced tilled.	Fertilizer use will increase slightly. Herbicide use is up (0.6%) and insecticide use increases by 8.6%. An estimated 5 million acres of land will be shifted to crop production with reduced and no-tillage methods. Labor output will decrease. Energy to plant crops decreases, but increased energy will be used in manufacture of increased fertilizers and insecticides. Some soil moisture will be conserved with reduced tillage.	Sediment will be reduced an average of 14%. Reduced tillage is less effective than no-till in controlling soil loss.	There will probably be reductions in total nutrient loss to surface water, but reduction will not be proportional to reductions in soil loss (14%).	Effect of reduced tillage on pesticide loss is not well documented. Loss to surface water is greater when a pesticide is surface applied and total pesticide use is 9% greater for reduced till. While reductions of pesticides in surface water could occur, there is not enough research data to support this.	Nitrates in ground water will show no change to slight increases. Pesticide levels in ground water will not be significantly changed with reduced tillage.	Erosion losses decrease an estimated 14%. Wind erosion losses will also decrease slightly. Crop residues increase, which lead to increased N available to the soil for leaching and runoff. Residues on soil also increase the incidence of pests.	Surface applications of some pesticides types leads to increased volatilization losses. The vapor pressure, molecular weight, and other chemical properties of a pesticide will determine the extent of vaporization.

SOURCE: U.S. Environmental Protection Agency, Environmental Implications of Trends in Agriculture and Silviculture, Volume II: Environmental Effects of Trends, EPA-600/3-78-102, December 1978.

Table 34.—Ecological Effects of Agricultural Conservation Practices

Contour farming/contour stripcropping
Extensiveness of contouring in 1985 (over 1976 use) will be low, but will increase by 2010. Beneficial aquatic effects result from decreased turbidity and pesticide residues in surface water. Species diversity will also increase in the aquatic ecosystem. Decreased erosion and retention of soil nutrient cycles will have long-term beneficial terrestrial effects. Since pesticide residues at current levels in drinking water are not known to be a human health hazard, reduction of pesticide residues will have no significant human health effects. However, if pesticide residues are later determined to be dangerous at current levels, then human health effects would be beneficial.

Terraces and grass waterways
Terraces are more effective than contouring in reducing pollutants, but extensiveness of use is lower for terraces. Aquatic effects are decreased turbidity, increased species diversity, and decreased pesticide residues. Terrestrial effects are beneficial, resulting from increased vegetation on terraces and grass waterways, increased diversity of wildlife, and more pathways for animal populations to travel. Valuable topsoil will also be retained. Based on present knowledge, there is no known human health effect. Decreased sediment in water might result in an unpleasant taste or odor in drinking water.

Reduced tillage
Reduced tillage (with crop residues remaining) is less effective than no-till in reducing soil loss, but extensiveness of reduced tillage will be greater. Therefore, the intensity of ecological effects are comparable for the two practices. Sediment reductions will reduce turbidity and increase species diversity. However, the potential for increased pesticide residues in surface water could have adverse effects on the aquatic ecosystem. Crop residues remaining on the soil and decreased soil loss are beneficial to the terrestrial system, but increased pesticide use will have adverse effects on nontarget organisms. Human health effects will not be significant.

No-till
Aquatic and terrestrial effects are both beneficial and adverse. Aquatic systems will benefit from reduced turbidity and increased species diversity. However, pesticide residues in surface water could potentially be increased with no-till and create adverse effects in the aquatic ecosystem. Increased pesticide use can also have adverse effects on nontarget terrestrial life. Retention of crop residues and reductions in erosion will have beneficial terrestrial effects. Human health effects will not be significant since pesticide residue in surface water should still be within safety limits even if they increase slightly with no-till.

SOURCE: U.S. Environmental Protection Agency, *Environmental Implications of Trends in Agriculture and Siliviculture, Volume II Environmental Effects of Trends*, EPA-600/3-78-102, December 1978.

preparation methods, harvesting times, and other factors that may potentially affect impact. Some of the more important crop-determined factors are:

- *Annual or perennial.*—Perennial crops (trees, sugarcane, perennial grasses, etc.) offer a substantive environmental advantage over annuals because their roots and unharvested top growth protect the soil from erosion year round, while annuals offer protection only during the growing season and require seasonal tilling (unless no-till is used) and planting.

- *Row or close-grown crops.*—Row crop cultivation is generally more erosive than cultivation of close-grown crops. For example, the average erosion rates of close-grown crops are significantly lower than those of row crops in every land capability class and subclass shown in table 34. In general, the rates of the close-grown crops appear to be about half those of the row crops.

In the previous calculation of the expected average erosion rates from new biomass production, a mix of row and close-grown biomass crops (in the same proportion as existed in 1977 food produc-

tion) would be expected to have an average (sheet and gully) erosion rate of about 7.5 ton/acre-yr compared with about 6.3 ton/acre-yr for food production. If the entire biomass crop were a row crop (e.g., corn for large-scale alcohol production), the average erosion rate from the biomass acreage is estimated to be 9.3 ton/acre-yr—almost 50 percent higher than the erosion rate from food production.

- *Water requirements.*—High irrigation water use means greater competition for water among competing uses, greater drawdown of streams and consequent loss of assimilative capacity, potential for entry of more salts into surface and ground waters, depletion of aquifers (ground water mining), and energy use for pumping. There are substantial differences in water consumption among different crops. For example, irrigation requirements for crops in Arizona during a dry year[58] are:

Water use, acre-ft/ton of crop
Wheat. 0.9
Oats 1.6
Barley. 1.3
Alfalfa 0.7

[58]*Conservation Needs Inventory*, op. cit.

Most discussions of biomass energy assume that irrigation generally will not be used in growing feedstocks. However, an extension of the types of irrigation water subsidies now available to Western farmers, however unlikely, could lead to such use.

• *Soil requirements.* —The ability to utilize marginal lands can avoid the problem of competition with food production that is a major environmental and social/economic issue in evaluating biomass fuels. As discussed elsewhere in this chapter, however, the potential for high biomass yields under marginal soil, temperature, and water conditions has been exaggerated.

• *Pesticide requirements.* —The importance of reducing pesticide applications is a matter of considerable controversy. However, crops that have low pesticide requirements will be perceived as more environmentally benign. In some instances, present pesticide use may be a poor indicator of future requirements for energy crops because cropping practices and land characteristics may be altered significantly in going to a crops-for-energy system. For example, regulatory restrictions on soil erosion could force virtually universal use of conservation tillage and consequent increases in herbicide and (to a lesser extent) insecticide applications. The lack of esthetic requirements for biomass feedstocks might also lead to some decrease in pesticide requirements, but this effect may be small because minor insect damage can lead to further damage by fungal and viral infections (especially during storage). Finally, although pesticide requirements for grasslands currently are very low, pest problems conceivably may accelerate if productivity is pushed by expanded use of fertilizers.

• *Fertilizer requirements.* —In general, high fertilizer requirements are an environmental cost because of the energy used to produce the fertilizer and the nutrient runoff that results from applications. However,

crop requirements for very high levels of nitrogen may be an environmental advantage; some high-nitrogen crops are compatible with land disposal of sewage sludge and effluents and thus can be an important component of urban sewage treatment strategy.

• *Yield.* — Because yield per acre determines the amount of land necessary to produce a unit of energy, it is one of the most important factors determining impact. Measurements of input requirements (water, fertilizer, pesticides, etc.) and measurable damages (such as erosion) on a "per acre" basis are inadequate measures of relative environmental impact because of the large variation in biomass yields from crop to crop. For example, corn is widely perceived as an extremely energy- and water-intensive crop, but its very high yields essentially cancel its high "per acre" fertilizer, pesticide, and water needs; it is, in fact, a relatively average crop on an "energy per ton of product" basis.

The importance of these factors in determining environmental impacts is extremely site and region specific. For example, water requirements clearly are more important in the arid West than in the wet Southeast, while factors affecting sheet and rill erosion potential are more or less important in the reverse order. Much of the data needed to assess the different potential crops are not available, and thus it is premature to suggest which crops would be the most environmentally benign in each region or subregion. There are sufficient data, however, to draw some rough sketches of some of the possible advantages or disadvantages of several of the suggested biomass crops.

Corn has been most often mentioned as the primary candidate for an ethanol feedstock. It is an annual row crop and thus a major contributor to erosion, but much of the land on which it is grown is relatively flat, a factor that limits the erosion rate. Corn's high yield rate—currently about 100 bu/acre, or about 260 gal/acre of ethanol—will minimize the land use impact of additional production, although yields on new lands will not be as high as the current average, and the land displaced would be of high quality.

Because the protein-rich residue from the fermentation (ethanol producing) process is a substitute (although not necessarily a perfect one) for soybean meal in cattle feed, switching existing cropland from soybean to corn production may allow large quantities of ethanol to be produced using far less acreage than would be needed if corn for ethanol production were planted only on new acreage. As discussed in the section on "Energy Potential From Conventional Crops," corn's effective yield per acre of new land could grow by over 300 percent (i.e., about three-fourths of the corn used for ethanol would be grown on land formerly planted in soybeans with no loss in national food and feed values) as long as the soybean meal market remained unsaturated. Significant uncertainties concerning the corn residue's nutritive value, potential corn yields on soybean land, soybean market response, and other factors must be overcome, however, before this crop-switching scenario can be accepted as valid. In the absence of the necessary research, the higher estimate of new land required for each gallon of ethanol produced should be used as a pessimistic measure of potential impact. At low levels of production, the more optimistic, lower acreage requirements are likely to be accurate, but the requirements may increase as production increases. Above 2 billion to 7 billion gal of ethanol produced annually, feed markets would be saturated even under the most optimistic assumptions and additional ethanol production would require cropland conversion at the higher rate.

Sweet sorghum has been praised as a crop of high biomass potential for fermentation and alcohol production. Although ethanol yields of 260 to 530 gal/acre have been projected, these projections are based on minimal—and clearly inadequate—experience. However, these high yields, if confirmed, would limit displacement of alternative land uses. Sweet sorghum may be more tolerant of marginal growing conditions than corn, which could lead to a lower

level of displacement of the most productive ecosystems.

Sugarcane has been suggested as a biomass crop for alcohol production in Hawaii and the Gulf Coast. Because its cellulosic content is high enough to supply all of the heat energy necessary to ferment the sugar and distill alcohol from it, no coal or other fossil fuel use would be necessary to power the system. Sugarcane requires high-quality land and thus may displace particularly valuable alternative land uses.

Perennial grasses can be supplied in large quantities by increasing yields on present acreage with more intensive harvesting and fertilization; the present average yield is 1½ to 2 ton/acre, and this can be increased to 3 to 5 tons. Because perennials provide excellent erosion control, and because no additional acreage would have to be converted from alternative uses, the environmental impact of a grass-based biomass strategy should be far less than that of a strategy based on annual crops. Environmental impacts of some significance could occur because of the expanded use of fertilizer (150 lb N, 30 to 50 lb P_2O_5, 80 to 150 lb K_2 for an incremental production of 78 gal of ethanol on each acre) and pesticides. Recovery of added fertilizer is very high for grasses, however, so the potential for water pollution will be less than for annual crops. Also, there is uncertainty about changes in susceptibility to disease and insect damage because of the intensification of production, and substantial new use of pesticides conceivably could be required. Finally, the frequent harvesting and greater use of chemicals may disrupt the populations of wildlife that now flourish in the less intensively maintained grasslands.

Trees may be grown plantation-style and harvested by coppicing to supply significant quantities of biomass. A carefully designed tree plantation should have few problems of erosion unless cultivation is practiced (which appears unlikely); however, harvesting may conceivably create an erosion problem unless low-bearing-pressure machines are used to avoid damaging the soil. Tree plantations present basically the same ecological problems as do agricultural monocultures—higher potential for disease attack and displacement of alternative ecosystems. The spacing necessary for tree growth may also allow greater competition from weeds— and consequently larger herbicide requirements—but the sheltering effect of the tree canopy and the greater ability of some tree species to compete for water may counterbalance this effect.

Environmental Impacts of Harvesting Agricultural Residues

The residues from agricultural production have a number of significant effects—beneficial or otherwise—when left on the land. Understanding these effects is critical to understanding the potential environmental impacts of the collection and use of these residues as an energy feedstock.

The effects of residues left on the land include (table 35):

- Control of *wind and water erosion.*— Retention of residues as a surface cover is a major erosion control mechanism on erosion-prone lands. For example, residue retention on land that is conventionally tilled (i.e., plow-disk-harrow) can cut erosion in half.[59]
- Retention of plant *nutrients.*—Residues from the nine leading crops in the United States contain about 40, 10, and 80 percent as much nitrogen, phosphorus, and potassium, respectively, as in total fertilizer use in U.S. agriculture.[60]
- *Enhanced retention of water* by soils and maintenance of ability of soil surfaces to allow water infiltration.
- Maintenance of *organic matter* levels (necessary to maintain soil structure, ion exchange capacity, water retention properties) in soils.—Croplands in the United States have lost major portions of their organic content. Reductions (in North Central and Great Plains soils) of one-half to two-thirds of what was present under native grassland have been cited.[61] Retention of crop residues is a critical factor in maintaining organic matter levels.

[59]W. E. Larson, et al., "Residues for Soil Conservation," paper No. 9818, Science Journal Series, ARS-USDA, 1978.
[60]Ibid.
[61]Ibid.

Table 35.—Environmental Impacts of Plant Residue Removal

Water
- Increased *erosion and flow of sediments* into surface waters if restrictions on removal are not observed, causing increased turbidity, obstruction of streams, filling of reservoirs, destruction of aquatic habitat, increase of flood potential; under circumstances where conservation tillage is encouraged by removal of a portion of the residues, erosion and its consequences will decrease.
- Increased *use of herbicides* and possible increased flow into surface and ground waters if conservation tillage is required for erosion control; in some situations, removal of a portion of the residues would increase herbicide efficiency and greater use may not occur.
- Increased *flow of nutrients* if more runoff results from decreased water retention of soil and greater erosivity of soil; if more fertilizer is applied to compensate for nutrient loss, flow of nutrients will change but the net affect is not certain.

Air
- *Dust* from decreased cover on land, operation of residue harvesting equipment (unless integrated operation).
- Added *herbicides* from aerial spraying or as a component of dust.
- *Decreased insecticides, fungicides.*
- *Reduction in pollution from open-burning* of residues, where formerly practiced.

Land
- *Erosion and loss of topsoil, degrading of productivity* if restrictions on removal are not observed; the opposite, positive effect if conservation tillage is encouraged by residue removal.
- Decrease in *water retention capabilities* of land, increased flooding potential if restrictions are not observed.
- *Depletion of nutrients and organic matter from soil* (nutrients may easily be replaced).

Other
- *Reduction in plant diseases and pests* (if lowering of soil organic matter does not adversely affect this factor) because residues can harbor plant pathogens.

SOURCE: Office of Technology Assessment.

- A number of negative effects depending on the type of crop and amount of residue—"poor seed germination, stand reduction, phytotoxic effects, nonuniform moisture distribution, immobilization of nitrogen in a form unavailable to plants, and increased insect and weed problems."[62] In all cases, the residues harbor crop pests; this can be a particularly sig-

[62]*Improving Soils With Organic Wastes,* op. cit.

nificant problem if single cropping is practiced (the same crop is grown in consecutive years). Because the residues shield the soil, they may hinder soil warming and delay spring planting (causing reduced yields in corn).

When the problems associated with crop residues outweigh the benefits, farmers will physically remove the residue (this practice is necessary in rice cultivation) or plow it under in the fall (a common practice in the Corn Belt). The collected residues may be burned, although they have alternative uses such as livestock bedding. Where removal is normally practiced, use of the residues as an energy feedstock is at worst environmentally benign and possibly beneficial (if air pollution from open burning is prevented). Because fall plowing negates much of the residues' value as an erosion control, collection of a portion of the residue is usually considered benign (full removal may affect soil organic content, the importance of which is somewhat in debate). Because an excess of residue may inhibit the effectiveness of herbicide treatments—especially preemergence and preplant treatments—and also leave large numbers of weed seeds near the soil surface, removal of a portion of the residues on land where they are in excess may promote the use of reduced tillage by allowing more effective chemical weed control, and thus be considered environmentally beneficial.

When residues are normally left on the soil surface as an erosion control, their removal potentially may be harmful. However, where substantial quantities of residue are produced on flat, nonerosive soils, a portion of these residues may be removed without significantly affecting erosion rates. SCS and the Science and Education Administration—Agricultural Research have sponsored extensive research designed to compute the effects of residue removal practices and other practices on soil erosion. USDA believes that it can identify the quantity of residues that can be safely removed from agricultural lands in all parts of the United States. Although controversy exists over the rate of creation of new topsoil, and

thus the erosion rate that will maintain productivity over the very long term, it seems likely that errors in these computations will not cause significant harm as long as SCS maintains its monitoring efforts at the current level.

The key to preventing significant environmental damage while harvesting large quantities of residues is for the agricultural system to act in accordance with USDA's knowledge The discussion of the impacts of U.S. agriculture presented previously seems to indicate a willingness among farmers to ignore warnings about using erosive practices or cultivating fragile land, in order to gain short-term benefits. In the absense of additional constraints, a significant number of farmers might be willing to remove their crop residues even when adverse erosion effects would occur. (Interestingly enough, some farmers may ignore USDA with the opposite effect—they may be reluctant to remove any residues because of their fear of erosion and other negative consequences). Under these circumstances, the establishment of a market for crop residues could result in additional erosion from croplands that cannot afford it and add to the already significant sediment burden on surface waters caused by current farming practices.

Although the negative effects of any increase in erosion are straightforward, other effects that have been associated with residue removal are more ambiguous. For example, the removal of plant nutrients in the residues may be compensated for by the return of the conversion process byproducts or by chemical fertilizers (both of which may have some adverse effects on water quality). The removal of organic content has been identified as a significant impact[63] and soil scientists have long thought that soil organic content is a critical variable of the health of the agricultural ecosystem (e.g., increasing the organic content of soils can stimulate the growth and activity of soil micro-organisms that compete with plant pathogens). However, despite a variety of papers in the agronomy literature that treat yield

---

[63]Pimentel, et al., op. cit.

as a function of soil carbon level, there is insufficient experimental evidence to establish that any significant effects on crop yields would occur. Also, the much higher yields of today's agriculture means that removal of half of the residue will leave the same amount of organic material as would have occurred 25 years ago if all of the residue had been left on the land. This is an area that clearly deserves further research.

R&D Needs

Considerable research has been and is directed at improving agriculture for food, feed, and materials production. While much of this research is applicable to energy production, the specific goal of producing various types of energy crops has not been adequately addressed. Changing the emphasis to energy or energy and food production and the environmental concerns with agriculture suggest several R&D problems. Some examples are listed below.

- A wide variety of crops that are not used as food or feed crops could, potentially, be good bioenergy crops. The promising varieties should be developed. From a theoretical point of view, grasses appear to be promising candidates for high biomass producers and on marginal cropland (see ch. 4): and arid land and saline tolerant crops may enable the economic use of lands and water supplies that are otherwise unsuited for agriculture.
- Food and feed crops are usually quite specific as to their use. Corn, for example, is not interchangeable with wheat. Many different types of crops, however, can produce the same or interchangeable biomass fuels. Consequently, extensive comparative studies between various crops are needed to determine the promising bioenergy crops for the various soil types and climates.
- If both the residues and the grain can be sold, then the optimum plant may not be the one that produces the most grain. Farming practices and hybrids that can change the relative proportions of grain to residue in the plant while maintaining a high overall yield should be investigated.

- Various crop-switching possibilities that involve fuel production should be investigated further to determine the extent to which they can provide fuels and the traditional products from agriculture without expanding the quantity of cropland cultivated. The extent to which the corn-soybean switch actually takes place should be studied, as should novel possibilities such as sugar beets used for animal fodder. Included in this should be investigations of the effect of substituting current feed rations with varying amounts of forage-distillers' grain, forage-corn gluten mixtures, and other feeds that may be involved in the crop-switching schemes.
- Large-scale biomass development will require the placement of millions of acres of land—now in low-intensity agriculture (e.g., pasture), forest, or other uses—into intensive production, coupled in many cases with very high rates of removal of organic matter. Environmental R&D that should accompany, and preferably precede, such development includes:
  —further investigation of long-term effects of reduction in soil organic matter,
  —determination of pesticide requirements for high-yield grasses in intensive production,
  —intensification of breeding programs for insect/disease-resistant strains of crops with high biomass potential,
  —determination of economically optimum strategies for minimization of soil erosion, and
  —development of effective/ programs to improve farmer (environmental) behavior.

Chapter 4
UNCONVENTIONAL BIOMASS PRODUCTION

Chapter 4.—UNCONVENTIONAL BIOMASS PRODUCTION

UNCONVENTIONAL BIOMASS PRODUCTION

Introduction

A number of unconventional approaches to biomass energy production have been proposed. Several nontraditional crops that produce vegetable oils, hydrocarbons, and other chemicals or cellulosic material are under investigation. Both freshwater and saltwater plants are being considered, and various other approaches to biomass fuel production are being examined. A common feature to all of these approaches is that the full potential of individual plants proposed as fuel-producers cannot be fully assessed without further R&D. A description of some general plant characteristics, however, can aid in comparing the various possible types of energy crops.

The general aspects of farming, plant growth, and the efficiency of photosynthesis are considered in chapter 3. Since future crop yields will depend on these factors and on the development of hybrids for energy production, the possibilities for genetic improvements are considered here. Following this, crop yields and various unconventional bioenergy crops and approaches to farming them are discussed.

Genetics

There are two major areas of genetics. The first, which plant breeders have used most effectively to date is the classical Mendelian approach (introduced by Gregor Mendel in the 19th century). It involves selecting and crossbreeding those plants with desired characteristics (e.g., biomass yield, grain yield, pest resistance). The process is continued through each succeeding generation until a hybrid, or particularly favorable strain, is isolated. Strains with unique and desirable properties are often crossbred to produce hybrids that outperform the parents. Hybrid corn is an example. This technique is limited, however, by the variability of characteristics that exist naturally in plants or mutations that occur spontaneously during breeding. One can isolate the best, but one cannot produce better than nature provides.

The second approach to genetics, molecular genetics, is a recent development that involves manipulating the genetic code more or less directly. Three types of potential advances from molecular genetics can be distinguished: 1) improvements in the efficiency or rate of biological conversion processes (e.g., fermentation, anaerobic digestion), 2) introducing specific characteristics into specialized cells such as the ability to produce insulin,[1] and 3) improvements in photosynthetic efficiency, plant growth, and crop yields. The complexity of the tasks increases greatly as one goes from 1) to 3), as described below.

The first type involves subjecting single cells to chemicals or radiation that cause mutations in the cells' genes. The way these mutations occur is not well understood and the effects are generally unpredictable. The result is to increase the diversity of cell types over what occurs naturally; and in favorable cases one may produce a cell that performs a particular function "better" than naturally occurring cells. This method has been applied successfully to the production of antibiotics, in biological conversion processes,[2] and in increasing the tolerance of plants to certain diseases; but it is generally a "hit and miss" proposition.

---

[1]A. Elrich, et al., *Science*, vol. 196, p. 1313, 1977.
[2]For example, see G. H. Emert and R. Katzen, "Chemicals From Biomass by Improved Enzyme Technology," presented at the *Biomass as a Non-Fossil Fuel Source*, ACS/CSJ Joint Chemical Congress, Honolulu, Hawaii, April 1979.

The second type involves identifying the genes responsible for a particular function in one cell and transferring these genes to another cell. This transfer does not always require a detailed knowledge of how the gene produces the desired characteristic. One can draw from the pool of naturally occurring characteristics, but the conceptual link between the gene and the characteristic must be relatively direct.

The third type probably would involve altering a complex set of interdependent processes in the plant. Although some plant physiologists believe that some improvements in photosynthetic yield can be achieved by suppressing processes like photorespiration (a type of plant respiration that occurs only in the presence of light), this belief is highly controversial among specialists in the field. It is generally believed that the processes involved in plant growth and photosynthesis and their relation to specific genes are too subtle and poorly understood at present to know what biochemical processes should or can be altered to improve plant growth and crop yields.

Some additional near- to mid-term advances are likely in the area of biological conversion processes and with gene transfers in the area of synthesizing high-value chemicals, like insulin, that would be either impractical or impossible to synthesize by other means. The complexity of plant growth and photosynthetic efficiency, however, reduces the chances of improving these characteristics in plants through molecular genetics in the near future. Although the possibility cannot be precluded that a scientist will alter a crucial process in plant growth despite the lack of knowledge, there are few grounds for predicting that this will occur before the fundamental biochemical processes involved in plant growth and photosynthesis and the way that environmental factors limit them are better understood. There is a great deal of controversy surrounding this subject, but most arguments—both pro and con—are based on intuition rather than demonstrated fact.

Crop Yields

Current knowledge and theories of plant growth do not enable one to predict the crop yields that can be achieved with unconventional crops. Nevertheless, because of the importance of biomass yields in determining the economics of production, it is important to have an idea of the approximate magnitude of the yields of various options that may be possible.

To this end, corn—a highly successful example of crop development—is used as the basis for these estimates. Corn has the highest photosynthetic efficiency of any plant cultivated over large areas of the United States. As discussed in chapter 3, an optimistic estimate for average corn grain yields would be about 140 bu/acre (3.9 tons of grain/acre) by 2000. Many farmers routinely exceed this yield, as do experimental plot yields. This number, however, is quite optimistic for average yields from cultivation on millions of acres of average U.S.

cropland. Furthermore, **since cropland that could be devoted to energy crops is generally of poorer quality than average cropland, using this as a basis for estimates may be overstating the potential.**

A yield of 140 bu/acre for corn corresponds to a photosynthetic efficiency of about 1.2 percent over its 120-day growing season. Perennial crops, however, probably will have somewhat lower efficiencies during the cold weather at the beginning and end of their growing seasons. Consequently, it is assumed that perennials can achieve an average photosynthetic efficiency of 1.0 percent. With these assumptions, and the others stated below, the following yields may be possible.

• **Dry matter yield.**—With an 8-month growing season in the Midwest, biomass production could yield 15 ton/acre-yr of dry plant matter. For the Gulf Coast (12-month growing season), the yields could reach 21 ton/acre-yr.

- **Grain yields.** — Based on corn yields, average grain production from some plants could yield 3.9 ton/acre-yr.

- **Sugar yields.** — Good sugar crops are 40- to 45-percent sugar on a dry weight basis (e.g., sugarcane, sweet sorghum). In the Midwest, sugar crops will probably be annual crops leading to possible yields of 4 tons of sugar/acre-yr. Along the Gulf Coast, there is a longer season. Current average sugar yields are 4 ton/acre-yr. As with corn, the yields could conceivably be increased by 40 percent to 5.6 ton/acre-yr.

- **Aquatic plant yields.** — Estimates for water-based plants are more difficult to derive, since there is considerably less experience and applicable information. Water plants have a continuous supply of water and are never water stressed. For maximum productivity, nutrients and carbon dioxide (CO_2) (for submerged plants) would be added to the water and could be available continuously at near-optimum levels. The water would prevent rapid changes in temperature. All of these factors favor plant growth, and if other problems with cultivating aquatic plants can be solved, yields may be quite high (see "Aquaculture" and "Mariculture"). Nevertheless the uncertainty is too great to make a meaningful comparison with the land-based plants. As with other plants, experimental yields will probably not be representative of commercially achievable yields.

- **Yields in greenhouses.** — Yields in greenhouses are also very uncertain, due to a lack of sufficient data and potential problems such as fungal attacks on plants, root rot, and other problems with extremely humid environments. If these and other problems are solved, then crop yields approaching those estimated for the milder climates may be achieved.

- **Vegetable oil or hydrocarbon yields.** — In addition to solid material, plant biomass includes oils. New seed oil crops typically contain 10-to 15-percent vegetable oil, and in sunflowers the oil comprises up to 50 per-

cent of the seed weight.[3] Assuming that plants which are 50-percent seed contain seeds that are 50-percent oil, the oil content may reach 25 percent of the total plant weight.

Assuming the biochemical reaction producing the oil is 75 percent as efficient as that which produces cellulose, then for 1-percent photosynthetic efficiency the oil production would be 16 bbl/acre-yr for a plant that is 25-percent oil. For an oil-producing reaction that is 50 percent as efficient as the reaction that results in cellulose, the yield would be 12 bbl/acre-yr.

Plant material stored as hydrocarbons has also been proposed as a source of liquid fuels. Eucalyptus trees and milkweed, for example, contain up to 12-percent hydrocarbons. Assuming that this content could be doubled, the same yields as for oil crops would apply.

- **Arid land crop yields.** — Another important and sometimes limiting factor in biomass production is water. Generally plants will grow well without irrigation in areas of the United States where the rainfall is 20 to 30 inches or more. For high biomass-producing crops in relatively humid climates (like the Midwest), the minimum water necessary for plant growth in open fields is about 200 weights of water for 1 weight of plant growth. There has been interest, however, in plants that can grow under more arid conditions. In desert regions with very low humidities, requirements are more typically 1,000 weights or more of water per 1 weight of plant growth. (Some plants survive for long periods of time without water, but they do not grow.) Assuming the 1,000:1 figure, the maximum plant growth that could be expected in a region with 5 inches of rain and no irrigation is 0.6 ton/acre-yr. Oil yields would be correspondingly low or less than 1 bbl/acre-yr.

- **Natural systems.** — In addition to agriculture, there has also been interest in using biomass produced by plants in their natural state. In

[3]D. Gilpin, S. Schwartzkopf, J. Norlyn, and R. M. Sachs, "Energy From Agriculture—Unconventional Crops," University of California at Davis, contractor report to OTA, March 1979.

the natural state, most of the nutrients are returned to the soil as the leaves drop off or the plant dies and decays. Harvesting of some of the biomass removes some of the nutrients, although animal excretions and the natural breakdown of minerals in the soil provide new nutrients. The rate of replenishment varies considerably from area to area, however, and this determines the rate that biomass could be removed from natural systems without depleting the soil.

The potential growth of biomass in continuously harvested natural systems has apparently not been studied. (Forestlands are an exception, although the emphasis there has been on the production of commercial timber rather than on total biomass.) It has been estimated, however, that some natural wetlands produce more than 5 ton/acre-yr of growth, and that 11.4 million acres of rangeland produce more than 2.5 ton/acre-yr.[4] While no estimates for the production of natural systems can be given, they will certainly permit less harvestable growth than intensively managed systems on comparable soil.

In evaluating the possible yields for biomass production, all of the yield estimates here should be treated with extreme caution. None of these yields has been achieved under large-scale cultivation (i.e., millions of acres) and the estimates for oil-producing plants are particularly uncertain. Experimental plot yields, on the other hand, exceed these yields for many plants.

Moreover, several factors operate to prevent average yields from reaching these estimates for large-scale production of biomass. The most important are the less than ideal soils of most potential cropland and the fundamental limitations of plant genetics with current knowledge. On the other hand, management practices improve with time and increased costs for farm products may eventually justify more extensive management practices, such as additional fertilizers, extensive soil treatment, and expanded irrigation.*

[4]"An Assessment of the Forest and Range Land Situation in the United States," Forest Service, U.S. Department of Agriculture, review draft, 1979.

*It is unclear whether irrigation will be socially acceptable for

Each plant is, to a certain extent, a special case. The experience with large-scale cultivation of crops is limited to a few food, animal feed, fiber, and chemical crops. Many plant scientists argue that maximum food production implies maximum biomass production. However, few genetic and development programs have been specifically aimed at maximizing biomss output for crops suitable to large areas of the United States.

These contradictory factors mean that the potential for biomass production is uncertain. And the uncertainty of the estimated yields is judged to be at least 50 percent. Consequently, the yields could easily vary anywhere from 0.5 to 1.5 times the numbers reported.

It is highly unlikely, however, that average U.S. yields for corn will exceed 140 bu/acre before 2000, and perhaps not after then. And corn is one of the best biomass producers for the U.S. climate known to man. Consequently, the numbers reported represent reasonable limits in terms of what is known today. Any large-scale production of biomass that significantly exceeds these yields would represent a major breakthrough. Estimates that are based on projected yields significantly exceeding those in table 36 either: 1) are limited to the relatively small acreage of the best U.S. soils,

Table 36.—Optimistic Future Average Crop Yields for Plants Under Large-Scale Production[a]

Region	Product	Plausible average yield[b] (ton/acre-yr)
Midwest	Dry plant matter	15
Gulf Coast	Dry plant matter	21
Midwest	Sugar	4
Gulf Coast	Sugar	5.6
Midwest	Grain	3.9
Midwest	Vegetable oil or hydrocarbons	1.7 - 2.2 (12 - 16 bbl)
Area with 5 inches rainfall per year and no irrigation..........	Dry plant matter	0.6
Area with 5 inches rainfall per year and no irrigation..........	Vegetable oil or hydrocarbons	0.1 (0.7 bbl)

[a]In this context, large-scale production means cultivation on millions of acres of average U.S. cropland.
[b]Estimated uncertainty ± 50 percent.

SOURCE: Office of Technology Assessment.

energy production or whether the necessary water will be available.

2) rely on technologies that do not now exist and are not anticipated in the near future, or 3) require extensive management practices that are not likely to be cost effective unless there are dramatic increases in the prices for farm commodities.

Unconventional Land-Based Crops

A large number of plants not now grown commercially in the United States are potentially energy crop candidates. Some are relatively high biomass producers and others could provide a source of a variety of chemicals that could be used as fuel or as chemical feedstocks. Unlike conventional crops, these crops could be considered primarily for their value as fuel. (However, see also ch. 10.)

Assessing and comparing potential yields for the unconventional crops from literature reports are extremely difficult, since these reports often do not give dry yields, the plants often are grown on unspecified soils and in different climates, and the water and nutrient inputs often are not given. Furthermore, it is a well-known fact that experimental plot yields are larger than those achieved with large-scale commercial cultivation. For these reasons, the yields reported below should be treated with extreme skepticism. Comparative cultivation experiments and crop development will be needed in the various regions and soil types in order to establish which crops are, in fact, suitable or superior for bioenergy production. In broad terms, the categories include: 1) lignocellulose, 2) vegetable oil and hydrocarbon, and 3) starch and sugar crops. Each group is considered briefly below, and an incomplete list of candidate bioenergy crops is shown in table 37.

Lignocellulose Crops

Various species of hardwood trees (e.g., red alder, hybrid poplar) and grasses (e.g., kenaf, Bermuda grass, Sudan grass, big bluestem) are candidates for crops grown primarily for their high dry matter yields (lignocellulose crops).

Theoretically, one would expect perennial crops (like trees and some grasses) to be superior biomass producers to annual crops,

Table 37.—Incomplete List of Candidate Unconventional Bioenergy Crops[a]

Lignocellulose crops	
American sycamore	Red alder
Bermuda grass	Russian thistle
Big bluestem	Salt cedar
Gum tree (eucalyptus)	Sudan grass
Kenaf	Switchgrass
Napier grass	Tamarix
Poplar	Tall fescue
Reed canarygrass	

Vegetable oil and hydrocarbon crops	
Crambe	Milkweed
Guayule	Mole plant (euphorbia)
Gum tree (eucalyptus)	Safflower
Jojoba	Turnip rape

Starch and sugar crops	
Buffalo gourd	Kudzu vine
Chicory	Sweet potatoes
Fodderbeets	Sweet sorghum
Jerusalem artichoke	

[a]Some of these crops are produced commercially today on a limited scale, but not for their energy value.

SOURCE: D. Gilpin, S. Schwartzkopf, J. Norlyn, and R. M. Sachs, "Energy From Agriculture— Unconventional Crops," University of California at Davis, contractor report to OTA, March 1979; S. Barber, et al., "The Potential of Producing Energy From Agriculture," Purdue University, contractor report to OTA; and J. S. Bethel, et al., "Energy From Wood," University of Washington, contractor report to OTA.

since the perennials develop their leaf cover sooner in the spring and do not need to generate a complete root system each year. One would also expect grasses to be superior biomass producers to trees because of their larger leaf area per acre of ground,* but considerable attention has been focused on trees, since the technologies for using wood are more advanced. Experimental plot yields for short-rotation trees are 5 to 20 ton/acre-yr.[5] Yields of as much as 10 to 15 ton/acre-yr may be achieved for large-scale cultivation of some of these crops in good soil (see "Crop Yields" section) but are likely to be 6 to 10 ton/acre-yr in poorer soils. Since farming costs could be similar to corn, this could result in biomass for about $20 to $50/ton.

---

*Thereby reducing light saturation, which lowers photosynthetic efficiency.

[5]Gilpin, et al., op. cit.

The trees would typically be grown for 6 to 10 years before harvest, while the grasses would be harvested several times a year. With fewer harvests for the trees, each harvest could be considerably more expensive and consume more energy than grass harvests without unfavorably affecting the economics or net energy balance. However, tree crops would require that the land be dedicated to the crop for several years and converting the land to other uses would be more expensive, due to the developed root system. Also, if a disease were to kill the crop, reestablishing a tree crop would be more expensive. Both trees and some grasses are perennial crops and, consequently, would require fewer herbicides and would reduce erosion on erosion-prone land as compared to annual crops. Grasses, having a more complete soil cover, would be more effective in preventing soil erosion.

Vegetable Oil and Hydrocarbon Crops

Vegetable oils and hydrocarbons are chemically quite different from petroleum oil. Nevertheless, most vegetable oils and hydrocarbons can be burned and might prove to be a substitute for fuel oils or, with refining, for other liquid fuels. However, appropriate methods for extracting the oil from the plant and for refining the oil are not well defined at present.

A number of edible and inedible vegetable oils are currently produced commercially.[6] In addition, unconventional crops such as gum tree (eucalyptus), mole plant (euphorbia), guayule, milkweed, and others could be used as vegetable oil and hydrocarbon crops (or for natural rubber). The maximum current yields of commercial oil plants are in the range of 100 to 200 gal/acre (2.5 to 5 bbl) of vegetable oil and/or hydrocarbon. Reports of 10 bbl/acre (420 gal) for euphorbia were apparently based on measurements of plants on the edge of a field, which were 1.5 times larger than interior plants. Also, in some cases, 16 months of growth were used to obtain "annual" yields.[7]

The theory of hydrocarbon and vegetable oil production in plants is not adequate to predict possible yields. However, from other considerations ("Crop Yields" section) there may be a significant potential for improvement. Furthermore, some of these crops (e.g., guayule) may do well on land where there is slightly less water available than would be needed for conventional crops.[8] Others, such as milkweed, can be grown with brackish water which would be unusable for conventional food crops.[9] Comparative tests under comparable conditions will be necesssary to determine which plants show the most promise for energy production.

Because of the higher prices that can be paid for chemicals and natural rubber, the fact that these products are economic in some cases does not in any way imply that energy production from vegetable oil and hydrocarbon plants will be economic. Some proponents of hydrocarbon plant development have failed to distinguish between these end uses, a fact that has led to considerable confusion and misunderstanding.

Critics of the development of vegetable oil and hydrocarbon plants for energy argue that the production of these products by plants is considerably less efficient than normal chemical synthesis (e.g., to produce methanol or ethanol from dry plant matter). They also point out that the plant often must be subjected to stress (drought or cold) to produce hydrocarbons, and this lowers the photosynthetic efficiency. Consequently, they contend that the high yields being predicted (e.g., 26 bbl/acre[10]), will not be achieved in the foreseeable future.

At present, however, the theory of and experience with these types of plants is inadequate to make a meaningful judgment.

---

[6]*Agricultural Statistics* (Washington, D.C.: U.S. Department of Agriculture, 1978).

[7]Gilpin, et al., op. cit.

[8]K. E. Foster, et al., "A Sociotechnical Survey of Guayule Rubber Commercialization," Office of Arid Land Studies, University of Arizona, Tucson, Ariz., prepared for the National Science Foundation under grant PRA 78-11632, April 1979.

[9]W. H. Bollinger, Plant Resources Institute, Salt Lake City, Utah, private communication, 1980.

[10]J. D. Johnson and C. W. Hinman, "Oils and Rubber From Arid Land Plants," *Science,* vol. 280, p. 460, 1980.

Starch and Sugar Crops

Starch and sugar crops are of interest since they can be used to produce ethanol with commercial technology. Current corn grain yields can be processed into about 260 gal of ethanol per acre cultivated and sugar beets (usually irrigated) can produce about 350 gal/acre, on the average. Irrigated corn, however, would match the sugar beet yield. Furthermore, experimental plot yields for corn produce about 430 gal/acre-yr and record yields exceed 850 gal/acre. In addition to the conventional starch and sugar crops, several other plants have been proposed as ethanol feedstocks including sweet sorghum and Jerusalem artichokes.

Experimental plot yields for sweet sorghum could be processed into about 400 gal of ethanol per acre year. Furthermore, this crop produces large quantities of residues that are suitable for use as a distillery boiler fuel. The yields for large-scale cultivation, however, are still unknown, and concern has been expressed that droughts during parts of the growing season could reduce sugar yields significantly.

Experimental plot yields for Jerusalem artichokes have produced about twice the sugar yields of sugar beets under the same growing conditions in Canada.[11] Whether this result can be applied to other regions is not known. Jerusalem artichoke, like the sugar beet, is a root crop. Harvesting it, therefore, causes extensive soil disturbance which increases the chances of soil erosion.

[11]Gilpin, et al., op. cit.

Other plants such as fodderbeets, sweet potatoes, and Kudzu vine are also potential ethanol crops. Comparative studies are necessary to determine which crops are best in each soil type and region of the United States. As was emphasized in chapter 3, this comparison should include the displacement of other crops that can be achieved by the byproducts of ethanol production. This factor tends to favor grains, but other possibilities do exist.[12]

General Aspects

Intensive cultivation of unconventional crops may cost about the same as corn, or $200 to $400/acre-yr in the Midwest. These costs, together with the yield estimates given above, allow an approximate comparison of the costs for various unconventional land crops, which is shown in table 38. Since the exact cultivation needs have not been established, a more detailed comparison is not warranted at this time. These costs estimates, however, show that unconventional crops may be economic energy sources. The ultimate costs will depend to a large extent on the yields that can actually be attained with intensive management and the success of developing crops that can be cultivated on land that is poorly suited to food production.

The crops that are now grown in U.S. agriculture were selected for properties that are unre-

[12]R. Carlson, B. Commoner, D. Freedman, and R. Scott, "Interim Report on Possible Energy Production Alternatives in Crop-Livestock Agriculture," Center for the Biology of Natural Systems, Washington University, St. Louis, Mo., Jan. 4, 1979.

Table 38.—Optimistic Cost Estimates for Unconventional Crops[a]

Product	Ultimate fuel	Yield of ultimate fuel per acre cultivated		Contribution of feedstock to fuel cost[b]	
Dry plant matter	Combustible dry matter	15 ton	(195×10^6 Btu)	$20/ton	($1.53/$10^6$ Btu)
Dry plant matter	Methanol	1,500 gal[c]	(95×10^6 Btu)	$0.20/gal	($3.15/$10^6$ Btu)
Dry plant matter	Ethanol	1,300 gal[d]	(107×10^6 Btu)	$0.23/gal	($2.80/$10^6$ Btu)
Grain	Ethanol	364 gal	(31×10^6 Btu)	$0.82/gal[e]	($9.89/$10^6$ Btu)
Sugar (Midwest)	Ethanol	540 gal	(46×10^6 Btu)	$0.56/gal	($6.50/$10^6$ Btu)
Vegetable oil or hydrocarbon crop	Vegetable oil or hydrocarbon	504-670 gal	($63-84 \times 10^6$ Btu)	$0.45-$0.60/gal	($3.60-$4.70/10^6 Btu)

[a]Based on yields in table 36.
[b]Assuming $300/acre cultivation and harvest costs; does not include conversion costs.
[c]Assuming yields of 100 gal/ton of biomass.
[d]Assumes yields of 85 gal/ton of biomass.
[e]Does not include byproduct credit for distillers' grain. If byproduct credits included, the situation becomes more complex as described in ch. 3, in the section on "Energy Potential From Conventional Crops." The byproduct credit would reduce the costs by roughly one-third.

SOURCE: Office of Technology Assessment.

lated to energy. It is likely, therefore, that other plants will prove to be superior to conventional crops for energy production. Beyond the yields of these crops, properties like insensitivity to poor soils, multiple products (e.g., vegetable oil, sugar, and/or starch plus dry plant herbage) displacement of other crops with crop byproducts (e.g., corn distillery byproduct), the energy requirements to cultivate the crop, the energy needed to convert it into a form that can be stored (especially for sugar and starch crops), tolerance to adverse weather conditions, ease of harvesting and conversion to fuels, and the environmental impacts of growing the crop are all factors that should be considered when choosing energy crops. In short, analyses of the net premium fuels displacement per new acre cultivated (as was done for various conventional crops in ch. 3), the cost, and the environmental impacts are needed to compare the options. Due to the diversity of U.S. soils and climates, different crops will no doubt prove to be superior in different regions. Many of the possible unconventional crops appear promising, but the ultimate decisions will have to come from experiment and experience. (Typically it requires 10 to 20 years to develop a new crop.) Nevertheless, some general aspects of plants can be expected to hold for the unconventional crops.

Root plants (e.g., Jerusalem artichokes, sugar beets, potatoes, sweet potatoes) will cause the most soil erosion. Annual crops will be next, and perennial grasses can virtually eliminate soil erosion.

Soil structure and climate are dominant features controlling plant growth and these can be controlled by man only to a very limited extent. Plants vary as to their sensitivity to these factors and to the presence of nutrient solubilizing mycorrhizae in the soil,[13] but yields will decrease on going to poor soils and climates. Crops grown in arid climates without irrigation or an underground supply of water will give low yields; and social resistance to using water for energy production in the West could preclude irrigated energy crops, although some people maintain that this resistance will not extend to crops.

Finally, any crop that grows very well in an area without inputs from man is likely to spread and become a weed problem.

---

[13]J. M. Trappe and R. D. Fogel, "Ecosystematic Functions of Mycorrhizae," reproduced from *Range Sci. Dep Sci.*, series No. 26, Colorado State University, Fort Collins, by U.S. Department of Agriculture.

Aquaculture

Aquatic plants comprise a diversity of types, from the single-celled microalgae to the large marsh plants such as cattails and even some trees such as mangroves. Considerable interest exists in the cultivation of many different aquatic plants as energy sources. Examples are the production of cattails in the extensive marshes of Minnesota, the cultivation of water hyacinths on wastewaters in Mississippi or Florida, and the establishment in the Southwest of large-scale brackish water ponds for microalgal production of chemical feedstocks. OTA has prepared a detailed review of the potential of fresh and brackish water aquaculture

systems for energy production.[14] The general conclusions were that the production of aquatic biomass has near-term potential in conjunction with wastewater treatment and high-value chemicals production. However, the development of large-scale "energy farms" based on aquatic plants is less promising at present, from both an economic or a resource potential viewpoint. Nevertheless, aquatic plants have certain unique attributes, the key one being high achievable biomass production rates

---

[14]J. Benemann, "Energy From Aquaculture Biomass Systems: Fresh and Brackish Water Aquatic Plants," Ecoenergetics, Inc., Vacaville, Calif., contractor report to OTA, April 1979.

which justify continued research on a variety of approaches to the development of aquaculture energy systems.

Higher aquatic plants growing in or on water are not, as a rule, water limited—a common and natural state of land plants. Thus, they are capable of higher rates of photosynthesis by keeping their stomata (plant pores) open longer than land plants* thereby, increasing CO_2 absorption. Thus, plants such as the water hyacinth and cattails exhibit very high rates of biomass production, often exceeding 20 ton/acre-yr.[15] This high productivity is achieved, however, by evaporation of large amounts of water, exceeding by a factor of two to four that transpired by land plants. Thus, cultivation of water plants can only be considered where ample supplies of water exist or where the systems are covered, such as in greenhouse structures.

Some aquatic plants, however, do not exhibit very high biomass production rates. For example, the common duckweed (Lemna sp.) covers a water surface very rapidly; however, once this is achieved, further growth in the vertical direction is minimal. Thus, the productivity of such plants is relatively low when compared to plants such as water hyacinths and marsh plants which extend their shoots up to several feet into the air. Indeed, the high leaf area index (the ratio of the total leaf area to the ground area), sometimes exceeding 10, of these plants, accounts, along with high transpiration rates, for their high productivity.

Another type of aquatic plant that exhibits relatively low productivity is the salt marsh plant Spartina, which does not produce as much biomass as its freshwater analogues such as Typha (cattails) or Phragmites (bullrush). The high salt concentration tolerated by Spartina also results in a decrease of transpiration and productivity. Even among the freshwater marsh plants, biomass productivities are limited by both the seasonal growth patterns of the plants in the temperate climate of the

United States and the large fraction of biomass present in the root system which may be difficult to recover. The submerged aquatic plants such as the notorious weed Hydrilla, are also not remarkable for their biomass productivity. Adaptation to the light-poor environment frequently encountered below the water surface has made these plants poor performers at the high light intensities that would be the norm in a biomass production system.

Finally, the case of the microalgae must be considered. Being completely submerged they also are subject to significant light losses by reflection from the water surface (at low solar angles) and scattering of light. More importantly, in a mixed algal pond, the cells near the surface tend to absorb more light than they can use in photosynthesis, resulting in a significant waste of solar energy. However, if a microalgal production system is designed to enhance mixing, then rapid adjustment by the algae occurs, thus overcoming, to some extent, the handicap inherent in inefficient sunlight absorption by microalgal cultures. Therefore, microalgal cultures could be considered in a biomass production system. A review of the rather sparse productivity data available, together with consideration of the basic photosynthetic processes involved, suggest that green algae and diatoms are promising candidates for mass cultivation, probably with achievable production rates of at least 20 ton/acre-yr, with blue-green algae, particularly the nitrogen-fixing species, considerably less productive.

It must be noted that the available data on aquatic plant productivity are too limited to allow confident extrapolations to large-scale systems. Most available data are based on natural systems where nutrient limitations may have depressed productivity or small-scale, short-term experimental systems where edge effects and other errors may have increased productivity. Actual yearly biomass production rates in sufficiently large-scale managed systems must be considered uncertain for any aquatic plant, particularly if factors such as stand establishment, pest control, optimal fertilizer supply, and harvesting strategy are concerned. Thus, to a considerable extent, assess-

*Somata are closed to conserve water, but this also prevents carbon dioxide from entering the leaf.

[15]Ibid.

ing the potential of aquatic plants in energy farming, like that of other unconventional crops, involves more uncertainty than specific detailed knowledge.

Among the uncertainties are the economics of the production system, including the harvesting of the plants. Detailed economic analyses are not available; those that have been carried out are based on too many optimistic assumptions to be credible or useful. Of course each type of plant will require a different cultivation and harvesting system. However, in all cases, these appear to be significantly more expensive per acre in both capital and operations than the costs of terrestrial plants. This increased cost per acre can only be justified by an increased biomass production rate or a specific, higher valued product. Because the productivity and economics of aquatic plants are, to a large degree, unknown, the potential for aquatic plant biomass energy farming is in doubt.

One approach to improve the economics of such systems is to combine the biomass energy system with a wastewater treatment function. As aquatic plants are in intimate contact with water, they can perform a number of very important waste treatment functions—oxygen production (by microalgae) which allows bacterial breakdown of wastes, settling and filtration of suspended solids, uptake of organics and heavy metals, and, perhaps most importantly, uptake of the key nutrients that cause pollution. The relatively high concentrations of nitrogen and phosphorus in aquatic plants (e.g., about 10 percent nitrogen and 1 percent phosphorus in microalgae and 3 percent nitrogen and 0.3 percent phosphorus in water hyacinths), makes these plants particularly useful in nutrient removal from wastewaters. Research in wastewater aquaculture is well advanced, although some critical problems remain to be elucidated, and several large demonstration projects are being initiated throughout the United States. For example, water hyacinths are being used in wastewater treatment plants in Coral Gables and Walt Disney World, Fla., in projects which involve fuel recovery by anaerobic digestion of the biomass. Microalgal

ponds have been used for several decades in many wastewater treatment systems throughout the United States. More stringent water quality standards are resulting in a need for better microalgal harvesting technology and presenting an opportunity for fuel recovery from the harvested microalgae. Several projects throughout the United States have demonstrated the beneficial effects of marsh plants in wastewater treatment. In all cases, wastewater aquaculture appears more economical and less energy intensive than conventional technologies.[16] However, the total potential impact of wastewater aquaculture on U.S. energy supplies, even when making favorable market penetration assumptions, is minimal— about 0.05 to 0.10 Quad/yr.[17]

For aquatic plants to make a more significant contribution to U.S. energy resources, other types of aquatic biomass energy systems must be developed. One alternative is the conversion to fuel of aquatic plants already harvested from natural, unmanaged stands. Examples are water hyacinth weeds removed by mechanical harvesters from channels in Florida and other southern States and cattails or bullrushes cut periodically in natural marshes in Minnesota or South Carolina to improve wildfowl habitats.[18] However, the infrequent occurrence of such harvests, the small biomass quantities involved, and transportation difficulties make energy recovery from such sources essentially impractical. The conversion, if practiced, of natural marsh systems to large-scale managed (planted, fertilized, harvested) plantations will present significant ecological problems and, even if these are ameliorated or overcome, opposition by environmental groups. Nonetheless, large areas of marshes do exist in the United States and they, in the long term, may become resources that could be exploited on a multipurpose and sustained yield basis like the national forests. In the near term, however, the technology for aquatic plant biomass energy systems must be developed with presently unused or "margin-

---
[16]Ibid.
[17]Ibid.
[18]Ibid.

al'' land and water resources. In addition, relatively high-value biomass energy products, specifically chemicals and liquid fuels, should be produced by such systems. Examples of such systems include the production of alcohol fuels from cattails (either by hydrolysis of the areal parts or directly from the starches stored in the roots) or the production of hydrocarbon fuels and specialty chemicals from microalgae.

Microalgae are known to produce a variety of useful chemicals. However, the development of such production technology is only just now beginning, and the potential resource base (land, water, nutrients) available for such systems is not yet quantified. Thus, the future contribution to U.S. fuel supplies of aquatic plant biomass energy systems cannot be predicted. However, sufficient possibilities and promise exist to warrant further R&D efforts.

Mariculture

This section describes problems and opportunities associated with developing future ocean farms which might use the giant kelp (macrocysts) as a future biomass energy source. Other macroalgae have also been proposed as potential marine biomass crops. By examining the possibilities of kelp and also noting other proposals, OTA hopes to illustrate the status of this technology in general, its future potential, the problems involved, and the Federal role in this segment of alternative energy research.

Macroalgae are harvested around the world. About 2 million wet metric tonnes are now cut annually, and estimates are that the total potential worldwide crop is 10 times this much—about 20 million wet tonnes.[19]

In recent decades seaweed cultivation has rapidly become more successful and has substantially added to annual harvest figures. For example, as of 1970 there were 130,000 acres of sea surface under cultivation in Japan, about 25,000 acres in The Peoples Republic of China, and additional acreage in Taiwan, Korea, the Philippines, and elsewhere. None of the current annual world harvest is being used for energy production.

In the United States, where wild seaweed beds have been harvested for many years, the possibility is beginning to be studied of increasing production through ocean farm cultivation techniques. A small test farm has been installed along the California coast.[20]

Large ocean kelp farms could theoretically supply significant quantities of natural gas (methane). Linked to a methane production system, for example, and assuming serious technical problems are solved, a 1-million-acre kelp farm could produce enough gas to supply 1 percent of current U.S. gas needs.

It would be no easy matter to farm such vast tracts of ocean. Much still needs to be learned about macroalgae cultivation. But serious research is reducing the areas of ignorance and seaweed may some day become a biomass producer.

Algae are among the simplest and most primitive of plants. The larger macroscopic algae are commonly referred to as seaweeds or macroalgae. Large seaweeds are the dominant plant in most shallow coastal waters including those off California and Mexico, where they attach themselves to rocks or some other hard substrate under water.

To date, the seaweeds apparently most adaptable to human cultivation are the red and the brown algae. People have eaten red algae varieties for thousands of years, especially in countries such as Japan and China.

---

[19]G. Michanek, *Seaweed Resources of the Ocean,* U.N. Food and Agriculture Organization, Rome, 1975.

[20]A. Flowers, statement before the House of Representatives Committee on Merchant Marines and Fisheries, Subcommittee on Oceanography, p. 18, committee report serial No. 95-4, June 7, 1978.

The brown algae group includes the giant kelp *Macrocystis* (figure 13), already harvested

Figure 13.—Macrocystis Pyrifera

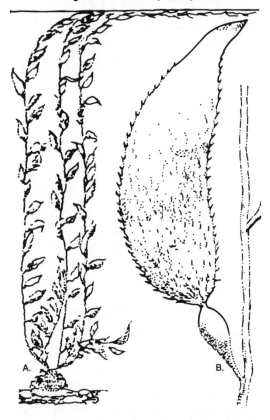

(A: 1/64 natural size; B: 1/4 natural size.) The Giant Kelp is shown in the left part of the plate in a natural pose with the long leafy stipes rising to the sea surface from the massive holdfast. On the right is one of the leaf-like fronds showing the gas-filled float bladder at its base and the distinctive teeth along the margin (Anon. 1954).

SOURCE: Velco, Inc.

in the United States from wild and semiculti-vated beds and considered at present as the best candidate for intensive cultivation off California and as a possible fuel producer.[21]

Kelp may grow in length as much as 2 ft/day or increase its weight by 5 percent per day un-der optimum conditions. The plants form natu-ral beds up to 3 miles wide and several miles long in southern California. This kelp is now harvested and put to a variety of uses, prin-cipally in the food-processing industry. Fuels have never been produced from kelp except in minute quantities as part of research testing.[22]

Unfortunately, there is no consensus among the experts who have made projections as to the potential of ocean energy farms. Their estimated costs vary widely and are based on such very sparse data that they cannot be used to either support or reject ocean farm propos-als. Estimates of production rates vary by fac-tors of as much as 100. Better experimental data and more complete biological engineer-ing tests will allow for better estimates in the future. The estimates used here lie approx-imately in the middle of responsible optimistic and pessimistic projections for a 400,000-hec-tare (1 million acre) ocean kelp farm:

- average productivity = 20 dry ash free (DAF) tons per acre per year, and
- average annual energy produced = 0.2 Quad (1 percent of U.S. gas consumption of 20 Quads/yr).

Such a system, if built, would provide the equivalent in energy supply of one large LNG-importing plant such as the one located at Cove Point, Md. It would, of course, be a do-mestic rather than an imported fuel, however.

Experiments are underway into the best laboratory-reared seaweed farms. Eventually, some researchers hope to produce a "pedi-greed" kelp bred specifically for high methane production, fast growth, and hardiness.

A key problem faced by potential ocean kelp farmers is to deliver enough nutrients to the plants to fertilize them. This is because, while the deep waters of the ocean contain many necessary nutrients, surface water is often as devoid of nourishment as a desert is

[21]Neushal, et al., "Biomass Production Through the Cultiva-tion of Macroalgae in the Sea," p. 100, Neushul Mariculture, Inc., for OTA, Oct. 6, 1978.

[22]M. Neushal, "The Domestication of the Giant Kelp, *Macro-cystis* as a Marine Plant Biomass Producer," presented at the *Ma-rine Biomass of the Pacific Northwest Coast Symposium*, Oregon State University, Mar. 3, 1977.

devoid of water. One fertilizing technique being tried is artificial upwelling of seawater, which involves pumping nutrient-rich, deep ocean water to the surface to benefit the kelp plants.[23]

Current research on marine plants can be divided into two categories.

The first category, funded by several Federal agencies to a total of about $1 million in 1979, generally includes research projects aimed at a better understanding of marine plants, their cultivation, and potential new uses of the plants.

The other "category" is actually just one project: the Marine Biomass Research Program jointly funded by the Gas Research Institute (GRI) and the Department of Energy (DOE), which has funded over $9 million of directed research as of 1979.

This ongoing marine biomass project includes a test farm off California. The farm began artificial upwelling experiments late in 1978, but was forced to suspend operations in early 1979 due to storm damage. This prototype is meant to provide biological information and research clues needed to operate much larger culture farms. It also aims at experimental work into cultivation of giant kelp on moored structures in the open ocean. The test farm, may lead to the actual operation of a full-scale ocean farm.

There is considerable difficulty at this time in evaluating the appropriateness of the Marine Biomass Research Program because little has been produced. It is important that research results on the cultivation of kelp on ocean farms be reported in a comprehensive way and subjected to critical review if a future large program is to be justified.

Kelp and other seaweeds are potentially a highly productive source of biomass for fuels. Estimates can vary drastically as to what may be possible for future large ocean farms, but OTA's evaluation of a hypothetical ocean kelp farm indicates productiveness could range from a low value of 6 DAF ton/acre-yr to a high value of 30. In comparison, this country's aver-age corn harvest is 6 DAF ton/acre-yr and Hawaiian sugarcane averages 14 DAF ton/acre-yr.

OTA estimates that if about 1 million acres were ever farmed, the gross energy production could amount to 0.2 Quad. This is equal to approximately 1 percent of current U.S. natural gas consumption. These production estimates should be treated with caution since there are no ocean kelp energy farms and nobody has ever planted and harvested a macroalgae energy crop.

Actual gross energy production from such a huge hypothetical ocean kelp plantation has been projected by other researchers to range from 10 times OTA's estimate to only one-tenth that figure. The entire project might simply prove impossible, others caution. Years of experiments will be necessary before any projections can be confirmed.

There is even less data to draw on in estimatint net energy possibilities. In a report prepared for DOE by the Dynatech R&D Corp., net energy outputs were estimated to range anywhere from a negative number to about 70 percent of crop energy.[24]

Much of the technology to construct present concepts of open ocean farms is already well known. Similar platforms, structures, and moorings have been built for the offshore oil industry and the existing seaweed industry uses mechanical harvesting techniques.

Less certain areas of ocean farm engineering at present include nutrient distribution, dispersion characteristics of upwelled water, and specific configuration of the structure to which kelp plants will be attached. A major problem for cultivated kelp beds may be to supply an ocean farm with proper nutrients in correct quantities. The extreme difficulties of noting the delicate balance of nutrients found in a natural environment and reproducing this in a cultivated one are well known to researchers.

[24]Dynatech R&D Co., "Cost Analysis of Aquatic Biomass Systems," prepared for the Department of Energy, contract No. HCP/ET-400-78/1.

Test farms will upwell deep ocean water to supply kelp with proper nutrients. Reservations about this procedure are twofold among skeptics. They worry that deep water could become stagnant under the farms, or that, once upwelled, the water would dilute too rapidly and sink again.

As previously mentioned, this country's major ocean biomass project is jointly supported by GRI and DOE. The project may become the most heavily funded biomass program of the 1980's, with grants projected to grow to over $50 million yearly by 1983. Plans for this project have been developed mainly by GRI, although regular DOE approval for phases of the project is mandatory.[25]

GRI estimates imply that ocean kelp farming could be a commercially viable project for this country. The Institute's fuel production cost estimates for methane generation from kelp range from $3 to $6/million Btu.

The previously mentioned Dynatech report on fuels from marine biomass comes to a different conclusion. Its estimates range from $7/million Btu up to several hundreds of dollars per million Btu, should productivity prove low and design costs high.

Some critics of the GRI marine biomass program contend that there is not enough data available to justify the level of expenditures for the biological test farm.

Critics have stated that the open ocean test farm is an inappropriate and perhaps premature step in a long, logical process of developing future deep sea operations. Considerations which may be overlooked by this test farm approach include:

- the need for better information on kelp growth and productivity and limiting factors in natural beds;
- the need for additional basic research into nutrients and productivity (much research

is also needed on plant diseases, predators, and water movement and quality);
- the possibility of developing shallow water kelp farms either in areas of natural upwelling or in conjunction with other fertilizing techniques (see ch. 10);
- hard data on net energy expectations is lacking; and
- no plans are being readied at present as alternatives to fertilization by upwelling.

Since plans for future ocean biomass farms call for the use of millions of acres of ocean surface, there will be conflicts with other traditional users. The dedication of large areas of open ocean surface for a single commercial purpose such as this is unprecedented. It would require complex, special regulation after review of current local, national, and international laws.

Even though the ocean space within the 200-mile zone surrounding the United States is 1½ billion acres, conflicts can be expected with such traditional ocean users as commercial shipping, the navy, commercial and sport fishing, offshore oil and gas operation, and recreational boating. To date, no detailed investigation of legal or institutional approaches to resolving conflicts has been accomplished. This issue will need analysis prior to any large-scale initiative in ocean farming, and will have a major impact on feasibility, productivity, and cost of marine biomass in the future. Analyses of specific sites and siting problems will be crucial to the ocean question.

OTA has found that Federal research programs directed toward energy problems have not been adequately coordinated with similar research directed toward production of food, chemicals, or other products.

Much research is needed to develop any suitable marine plant culture regardless of whether the end product is food or fuel. Such basic research could be better supported and coordinated by all interested agencies. Programs supported by Sea Grant and the National Science Foundation have tended to focus on basic biological efforts or food production goals while DOE programs are focused on pri-

[25]General Electric Co., briefing, "Energy From Marine Biomass Project, Program Review," for the Gas Research Institute, Newport Beach, Calif., March 1978.

mary fuel production. Since DOE now has the major funds available for seaweed research, the tendency has been to create programs focused on fuel production. The encouragement of further diversity in existing seaweed research efforts is essential to a long-term improvement in the knowledge and capability of developing future marine plant culture programs.

One approach to conducting a systematic program for developing ocean farms would be to expand research in natural seaweed beds and shallow water farms prior to experiments in deepwater, open ocean farms. This possibility would mean coordination of several existing research efforts; expanding some, developing some new ones, and generally integrating many efforts focused on basic biological questions and food production as well as energy production.

Other Unconventional Approaches

There are several other unconventional approaches to biomass production. Because of the complexity of plant growth, it is likely that many approaches will be tried and fail. However, this complexity also gives rise to significant possibilities. While all unconventional approaches cannot be covered here, a few are discussed below.

Multiple Cropping

Multiple cropping consists of growing two or more crops on the same acreage in a year. Growing winter wheat on land that produced a summer crop is one example. The winter wheat can delay spring planting, so its use is applicable only for land where certain summer crops are to be grown. However, this is basically a conventional approach.

The unconventional multiple cropping consists of growing more than one summer crop on an acreage by harvesting the first crop before it matures or developing species that mature rapidly. Since starches, sugars, vegetable oils, and hydrocarbons are generally produced in the greatest quantities in mature plants, this approach would probably reduce the overall yields of these products. Also, the time between the harvest of the first crop and the development of a full leaf cover in the second crop will be a time when sunlight is not captured by the plants as effectively as it could. Consequently, this approach would also be expected to reduce the total biomass yields.

Chemical Inoculation

By subjecting some plants to herbicides like paraquat or 2,4-D, hydrocarbon or vegetable oil production can sometimes be increased. These chemicals block certain biochemical pathways, thus promoting greater production of other products. Preliminary results with guayule, for example, indicate that 2,4-D may cause a doubling of the natural rubber content of this plant.[26] While it is too soon to assess this approach, it may prove to be an effective way of improving yields of these products.

Energy Farms

Energy farms have been proposed[27] as a means of providing a reliable supply of large quantities of biomass for large conversion facilities located on or near the farm. The basic idea is to have a large tract of land (tens

---

[26]Gilpin, et al., op. cit.
[27]See, e.g., G. Szego, "Design, Operation, and Economics of the Energy Plantation," *Proceedings Conference on Capturing the Sun Through Bioconversion* (Washington Center for Metropolitan Studies, 1976); G. C. Szego and C. C. Kemp, "Energy Forests and Fuel Plantations," *Chemtech*, p. 275, May 1973; and *Silviculture Biomass Farms* (McLean, Va.: MITRE Corp., 1977).

of thousands of acres) dedicated to growing the biomass feedstock for a nearby conversion facility. Although this is technically possible, a number of practical and economic considerations probably will limit investment in energy farms. Moreover, this approach ignores the effect that bioenergy production has on related sectors. Some of the more important of these points are:

- **Land.**—The land available for energy farms has often been estimated to be several hundred million acres.[28][29] OTA's analysis, however, indicates that considerably less land is available for biomass production (see ch. 3). Furthermore, there would be practical difficulties with buying large contiguous tracts of the size needed for large conversion facilities (tens to hundreds of thousands of acres).

  If cultivation on very poor or arid land proves to be feasible or if irrigation for energy production is socially acceptable and the water is available, then these limitations could be somewhat less severe than they appear to be at present.
- **Crop yields.**—Estimates of future yields from short-rotation tree farms have been as high as 30 ton/acre-yr,[30] which OTA considers to be highly unrealistic. Yields of 6 to 10 ton/acre-yr are more realistic for the poorer soil that could be available for energy farms.
- **Initial investment.**—If short-rotation trees are used as the energy crop, there would be a 6- to 10-year leadtime before the first harvest could be made. This would be prohibitively long for many investors. Grasses, however, would reduce the leadtime to a fraction of a year. In either case, the cost of acquiring the land would increase the initial investment substantially.
- **Risk.**—Using short-rotation trees as the energy crop would give yields that are less sensitive to weather than grass because the growth would be averaged over sev-

eral years. A pest infestation, however, could destroy the entire crop in which an average of 3 to 5 years' cultivation had been invested, and this could be financially disasterous. If grass is the energy crop, or the time between tree harvests is reduced, the loss from a pest infestation would be considerably less, but the yields would fluctuate more from year to year, making it necessary to rely on outside sources of biomass in years with low harvests or to sell surpluses in years with bumper harvests.

- **Competition with other uses for the land.**—Because of the uncertainty about future cropland needs for food production, it would be unwise to assume that tens of millions of acres could be devoted to a conversion facility for 30 years without affecting the price of farmland and thus food.
- **Preclusion of nonenergy benefits.**—OTA's analysis indicates that bioenergy harvests, if properly integrated into nonenergy sectors, can provide benefits beyond the energy, such as increased growth of timber suitable for paper and lumber. Attempting to isolate bioenergy production from these other sectors would preclude some of the potential benefits.

Although none of these factors is insurmountable, taken together they make energy farms appear considerably less attractive than numerous other bioenergy options. Particularly because of the risk and the initial investment, it is more likely that bioenergy crops will be grown as one of the many crop choices available to farmers, rather than on large tracts dedicated solely to energy production. There is, however, no technical reason why energy farms cannot be constructed.

Biophotolysis

Biophotolysis is generally defined as the process by which certain microscopic algae can produce hydrogen (and oxygen) from water and sunlight. Two distinct mechanisms are known by which microalgae can carry out biophotolysis: 1) through a "hydrogenase" en-

---

[28]Szego, op. cit.

[29]*Silviculture Biomass Farms,* op. cit.

[30]J. A. Allich, Jr., and R. E. Inman, "Effective Utilization of Solar Energy to Produce Clear Fuel," Stanford Research Institute, final report No. NSF/RANN/SE/GI 38723/FR/2., 1974.

zyme (biological catalyst) which is activated or induced by keeping the microalgae in the dark without oxygen for a period of time; or 2) through the "nitrogenase" (nitrogen-fixing) enzyme which normally allows some types of microalgae (the "blue-green" algae) to fix atmospheric nitrogen to ammonia but which also can be used to produce hydrogen by keeping the algae under an inert atmosphere such as argon gas.

In the case of biophotolysis with hydrogenase the key problem is that when simultaneous oxygen production occurs, the hydrogenase enzyme reaction is strongly inhibited and the enzyme itself inactivated. Although it was recently demonstrated that simultaneous production of oxygen and hydrogen does occur in such algae,[31] it is uncertain whether it will be possible to sustain such a reaction in a practical system. This difficulty has led to proposals for separation of the reactions either by developing an algal system which alternates oxygen and hydrogen production, (possibly on a day-night cycle) or by developing a two-stage process. Such systems are still at the conceptual stage, although considerable knowledge exists about the basic mechanisms involved.

Somewhat better developed is a biophotolysis process based on nitrogen-fixing blue-green algae. In these algae the oxygen-evolving reactions of photosynthesis are separated from the oxygen-sensitive nitrogenase reaction by their segregation into two cell types—the photosynthetic vegetative cells and the nitrogen-fixing heterocysts. Heterocysts receive the chemicals necessary to produce hydrogen from vegetative cells but are protected from oxygen by their heavy cell wall and active respiration. Using cultures of such algae from which nitrogen gas was removed, a sustained biophotolysis reaction was demonstrated: about 0.2 to 0.5 percent of incident solar energy was converted to hydrogen gas over a 1-month period. However, significant problems still exist in the development of a practical system—10 times higher conversion efficiencies must be achieved, a goal which may not be reached

due to the high energy consumption of the nitrogenase reaction. Also, the mixture of hydrogen and oxygen generated by such a system may be expensive to separate.

Whichever biophotolysis mechanisms or processes are eventually demonstrated to be capable of efficient and sustained solar energy conversion to hydrogen fuel from water, they must take place in a very low-cost conversion system. The development of an engineered biophotolysis conversion unit must meet stringent capital and operational cost goals. As biophotolysis will be limited by the basic processes of photosynthesis—probably no more than 3 to 4 percent of total solar energy conversion to hydrogen fuel—this sets an upper limit to the allowable costs of the conversion unit. In principle, the algal culture—the catalyst which converts sunlight and water to hydrogen and oxygen—can be produced very cheaply; however, the required "hardware" to contain the algal culture and trap the hydrogen produced may be relatively expensive.

Biophotolysis is still in the early stages of development. No particular mechanism, converter design, or algal strain appears to be inherently superior at this stage. Claims that near-term practical applications are possible, that genetic engineering or strain selection can result in a "super" algae, or that biophotolysis is inherently more promising than other biomass energy options are presently not warranted. A relatively long-term (10 to 20 years) basic and applied research effort will be required before the practical possibilities of biophotolysis are established.

Inducing Nitrogen Fixation in Plants

The biological process of nitrogen fixation, the conversion of nitrogen gas (not a fertilizer) to ammonia (a fertilizer) has only been found to occur in bacteria and the related blue-green algae. These primitive organisms maintain the ecological nitrogen cycle by replacing nitrogen lost through various natural processes. The capability for nitrogen fixation expressed by many plants (soybeans, alfalfa, peas) is due solely to their ability to live in a symbiotic

[31]E. Greenbaum, *Bioengineering Biotechnology Symposium,* vol. 9, in press.

association with certain bacteria (of the genus *Rhizobia*), which form the characteristic "root nodules." A certain fraction of the photosynthetic products of these plants are transferred to the roots where they are used (as "fuel") by the bacteria to fix nitrogen to ammonia which is then sent (in bound form) to the protein synthesizing parts of the plant.

This process is, in principle, energy intensive, with each nitrogen atom (fixed) reducing the biomass production by several carbon atoms (about 2 to 3).[32] In practice, significant inefficiencies in the process are often noted, most particularly the recent discovery that some *Rhizobia* bacteria in root nodules waste a large fraction of the "fuel" supplied by the plant in the form of hydrogen gas.[33] By using *Rhizobia* strains that can effectively recycle the hydrogen gas, this loss may be overcome.

Although biological nitrogen fixation can substitute, to a large extent, for the fossil-fuel-derived nitrogen fertilizers currently used in agriculture, the tradeoff may be an overall reduction in biomass yields. In an era of decreasing fossil fuel availability, such a tradeoff may be desirable, particularly as the price of commercial fertilizers is a limiting economic factor in many biomass production proposals. However, nutrient recycling could be preferable to *de novo* production, as it probably would be less costly and energy intensive. Alternatively to biological nitrogen fixation, thermochemical conversions of biomass to synthesis gas and their catalytic conversion to ammonia are feasible. Whether this is more favorable both in terms of economics and energy efficiency is uncertain.

A number of scientists have proposed that, through genetic engineering, they could transfer the nitrogen-fixing genes directly to the plant. However, such proposals face technical barriers. For example, the nitrogen-fixing reaction is extremely oxygen sensitive and is unlikely to be able to take place in the highly oxygen-rich environment of a plant leaf. In principle, there would only be a relatively minor advantage for a plant to directly fix nitrogen rather than do so symbiotically. Much more basic knowledge in many areas of plant physiology, genetics, biochemistry, etc., as well as developments in genetic engineering and plant tissue culture will be required before the potential for practical applications of such concepts can be evaluated.

Greenhouses

It is well known that increasing the CO_2 concentration in the air results in significantly improved plant growth for some plants. Depending on the specific plant and the specific conditions of the experiments, a 50- to 200-percent increase in biomass production has been noted. Greenhouses have the additional advantages of providing a "controlled environment" where pest control, water supply, and fertilization can be better managed, resulting in potentially high yields. The higher temperature in greenhouses allows extended growing seasons in temperate climates. Greenhouse agriculture is rapidly expanding throughout the world to meet the demands of affluent countries for out-of-season vegetables and horticultural products. However, the high cost of greenhouse agriculture and its high energy consumption make production of staple crops unfeasible and proposals for biomass energy production unrealistic at present. Although significantly lower cost greenhouse technology is feasible in principle, biomass production costs in Arizona, for example, would still be 10 times as expensive as open-field biomass crops grown in the Midwest.[34] A significant inflation in farm commodity, farmland, and water prices could make greenhouse systems more attractive. At present and in the foreseeable future, however, greenhouses do not appear economically feasible for bioenergy production.

[32]K. T. Shanmugan, F. O Gara, K. Andersen, and P. C. Valentine, "Biological Nitrogen Fixation," *Ann. Rev. Plant Physiol. 29,* p. 263, 1978.
[33]Ibid.

[34]L. H. de Bivort, T. B. Taylor, and M. Fontes, "An Assessment of Controlled Environment Agriculture Technology," report by the International Research and Technology Corp. to the National Science Foundation, February 1978.

BIOMASS PROCESSING WASTES

Chapter 5.—BIOMASS PROCESSING WASTES

TABLES

FIGURES

BIOMASS PROCESSING WASTES

Introduction

There are a number of byproducts associated with growing biomass and processing it into finished products. The byproducts that are not generally collected in one place, such as logging or crop residues, are termed residues and are dealt with in chapters 2 and 3. The byproducts that are collected in one place are termed processing wastes for the purposes of this report and are considered in this chapter. The three main types of wastes considered are the primary and secondary manufacturing

wastes of the forest products industries, and the wastes associated with the processing of agricultural products and animal manures. Wastepaper, cardboard, and urban wood wastes are not considered in this report, since they fall into the category of municipal solid wastes, which is the subject of a previous OTA report.[1]

---

[1]*Materials and Energy From Municipal Waste* (Washington, D.C.: Office of Technology Assessment, July 1979), OTA-M-93.

Wood-Processing and Paper-Pulping Wastes

Based on published surveys and discussions with people familiar with the forest products industries, the fraction of wood feedstock that appears as residue was estimated for the various types of processes and regions of the country. These fractions and the U.S. Department of Agriculture's *Forest Statistics*[2] were used to estimate the quantities of residues generated by wood-processing and paper-pulping industries. There is, however, some uncertainty in these figures, since published data usually are reported in board feet or cubic feet (rather than dry tons) and often the bark is not counted. Furthermore, moisture loss during drying must be accounted for. Every effort was made to avoid these potential problems and adjust for the shrinkage.

Current data on the use of the manufacturing residues are not complete. In some cases data are available for only a few States or for some of the industries. In other published data, regional surveys are extrapolated to the entire country. The estimates presented here are based on several surveys,[3] but are nevertheless based on incomplete data.

---

[2]*Forest Statistics for the United States, 1977* (Washington, D.C.: Forest Service, U.S. Department of Agriculture, 1978).

[3]J. S. Bethel, et al., "Energy From Wood," contractor report to OTA, April 1979.

Figure 14 shows an approximate materials flow diagram for the harvested wood processed by the forest products industry. This is a national average diagram. There are, however,

Figure 14.—Material Flow Diagram for Forest Products Industry (in energy units, Quads/yr)

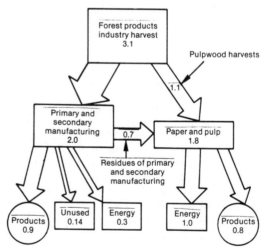

Total wastes used as energy: 1.2 to 1.3 Quads/yr
Unused waste   0.1 to 0.2 Quad/yr
Energy and unused   1.4 Quads/yr

SOURCE: Office of Technology Assessment.

significant variations between the regions, with the unused fraction being about twice as large in the East as in the West.

The largest user of biomass energy in the United States is the pulp and paper industry. This industry is currently 45- to 55-percent energy self-sufficient, up from 37 percent in 1967.[4] A major reason for the use of wood energy in the forest products industry is that the process used to recover the paper-pulping chemicals in most of the pulping processes involves burning the spent pulping liquor. This accounts for about 0.8 Quad/yr. The remaining 0.2 Quad/yr of bioenergy used in the pulp and paper industry comes from the bark of the harvested wood and reject woodchips.

The primary manufacturing industry produces lumber, plywood, poles, etc. The secondary industry produces furniture, prefabricated housing, etc. These industries are 20- to 40-percent energy self-sufficient.[5] About 50 percent (40 million dry ton/yr) of the primary

manufacturing wastes and 40 percent (4 million dry ton/yr) of the secondary manufacturing wastes go to paper pulping. Another 20 percent of each of these industries' residues goes to particle board and various other uses. About 20 million dry ton/yr (0.3 Quad/yr) of wood are used for energy; 9 million dry ton/yr (about 0.14 Quad/yr) are unused.

The main reasons that the unused portion is not used appear to be the very low quality of these wastes and a geographical mismatch between the source and potential users of the waste. However, either a strong wood energy market or cooperative agreements with electric utilities for cogeneration could bring these wastes into energy use.

There are alternative uses for some of the wastes other than for energy. If the demand for forest products increases and other fuels are available, then more of the primary and secondary manufacturing byproduct may be diverted from energy use to particle board and paper and pulp production. In addition, a small fraction of the spent pulping liquor could be used to produce ethanol and lignin products (as one Georgia Pacific Corp. plant does) instead of simply burning the spent liquor to recover the pulping chemicals.

---

[4]E. P. Gyftopoulos, L. J. Lazarides, and T. F. Widmer, *Potential Fuel Effectiveness in Industry* (Cambridge, Mass.: Ballinger Publications).

[5]S. H. Spurr, *Renewable Resources for Energy and Industrial Materials* (Austin, Tex.: LBJ School of Public Affairs, University of Texas, 1978).

Agricultural Wastes

With the exception of orchard prunings, agricultural waste byproducts are generally not collected at the place where the crops are grown. Rather, the wastes usually occur as byproducts to the agricultural product-processing industries. About 50 to 70 percent of these byproducts are sold as animal feed or for chemical production at prices that prohibit their use for energy.[6] The waste byproducts not being used for other purposes are considered in this section.

The various agricultural product-processing industries were surveyed[7] to determine the quantities and types of waste byproducts that

are produced. Table 39 shows the 10 major types of agricultural wastes and the energy potential of each. These 10 wastes represent over 95 percent of all agricultural wastes available for energy. Of these 10, about 90 percent are materials relatively low in moisture, and suitable for thermal conversion (combustion or gasification). The remaining 10 percent appear to be acceptable for anaerobic digestion or possible fermentation to ethanol in the case of fruit and vegetable wastes and cheese whey.

In addition, there is an unknown quantity of spoiled and substandard grain. One source[8] es-

---

[6]R. Hodam, "Agricultural Wastes," Hodam Associates, Sacramento, Calif., contractor report to OTA.

[7]Ibid.

[8]M. T. Danziger, M. P. Steinberg, and A. I. Nelson, "Storage of High Moisture Field Corn," Illinois Research, fall 1971.

Table 39.—The Ten Major Agricultural Wastes With
Potential to Produce Energy

Wastes	Btu/yr × 10^{12}
Orchard prunings[a]	30-61
Cotton gin trash[a]	20-31
Sugarcane bagasse[a]	4-8
Cheese whey[b]	4-8[c]
Tobacco (burley)[a]	2.3
Rice hulls[a]	2.2
Tomato pumice[b]	1.3-1.8
Potato peel and pulp[b]	1.0-1.1
Walnut shell[a]	0.9
Citrus rag and peel[b]	0.3-1.0
Total	66-117

[a]Suitable for combustion or gasification.
[b]Suitable for anaerobic digestion or fermentation.
[c]Based on starch content of milk and the volume of cheese production from *Agricultural Statistics* (Washington, D.C.: U.S. Department of Agriculture, 1978).

SOURCE: Office of Technology Assessment; and R. Hodam, ''Agricultural Wastes,'' Hodam Associates, contractor report to OTA, 1979.

timated corn spoilage from mold at 250 million bu/yr, but this number should be viewed as speculative. Furthermore, much of the spoiled grain may be accessible only as a supplement to existing distillery feedstocks because its occurrence is dispersed and unpredictable.

The four major sources of agricultural wastes are orchard prunings, cotton gin trash, sugarcane bagasse, and cheese whey. Most States have fruit or nut orchards, with the largest crops occurring in Arizona, California, Florida, Texas, New York, and Washington. Cotton gin trash is generally localized to the southern third of the United States and California. Sugarcane is processed primarily along the Gulf Coast, in Hawaii, and in New England. The majority of cheese whey is produced in Wisconsin, Minnesota, New York, Iowa, and California, but 30 States have some cheese production.

Orchard prunings are generally collected and burned onsite. A few growers disk whole prunings into the soil, although this is not a preferred practice for growers. With a strong energy market, much of this could be used for energy. The major expense is transporting the prunings to the place they are used.

Cotton gin trash is another potential source of energy. Texas cotton gins produce about five times as much energy in gin trash as they consume (mostly electricity). The major problems with using the trash for energy seem to be the difficulty of handling the trash, the seasonal nature of the ginning operations, and the difficulty in establishing cooperative ventures with the electric utilities. In addition, in the areas where the cotton plants are killed with arsenic acid prior to harvest, such as in much of Texas, special precautions will be necessary to burn the trash in an environmentally acceptable way.

Sugarcane bagasse is widely used in Hawaii as a source of energy. The sugar refineries have long-standing cooperative agreements with the electric utilities. Cogeneration is used to generate and export electricity to the utilities and to produce the process steam used by the sugar refineries. The electric generating facilities range in size from 1.5- to 33-MW electric. Most of the Hawaiian sugar refineries are 99- to 100-percent energy self-sufficient.

The New England and Southern sugar refineries should be analyzed in detail for the potential to duplicate the Hawaii experience, including the potential to purchase orchard prunings or wood wastes which are found in the same area in some cases.

OTA's analysis indicates that cheese whey is the largest source of food-processing waste suitable for conversion to ethanol, although other studies have indicated that citrus wastes are a larger source.[9] Based on total cheese production,[10] OTA estimates that 50 million to 100 million gal/yr of ethanol could be produced from cheese whey. Current production from this source is about 5 million gal/yr.

[9]*The Report of the Alcohol Fuels Policy Review* (Washington, D.C.: Department of Energy, June 1979), GPO stock No. 061-000-00313-4.
[10]*The Outlook for Timber in the U.S.* (Washington, D.C.: Forest Service, U.S. Department of Agriculture, 1974), report No. 24; and *Agricultural Statistics* (Washington, D.C.: U.S. Department of Agriculture, 1978).

Animal Manure

The major sources of animal manures suitable for energy are from dairy cows, cattle on feed, swine, chickens (broilers and layers), and turkeys. Only animals in confined animal operations are considered. However, it has been estimated that 48 percent of all manure voided from livestock (primarily sheep and cattle), is on open range.[11] This open range manure would require collection and, therefore, will not be economic in the foreseeable future.

The inventory of onfarm confined animals was derived from inventory numbers for animals that remain onfarm for more than a year and from sales numbers and the average time the animal spends on the farm for animals on farm for less than a year.[12] These inventory numbers were converted to the common basis of the number of animal units, or the equivalent of a 1,000-lb animal (defined in figure 15). The quantities of manure were calculated and, assuming that the manure is anaerobically digested to produce biogas (60 percent methane, 40 percent carbon dioxide), the energy equivalent was derived.

Table 40 shows the energy potential from each type of animal operation, and figure 15 shows the percent of this energy potential that is present on confined animal operations of various sizes (expressed in animal units). Currently most of this manure is used as nitrogen fertilizer and soil conditioner or is unused.

The total energy potential from manure produced in confined animal operations is about 0.3 Quad/yr. From one-third to one-half of this manure is currently allowed to wash away with rain or is allowed to dry which makes it unsuitable for anaerobic digestion. However, if it becomes economically attractive to digest the manure, then most of these operations can change their manure-handling techniques to accommodate anaerobic digestion.

Figure 15 shows that over 75 percent of the energy potential occurs on farms with less than

1,000 animal units and that about 45 percent of the potential is on farms with less than 100 animal units. Large feedlots (greater than 10,000 animal units) only account for about 15 percent of the total. Consequently, any technology development that is aimed at fully utilizing the potential for energy from animal manure will have to concentrate on relatively small-scale conversion units.

Figure 15.—Total Energy Available From Manure by Farm Size (confined animal operations)

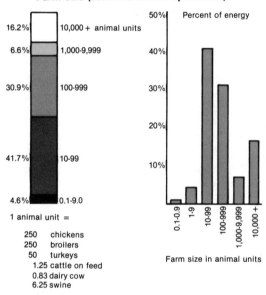

1 animal unit =

250	chickens
250	broilers
50	turkeys
1.25	cattle on feed
0.83	dairy cow
6.25	swine

SOURCE: K. D. Smith, J. Philbin, L. Kulik, and D. Inman, "Energy From Agriculture: Animal Wastes," contractor report to OTA, March 1979.

Table 40.—Energy Potential From Animal Manure on Confined Animal Operations

Type animal	Total energy potential Btu × 10¹²/yr	Percent or total
Dairy cattle	90	33
Cattle on feed	80	30
Swine	32	12
Chicken (broilers)	30	11
Chicken (layers)	25	9
Turkeys	18	6
Total energy potential from all manures	274	100

SOURCE: K. D. Smith, J. Philbin, L. Kulik, and D. Inman, "Energy From Agriculture: Animal Wastes," contractor report to OTA, March 1979.

[11]D. Van Dyne and C. Gilbertson, *Estimating U.S. Livestock and Poultry Manure and Nutrient Production* (Washington, D.C.: U.S. Department of Agriculture, 1978), ESCS-12.

[12]*1974 Census of Agriculture* (Washington, D.C.: Bureau of the Census, U.S. Department of Commerce), vol. 1-50.

Part II.

Conversion Technologies and End Use

Chapter 6
INTRODUCTION AND SUMMARY

Chapter 6
INTRODUCTION AND SUMMARY

Most bioenergy currently comes from direct combustion of solid biomass for space heating, process steam, and a small amount of electric generation. As chemically stored solar energy, biomass can be converted to a number of gaseous and liquid fuels, which can be used for a variety of energy purposes not suited to direct combustion.

Thermochemical conversion, or chemical processes induced by heat, is currently the most suitable process for the major biomass feedstocks—wood and plant herbage. Aside from direct combustion, these processes include gasification and liquid fuels production. Various types of gasifiers are being developed which could be used for process heat and retrofits of oil- and natural-gas-fired boilers. Suitable gasifiers may be commercially available in less than 5 years with adequate development support. Methanol synthesis is the near-term option for liquid fuels production. Wood-to-methanol plants can be constructed immediately, while herbage-to-methanol processes need to be demonstrated. Various other processes are being developed, and thermochemical conversion of biomass offers considerable promise for improved processes and new applications for fuel and chemical syntheses.

Fermentation is the biological process used to convert grains and sugar crops to ethanol—currently the only liquid fuel from biomass used in the United States. The byproduct of distillery grains can be used as an animal feed, thereby reducing the competition between food and fuel uses for the grain. Some farmers are producing ethanol onfarm, but with current technology the processes are not economic unless they are heavily subsidized or the onfarm production leads to increased grain prices—thereby enabling the farmer to earn more on the crops he/she sells for feed. However, process development could decrease the costs. Several processes for producing ethanol from wood and herbage are being developed, but the costs are highly uncertain.

Anaerobic digestion is a biological process, which produces a gas containing methane (the principal component of natural gas) and carbon dioxide. Suitable feedstocks include many wet forms of biomass, such as animal manure and some aquatic plants. For the near to mid term, digesters for onfarm production of gas from animal manure appear to hold the greatest promise. Not only can this technology serve as a waste disposal process, but it also could make most confined animal operations energy self-sufficient. There is a need to demonstrate a variety of digesters using different feedstocks to gain operating experience. Because the major cost is the initial investment, policies designed to lower capital charges will increase market penetration of the technology.

The alcohols most easily produced from biomass—ethanol and methanol—are not totally compatible with the existing liquid fuels system and automobile fleet. These alcohols can be used in gasoline blends or as standalone fuels, but methanol blends will have more problems than ethanol blends unless suitable additives are included with the methanol. All of the problems regarding the alcohols' incompatibility with the existing system have multiple solutions, but it is unclear which strategies will prove to be the most cost effective.

The energy balance for ethanol from grains and sugar crops has been the subject of considerable controversy, because the farming and processing energy consumption together are approximately the same as the energy contained in the ethanol. A net displacement of premium fuels—oil and natural gas—can be assured with ethanol, however, if: 1) distilleries do not use premium fuel for their boilers and 2) the ethanol is used as an octane-boosting additive to gasoline. Failure to fulfill either of these criteria could lead to ethanol production and use increasing the U.S. consumption of premium fuels, although there would be a small net displacement of premium fuels in most cases. Failure to comply with both criteria would almost certainly be counterproductive in terms of premium fuels displacement.

Methanol and ethanol can be produced from wood and plant herbage, although ethanol production is considerably more expensive with current technology. In each case, however, the biomass might be burned or gasified as a substitute for oil or natural gas. Liquid fuels production is considerably less efficient than combustion or gasification if the liquid is used as a standalone fuel. Using the liquid as an octane-boosting additive to gasoline, however, makes the options more comparable in terms of premium fuels displacement per ton of biomass. Future developments in refinery technology could change this conclusion.

Biomass already supplies substantial quantities of chemicals, and an expanded use of biomass chemicals is a widely discussed subject. Numerous plants produce potentially useful chemicals for industrial synthesis and as a source of natural rubber, mutant cells can produce highly specialized chemicals, and chemical synthesis from wood and plant herbage is developing or could be developed in a number of potentially very interesting directions. Because of the higher value of chemicals, as compared to fuel, the economic limitations on chemical production from biomass are considerably less severe than for energy production.

These topics and related aspects of conversion technologies and end use for bioenergy are presented in the following chapters.

Chapter 7
THERMOCHEMICAL CONVERSION

Chapter 7.—THERMOCHEMICAL CONVERSION

Chapter 7
THERMOCHEMICAL CONVERSION

Introduction

During the 1980's, the conversion processes with the greatest potential in terms of both the gross energy use and the largest possible displacement of oil and natural gas are the thermochemical processes, or processes involving heat-induced chemical reactions. Currently about 1.5 Quads/yr of biomass are combusted directly for process steam, electric generation (mostly cogeneration), and space heat. Intermediate-Btu gasifiers currently under development will be useful in retrofitting oil- and gas-fired boilers to biomass fuels and for crop drying and other process heat needs. Development of medium-Btu gasifiers is also underway and various processes for producing alcohols and other liquid fuels can be or are being developed. Also, methanol synthesis from wood can probably be accomplished with commercially available technology, while processes producing methanol from plant herbage can probably be demonstrated fairly rapidly. Moreover, there are good theoretical reasons for believing that the flexibility, efficiency, and usefulness of thermochemical processes can be significantly improved through basic and applied research into the thermochemistry of biomass.

Some generic aspects of biomass thermochemistry and generic reactor types are given first, followed by a discussion of the optimum size of some thermochemical conversion facilities and a more detailed consideration of select processes including densification, direct combustion, gasification, and direct and indirect liquefaction. Finally, the environmental impacts and research, development, and demonstration (RD&D) needs are presented.

Generic Aspects of Biomass Thermochemistry

Possible feedstocks for the thermochemical conversion processes include any relatively dry plant matter such as wood, grasses, and crop residues. Some conversion process designs accept a wide range of feedstocks, while others will be more suited to a specific feedstock. Although this is sometimes dependent on the chemical properties of the feedstock (e.g., manure), it more often depends on the physical properties of the material, such as its tendency to clog or bridge the reactor, the ease with which it can be reduced to a small particle size, and the materials' density.

Classification systems that provide information for assisting the designer of conversion equipment are not presently available for biomass feedstocks. Standard methods for biomass analysis or assays do not exist, although it is customary to use coal analyses (ultimate and proximate) for biomass. Some of the properties of some biomass materials using coal analyses are shown in tables 41 and 42. As a fuller understanding of biomass thermochemistry is developed, however, new classification schemes and methods of analysis are likely to be necessary.

Despite the differences in feedstocks, the generic thermochemical process consists of the following steps:

- moisture removal;
- heating the material to and through the temperature where it decomposes (about 400° to 800° F);
- decomposition to form gases, liquids, and solids; and
- secondary gas phase reactions.

The drying process absorbs the heat necessary to evaporate the water. This results in a decrease in the net usable heat from the feedstock as shown in figure 16. In this figure, the net heat content per pound of dry wood is

Table 41.—Proximate Analysis Data for Selected Solid Fuels and Biomass Materials (dry basis, weight percent)

	Volatile matter	Fixed carbon	Ash	Reference
Coals				
Pittsburgh seam coal.................	33.9	55.8	10.3	Bituminous Coal Research 1974
Wyoming Elkol coal...................	44.4	51.4	4.2	Bituminous Coal Research 1974
Lignite	43.0	46.6	10.4	Bituminous Coal Research 1974
Oven dry woods				
Western hemlock	84.8	15.0	0.2	Howlett and Gamache 1977
Douglas fir........................	86.2	13.7	0.1	Howlett and Gamache 1977
White fir	84.4	15.1	0.5	Howlett and Gamache 1977
Ponderosa pine....................	87.0	12.8	0.2	Howlett and Gamache 1977
Redwood	83.5	16.1	0.4	Howlett and Gamache 1977
Cedar	77.0	21.0	2.0	Howlett and Gamache 1977
Oven dry barks				
Western hemlock	74.3	24.0	1.7	Howlett and Gamache 1977
Douglas fir........................	70.6	27.2	2.2	Howlett and Gamache 1977
White fir	73.4	24.0	2.6	Howlett and Gamache 1977
Ponderosa pine....................	73.4	25.9	0.7	Howlett and Gamache 1977
Redwood	71.3	27.9	0.8	Howlett and Gamache 1977
Cedar	86.7	13.1	0.2	Howlett and Gamache 1977
Mill woodwaste samples				
−4 mesh redwood shavings	76.2	23.5	0.3	Boley and Landers 1969
−4 mesh Alabama oakchips	74.7	21.9	3.3	Boley and Landers 1969
Municipal refuse and major components				
National average waste	65.9	9.1	25.0	Klass and Ghosh 1973
Newspaper (9.4% of average waste)........	86.3	12.2	1.5	Klass and Ghosh 1973
Paper boxes (23.4%)	81.7	12.9	5.4	Klass and Ghosh 1973
Magazine paper (6.8%)	69.2	7.3	23.4	Klass and Ghosh 1973
Brown paper (5.6%)...................	89.1	9.8	1.1	Klass and Ghosh 1973
Pyrolysis chars				
Redwood (790° to 1,020° F)	30.0	67.7	2.3	Howlett and Gamache 1977
Redwood (800° to 1,725° F)	23.9	72.0	4.1	Howlett and Gamache 1977
Oak (820° to 1,185° F).................	25.8	59.3	14.9	Howlett and Gamache 1977
Oak (1,060° F).......................	27.1	55.6	17.3	Howlett and Gamache 1977

SOURCE: M. Graboski and R. Bain, ''Properties of Biomass Relevant to Gasification,'' in *A Survey of Biomass Gasification* (vol. II; Golden, Colo.: Solar Energy Research Institute, July 1979), TR-33-239.

Table 42.—Ultimate Analysis Data for Selected Solid Fuels and Biomass Materials (dry basis, weight percent)

Material	C	H	N	S	O	Ash	Higher heating value (Btu/lb)	Reference
Pittsburgh seam coal	75.5	5.0	1.2	3.1	4.9	10.3	13,650	Tillman 1978
West Kentucky No. 11 coal	74.4	5.1	1.5	3.8	7.9	7.3	13,460	Bituminous Coal Research 1974
Utah coal	77.9	6.0	1.5	0.6	9.9	4.1	14,170	Tillman 1978
Wyoming Elkol coal	71.5	5.3	1.2	0.9	16.9	4.2	12,710	Bituminous Coal Research 1974
Lignite	64.0	4.2	0.9	1.3	19.2	10.4	10,712	Bituminous Coal Research 1974
Charcoal...........................	80.3	3.1	0.2	0.0	11.3	3.4	13,370	Tillman 1978
Douglas fir	52.3	6.3	0.1	0.0	40.5	0.8	9,050	Tillman 1978
Douglas fir bark	56.2	5.9	0.0	0.0	36.7	1.2	9,500	Tillman 1978
Pine bark	52.3	5.8	0.2	0.0	38.8	2.9	8,780	Tillman 1978
Western hemlock.....................	50.4	5.8	0.1	0.1	41.4	2.2	8,620	Tillman 1978
Redwood...........................	53.5	5.9	0.1	0.0	40.3	0.2	9,040	Tillman 1978
Beech.............................	51.6	6.3	0.0	0.0	41.5	0.6	8,760	Tillman 1978
Hickory............................	49.7	6.5	0.0	0.0	43.1	0.7	8,670	Tillman 1978
Maple.............................	50.6	6.0	0.3	0.0	41.7	1.4	8,580	Tillman 1978
Poplar.............................	51.6	6.3	0.0	0.0	41.5	0.6	8,920	Tillman 1978
Rice hulls	38.5	5.7	0.5	0.0	39.8	15.5	6,610	Tillman 1978
Rice straw..........................	39.2	5.1	0.6	0.1	35.8	19.2	6,540	Tillman 1978
Sawdust pellets	47.2	6.5	0.0	0.0	45.4	1.0	8,814	Wen et al. 1974
Paper.............................	43.4	5.8	0.3	0.2	44.3	6.0	7,572	Bowerman 1969
Redwood wastewood	53.4	6.0	0.1	39.9	0.1	0.6	9,163	Boley and Landers 1969
Alabama oak woodwaste	49.5	5.7	0.2	0.0	41.3	3.3	8,266	Boley and Landers 1969
Animal waste........................	42.7	5.5	2.4	0.3	31.3	17.8	7,380	Tillman 1978
Municipal solid waste	47.6	6.0	1.2	0.3	32.9	12.0	8,546	Sanner et al. 1970

C = carbon   H = hydrogen   N = nitrogen   S = sulfur   O = oxygen

SOURCE: M. Graboski and R. Bain, ''Properties of Biomass Relevant to Gasification,'' in *A Survey of Biomass Gasification*, (vol. II; Golden, Colo.: Solar Energy Research Institute, July 1979), TR-33-239.

Figure 16.—Effect of Moisture on the Heat Content of Wood

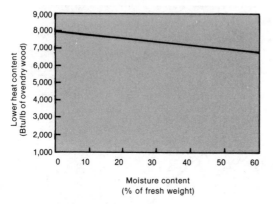

Moisture content
(% of fresh weight)

SOURCE: Office of Technology Assessment.

shown for various moisture contents. The heat content per pound of *moist* material, however, decreases much more rapidly with moisture content, due to the fact that part of each pound is water and not combustible material. (Nevertheless, the price of the moist feedstock will vary with the moisture content, so that $15/ton material at 50-percent moisture content is roughly equal to $30/ton of dry material. In this report, the feedstock costs are generally expressed as dollars per ton of dry material, so that variations in cost and heat content per ton are kept at a minimum.)

There is also a secondary effect of the moisture content of the feedstock. If moist feedstocks are combusted to produce steam, the boiler efficiency will usually drop if the feedstock moisture content is not that for which the boiler was designed. Aside from the heat lost in evaporating the water in the feedstock, high-moisture feedstocks have a lower flame temperature in direct combustion, which can result in particulate and creosote emissions (which escape without being completely combusted, if considerable excess combustion air is not used). (In poorly designed wood stoves or boilers, simply feeding excess air may not be sufficient to suppress these emissions.) In principle, a reactor can be designed to accommodate this excess combustion air, vaporized moisture, and lower flame temperature without a drop in efficiency, but in practice the efficiency is likely to drop.

A theoretical example of how the boiler efficiency drops with feedstock moisture content is shown in figure 17. Care should be exercised in applying these results to any given situation, since some factors which would vary with moisture content (e.g., excess air) are held constant in the calculations, but it does illustrate the point.

With gasification, the situation is slightly different. In this case the feedstock is decomposed into a fuel gas before combustion. The energy needed to vaporize the feedstock moisture is still lost, but the fuel gas can easily be mixed with the combustion air, so that excess air is not required, and the feedstock moisture is already vaporized, so the flame temperature can be high. Consequently, it may be possible to maintain the efficiency of gasification-combustion processes over a variety of feedstock moisture contents better than with direct combustion. Depending on reactor design, however, it may be necessary to limit the feedstock moisture in order to produce a flammable gas, and this point needs further investigation.

The rate that the biomass is heated to and through its decomposition temperature is a critical factor in determining the products. Many reactor designs are being developed to achieve high heating rates, as described below. (The heating rate is also determined by the particle size—small particles heating faster—and moisture content.) Depending on the products desired, however, one may want this heating rate to be slower.

The details of biomass decomposition are not well understood, but one can surmise the following. As the material is heated, the large biomass molecules (cellulose, hemicellulose, and lignin) begin to break down into intermediate-sized molecules. If the material stays in the heating zone long enough, the intermediate-size molecules decompose into still smaller molecules, such as hydrogen, methane, carbon monoxide (CO), ethane, ethylene, acetylene, and other chemicals. If the heating rate is too slow relative to the time the material is in the

Figure 17.—Effect of Feedstock Moisture Content on Boiler Efficiency

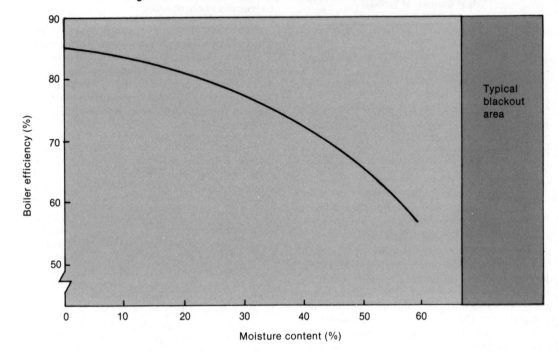

SOURCE: R. A. Arola, "Wood Fuels — How Do They Stack Up?" *Energy and the Wood Products Industry*, Forests Products Research Society, Proceeding No. 76-14, Nov. 15-17, 1976.

heating zone, the intermediate-sized molecules will escape and later condense as oils and tar. (This may also involve some intermediate reactions that are not well understood at present.) It also appears that a slow heating rate encourages the formation of char. Thus, a slow heating rate (either by design or due to excess moisture in the feedstock) will lead to the formation of varying amounts of char, tar, oil, and gas. With rapid heating, however, virtually the entire biomass goes to a gas with only the ash remaining.

Finally, the gases and vaporized tars and oils can react in the gas phase to form a new or modified set of products. Very little is understood about these secondary gas phase reactions, but they are of considerable importance in thermochemical processes. Depending on the oxygen and moisture content, the rate the biomass was decomposed, the temperature,

the pressure, and other variables not fully understood, the resultant gas can vary from almost pure carbon dioxide (CO_2) and water to gases with relatively high contents of materials such as hydrogen, methane, or ethylene (see ch. 12), or the gas can contain considerable quantities of particulates, various hydrocarbons, CO, and other pollutants.

Depending on the conditions chosen and the design of the reactor, the product(s) can be heat as in direct combustion, an intermediate- or medium-Btu gas suitable for oil- or gas-fired boilers and process heat, a gas suitable for chemical synthesis, oils, and/or char. But considerable research into the thermochemistry of biomass will be needed, before engineers will have the necessary information to design reactors that can achieve the full potential for the thermochemical conversion of biomass.

Reactor Type

Most commercial biomass reactors used for direct combustion or gasification are modifications of coal technology. The reactors proposed for direct liquefaction and densification, however, do not fall into this category and are considered in the sections dealing with these topics.

Although the technology for coal combustion and gasification is considerably more advanced than for biomass, it is generally agreed that grasses, wood, and crop residues are more readily gasified than coal or char. The biomass gasifies at a lower temperature and over a narrower temperature range than does coal, as illustrated in figures 18 and 19. Both of these properties favor rapid gasification. While these advantages of biomass over coal are partially offset by biomass' higher heat capacity (the amount of heat needed to raise the materi-

al's temperature a given amount)[1] coal gasification in advanced reactors will ultimately be limited by the rate that oxygen, CO_2, steam, etc., can diffuse to and into the surface of coal particles. Biomass gasification and decomposition, on the other hand, do not require the reaction of two or more separate species. Consequently, biomass gasification probably will be limited by the rate that heat can be transferred to the biomass.

In balance, these differences point to the conclusion that there is the potential for building biomass reactors that have considerably higher rates of throughput and thus lower costs than will be achieved with coal or has been achieved for either material so far. On the other hand, the most rapid heat transfer occurs when the feedstock particles are pulverized or of relatively small size. Most coals can readily be pulverized, but the fibrous nature of many types of biomass makes it difficult to reduce the particle size. Biomass densification (see below) makes it fairly easy to pulverize the biomass, but this and other pretreatments add to the costs. At present it is impossible to predict whether the difficulty and expense of reducing the biomass particle size or the inherent limitations in the rate that coal reacts will dominate the economic differences between the two types of fuel reactors. Nevertheless, it is clear that dramatic improvements in biomass reactors are possible and that achieving this full potential will require RD&D specifically aimed at addressing and exploiting the unique features of biomass. Furthermore, since biomass char is more like coal than wood, grasses, or crop residues, achieving this potential advantage of biomass will involve reactors that produce little or no char.

Generally, the biomass reactors are classified according to the way the feedstock is fed into them. Although there are numerous variations, the major types are moving grate, mov-

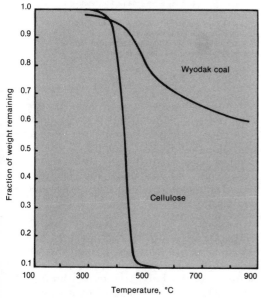

Figure 18.—A Comparison of Pyrolytic Weight Loss (on a mass fraction basis) v. Temperature for Coal and Cellulose

SOURCE: M. J. Antal, *Biomass Energy Enhancement—A Report to the President's Council on Environmental Quality* (Princeton, N.J.: Princeton University, July 1978).

[1]M. Graboski and R. Bain, "Properties of Biomass Relevant to Gasification," in *A Survey of Biomass Gasification* (vol. II, Golden, Colo.: Solar Energy Research Institute, July 1979), TR-33-239.

Figure 19.—Differential Loss of Weight Curves[a]

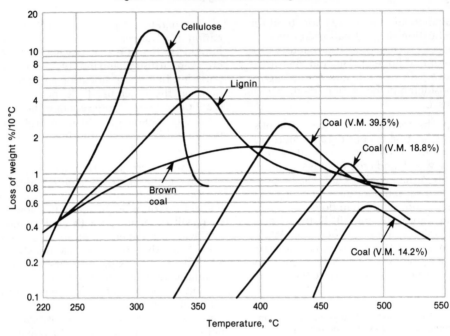

*These curves represent the derivative of curves similar to those given in figure 18. They were obtained by heating a small sample of solid material at a given rate and recording fractional weight loss v. temperature. The peak of each differential weight-loss curve (i.e., for cellulose the value is 15% per 10 °C at 315 °C) is indicative of the individual material's pyrolysis kinetics — a higher heating rate would displace all the curves to higher temperatures and would "sharpen" each peak. Thus the position of each peak is not related to "optimum" operating conditions. The curves simply show that biomass materials pyrolyze much more rapidly at much lower temperatures than coal.

SOURCE: H. H. Lowry, *Chemistry of Coal Utilization Supplementary Volume* (New York: John Wiley and Sons, Inc. 1963).

ing bed, fluidized bed, and entrained flow. The rate of heat transfer generally follows the order given, with the moving-grate reactors being the slowest. (There are, however, other classification schemes which can be useful as well.)

Moving-grate reactors consist of a grate that carries or moves the biomass through the zone where it is heated and decomposes. The heat transfer is relatively inefficient and slow, so an excess of heat must be generated to sustain the reaction. Therefore this type of reactor is generally best suited to direct combustion where the biomass is completely reacted and releases virtually all of its heat in the decomposition zone.

A slightly faster rate of heat transfer is achieved with moving-bed reactors. In these the bed, or clump of biomass, moves in a vertical direction as it is decomposed. Additional biomass is added at the top, which then gradually works its way down the reactor. Two types of moving-bed reactors exist: updraft and downdraft.

The updraft moving-bed reactors have a stream of air moving up through the bed of biomass. The hottest part of the bed is at the bottom. As the hot gases move through the bed, however, they cause relatively large amounts of tars and oils to form, which can condense causing maintenance problems and which may make it more difficult to burn the resultant gas without forming particulates.

The downdraft moving-bed reactors have a stream of air moving downward through the bed of biomass. Tars and oils are formed near the middle of the bed (where the air is injected) and subsequently move through a relatively large hot zone which gives them time to further decompose. The net result is a fuel gas with fewer tars and oils, thereby making gas cleanup easier and reducing the amount of particulates that form when the gas is burned.

Another type is the fluidized-bed reactor. In this case, gas is blown through the bed of solid fuel so rapidly that the bed of biomass levitates and churns as if it were fluid. In coal-fed fluidized-bed reactors, the fluid bed may contain limestone particles to react with and remove sulfur from the coal. Since biomass usually does not contain significant levels of sulfur, sand can be used as a fluidizing medium or one can rely solely on the biomass itself, with no separate fluidizing medium. Sand has

the advantage, however, of helping to retain heat in the bed, thereby increasing the rate that new pieces of fuel heat up in the bed.

Fluidized-bed reactors have a considerably faster heating rate than moving-bed or traveling-grate reactors. The churning in the bed, however, enables material at all stages of decomposition to be found throughout the bed. Consequently, there may be a tendency for oils and tars to escape from the heating zone before they can be fully decomposed.

The last type of reactor considered here is the suspension or entrained-flow reactor. In this type, small particles of feedstock are suspended in a stream of gas which moves rapidly into and through the decomposition zone. This type has the most rapid rate of heating, but the feedstock particles must be reduced to a relatively small size. As mentioned above, this would add to the total conversion costs and the details of this economic tradeoff are still uncertain.

Optimum Size for Thermochemical Conversion Facilities

Electric generating plants fueled with nuclear power or fossil fuels are generally quite large in order to take advantage of economies of scale. The same is true of most proposals for synthetic fuel plants. The optimum size of a biomass-fueled electric powerplant or synthetic fuels plant, on the other hand, is determined by a tradeoff between this economy of scale and the cost of transporting the feedstock to the conversion plant. Under favorable circumstances this optimum size could be several hundred megawatts electric (see app. A), and some paper-pulping mills do have wood inputs that would be sufficient for facilities of this size.[2] Under more common circumstances, however, the local availability of feedstock may limit the size of biomass conversion

facilities to the equivalent of 10- to 60-MW electric or less.

The economy of scale, however, is often matched by the cost savings associated with mass producing a large number of small units. Furthermore, in many industrial applications (e.g., process heat or steam boilers) the size is determined by the needs of that industrial plant rather than a potential economy of scale for the boiler or heat needs.

Large-scale facilities are technically feasible under some circumstances, particularly where the biomass arises as a waste byproduct in a large manufacturing plant. The number of sites where large quantities of biomass are available to a single plant on a continuing basis, however, may be limited. Consequently, the fullest utilization of the biomass resource for thermochemical conversion will require the development of small-scale, mass-produced units.

---

[2]Kip Howlett, Georgia Pacific Corp., private communication, 1979.

Biomass Densification

Freshly harvested biomass usually contains considerable moisture, has a relatively large volume per unit of energy (making it expensive to transport), and is fibrous (making it often difficult to reduce the particle size). These difficulties can be partially overcome by densifying the biomass.

There are several types of densification processes including pelletizing, cubing, briquetting, extrusion, and rolling-compressing. Pelletizing typifies the advantages and disadvantages of densification processes and is considered in more detail here.

Pelletization consists of drying the biomass, heating it until the lignin melts, and compressing the material into pellets. The pellets are denser than the biomass, more easily ground, and easier to handle and feed into reactors. Due to their lower moisture content, pellets usually burn more efficiently in boilers than does green biomass.

At present there are only commercial pelletization processes for wood. The lignin content in wood is generally high enough to bind the pellets so that no additional adhesives are required. Densified crop residues or grasses, however, may require the addition of adhesives to achieve the necessary binding strength to prevent the pellets from disintegrating to a powder; and the costs for this are uncertain.

The wood pelletization process has an energy efficiency* of about 90 percent if one starts with wood having 50-percent moisture content. Furthermore, wood pellets would burn in the boiler depicted in figure 17 to produce steam with an efficiency of about 83 percent as compared with an efficiency of 65 percent for woodchips with 50-percent moisture. Thus the overall efficiency (50-percent moisture woodchips to steam) is increased from 65 percent to perhaps 75 percent by including a pelletization process. This efficiency increase could also be achieved by predrying the wood-

*Efficiency is defined here as the lower heating value of the product divided by the lower heating value of the feedstock.

chips with heat escaping out the burner's chimney. The exact numbers will vary, however, depending on the specific boiler being considered. If the boiler is designed to accept high-moisture woodchips, then there may be no efficiency improvement with wood pellets.

Wood pellet costs are shown in table 43 for various feedstock costs. While the costs are

Table 43.—Cost of Pelletized Wood

	Wood feedstock cost (dollars/green ton)		
	$6.50	$10.00	$20.00
	Dollars/ton of pellets sold		
Wood...............	$14.39	$22.13	$44.26
Operation and maintenance........	7.95	7.95	7.95
Capital charges........ (30% of total investment per year)	5.14	5.14	5.14
Total	$27.48	$35.22	$57.35
Dollars/10⁶ Btu........	$1.72	$2.20	$3.58

Input: 540 ton/d of wood (50% moisture)
Output: 244 ton/d of pellets (10% moisture) for sale and 56 ton/d of pellets used to fuel the plant.
Load: 330 operating days per year

SOURCE: Office of Technology Assessment; and T. B. Reed, et al, "Technology and Economics of Close-Coupled Gasifiers for Retrofitting Gas/Oil Combustion Units to Biomass Feedstocks," in *Retrofit '79, Proceedings of a Workshop on Air Gasification*, sponsored by the Solar Energy Research Institute, Seattle, Wash., Feb. 2, 1979.

considerably higher than those for woodchips, the pellets' higher energy density allows them to be transported at a lower cost than green woodchips. This cost savings in the transportation pays for the pelletization process if the fuel is to be transported more than 50 to 150 miles depending on the transport and wood feedstock costs and the initial moisture content of the wood (see app. B for details of the calculation). However, this calculation does not include the added cost of transporting very bulky material such as plant herbage where the volume rather than the weight of the material determines the transport cost.

The most common and least expensive use of fuelwood, however, is likely to be in the region where it is harvested. Consequently, the use of densification processes may be limited. On the other hand, the increased ease of handling and burning pellets may make them attractive in applications where the process has to be extremely automated such as in very small industrial applications or where the feedstock is particularly unwieldy such as with plant herbage. In each application, the user will have to decide whether the higher fuel cost is justified in terms of the labor savings. For the remainder of this chapter, it is assumed that raw biomass rather than pellets are being used.

Direct Combustion of Biomass

Biomass can be burned together with coal (termed cocombustion) to produce process steam or electricity. Currently however, the largest amounts of energy produced from biomass come from the combustion of wood and food-processing wastes such as sugarcane bagasse by themselves. Another important use of direct combustion is in home heating. Each of these applications is considered below.

Cocombustion of Biomass

Currently, outdated—and therefore unusable—seed corn is being cocombusted with coal by the Logansport, Ind., Municipal Utility. Cocombustion of wood with coal has also been successfully demonstrated by the Grand Haven, Mich., Board of Light and Power.[3] And several assessments of the cocombustion of crop residues with coal concluded that it is technically feasible.[4]

Abdullah has estimated that the added costs at an electric powerplant needed to modify the boilers and handle the crop residues is $0.20 to $0.50/million Btu.[5] Consequently, for coal costing $1.50 to $2.00/million Btu ($30 to $45/ton), crop residues costing $13 to $24/ton would be economically cocombusted. Some crop residues may be available for these prices, but generally delivered crop residue prices are

likely to be higher. Higher coal prices, however, will make residue cocombustion more attractive.

Cocombustion can also be used to lower sulfur emissions somewhat. Since the biomass generally contains negligible amounts of sulfur, the quantity of sulfur being released in the combustion (per million Btu of heat) will decrease with the percentage of biomass, typically 20 to 30 percent. The economic savings associated with this will be highly site specific. The most advantageous situation would be where coal-fired boiler emissions are only marginally above the emissions standards without the use of sulfur removal equipment. Since the biomass costs, air pollution benefits, and feedstock availability are site specific, the economics of cocombustion will have to be determined through site-specific economic analyses. The principal determinant, however, will probably be the availability of a reliable supply of low-cost biomass feedstock.

Combustion of Biomass

Direct combustion of biomass for production of electricity or steam or for cogeneration (simultaneous production of steam and either electricity or mechanical shaft power) has commercially ready technology for wood, sugarcane bagasse, and many other feedstocks. There are also commercially available suspension burner retrofits for oil-fired boilers of 4.5 million Btu/hr or larger. The latter retrofitted boilers can return to oil if the biomass feedstock is temporarily unavailable, but they require a biomass feedstock that is quite dry (less than 15-percent moisture) and relatively small

[3]Pierre Heroux, *Supplemental Wood Fuel Experiment,* report to Grand Haven Board of Light and Power (Grant Haven, Mich.: J. B. Sims Generating Station, 1978).

[4]See, e.g., Wesley Buechele, *Direct Combustion of Crop Residues in Furnace Boilers* (Ames, Iowa: Agriculture and Home Economics Experiment Station), paper No. J8791.

[5]Mohammed Abdullah, "Economies of Corn Stover as a Coal Supplement in Steam Electric Power Plants in the North Central United States," ph. D. thesis, Agricultural Economics Department, Ohio State University, Columbus, Ohio, 1978.

in size (less than 1/8″ x 1/2″-3/4″).[6] A few types of biomass, however, involve special problems (e.g., the high silica content in rice hulls and residues) and boilers for these are not available.

In many applications today, feedstocks with 40- to 50-percent moisture content are used, resulting in boiler efficiencies of 65 to 70 percent. (The retrofit unit mentioned above, which is restricted to low-moisture feedstock, achieves an estimated 75-percent efficiency).[7] There has been little incentive to dry the feedstock in most current applications, since they usually involve relatively inexpensive waste products. As the use of biomass for direct combustion becomes more widespread and the average feedstock costs increase, however, predrying of the feedstock is likely to be more common.

As with cocombustion, the feedstock cost and availability of a reliable supply of the feedstock are major determinants of the economics of using biomass as a fuel. While these costs vary considerably from site to site, an average feedstock cost of $30/dry ton ($15/green ton) results in the costs of electric generation, cogeneration, and steam production shown in table 44. (More detailed cost calculations are given in app. B.) The costs for producing only electricity or only steam are also shown for various feedstock costs in figures 20 and 21.

---

[6]Peabody, Gordon-Piatt, Inc., Winfield, Kan., e.g., offers suspension burner retrofits to oil-fired boilers ranging from 4.5 million Btu/hr and up. The retrofit cost is slightly higher than for gasifiers, but where dry, small particle feedstock (e.g., sawdust) is available at low prices, the system is competitive with fuel oil. Private communication with Delvin Holdeman, Solid Fuels Marketing Division, Peabody, Gordon-Piatt, November 1979.

[7]Ibid.

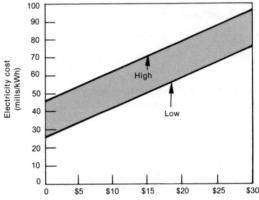

Figure 20.—Cost of Electricity From Wood for Various Wood Costs (field-erected generating station)

SOURCE: Office of Technology Assessment.

Obviously, where the feedstocks can be obtained inexpensively enough, biomass is competitive with coal for generating electricity and with oil for process steam. In the case of electricity, the investment costs are about the same as for coal-fired powerplants; but wood-fired boilers cost about three times that of oil or natural gas boilers.[8]

Wood Stoves and Fireplaces

Wood stoves and fireplaces have long been used as a means of space heating in residences, but fireplaces are more often used today for

---

[8]A Survey of Biomass Gasification (vol. I, Golden, Colo.: Solar Energy Research Institute, July 1979).

Table 44.—Cost of Electric Generation, Cogeneration, and Steam Production From Wood[a]

Product	Plant size	Wood cost (dollars/ green ton, delivered at 50% moisture)	Product cost
Electricity	60 MW (field erected)	15	50-70 mill/kWh
Steam	50,000 lb/hr (package boiler)	15	$3.50-$6.00/1,000 lb
Steam and electricity	390,000 lb/hr	15	$4-$6/1,000 lb[b]
	21.4 MW (field erected)		109-30 mills/kWh[b]

[a]See details in app. B.
[b]As the steam cost increases, the electric cost decreases.

SOURCE: Office of Technology Assessment.

Figure 21.—Cost of Process Steam From Wood for Various Wood Costs (package boiler)[a]

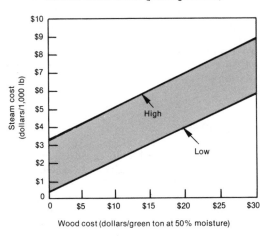

Wood cost (dollars/green ton at 50% moisture)

[a]Significant variations in installation costs can occur.

SOURCE: Office of Technology Assessment.

their recreational or esthetic value. Also many fireplaces are inefficient because excess air goes into the fireplace and up the chimney and this air often is drawn into the house through cracks in windows and doors. Consequently, while fireplaces do produce some local heating, the overall effect may be a net cooling of the house.

The efficiency of fireplaces can be improved (see table 45) through various methods of circulating room air past hot parts of the fire-

Table 45.—Small-Scale Heating Device Efficiency

Heat unit	Net efficiency (percent)
Fireplace	
Masonry	−10 -10%
Metal prefab, noncirculating	−10 - 10
Insert or retrofit, circulating	40 - 50
Metal prefab, circulating	10 - 30
Metal, freestanding	40
Stoves	
Franklin or fireplace stove	25 - 45
Cast iron airtight	50 - 65
Metal airtight	50 - 65
Box	25 - 45
Circulator, controlled airinlet	40 - 55
Furnace, convertor or adder	40 - 60

SOURCE: Auburn University, "Improving the Efficiency, Safety, and Utility of Wood Burning Units," DOE contract report DE-AS05-77ET-11288, 1979.

place, through tubes being heated by the fire, or by drawing the combustion air in from outside through tubes that are heated by the fire. Depending on the complexity of the arrangement, the cost can range from as little as $10 to $30 to over $1,000.

Wood stoves generally have a higher efficiency than most fireplaces, due to the greater degree of air circulation around and the radiation of heat from the hot stove. In the better wood stoves, the combustion efficiency (amount of heat liberated per pound of wood) is higher than in a fireplace. Often, however, wood stoves do not completely burn the wood gases, leading to deposits of creosote in the flue. The creosote deposits can present a fire hazard and, at best, need to be regularly cleaned from the flue. There is no fundamental reason, however, why these problems cannot be solved; and research into thermochemistry and development of advanced wood stoves are likely to lead to higher efficiencies, greater flexibility of operation, and fewer safety problems.

Wood furnaces for centralized heating of a home also have significantly better efficiencies than many fireplaces. Efficiencies as high as 80 percent have been reported under certain circumstances.[9] The possibility also exists of using wood furnaces as a backup to solar-heated houses. In this case, the heat storage system of the active solar heating system could be recharged in a few hours and thereby provide space heating for several days with low solar insolation. Hill has estimated that a wood furnace (300,000 Btu/hr) with hot water storage (500 gal) would cost about $3,000 installed.[10] This, however, should be treated as a rough estimate and additional work will be necessary to establish a more exact cost.

---

[9]Laatukattila Oy, Inc., Satamakatu 4, 33201 Tampere 20, Finland, sells a YR-60 furnace capable of burning either light fuel oil or wood. The Finnish Government Centre for Technical Research (Valtion Teknillinen Tutkimuskeskus) has rated this furnace at 79.3- and 78.8-percent efficiency at two-thirds and five-sixths full load, respectively, when using relatively dry birchwood as a fuel, according to information supplied to OTA by Laatukattila Oy, October 1979.

[10]R. C. Hill, University of Maine, Orono, Maine, private communication, Oct. 26, 1979.

In general wood stove and furnace heating require more labor than oil or natural gas heat. The fuel requires more handling, ashes must be removed, and the systems must be regularly serviced to maintain efficient and safe operation. This is less of a problem if wood pellets or well-dried wood is used or if the wood is used only as a solar backup. The use of wood heating exclusively, however, is likely to be limited to those people who consider this type of activity enjoyable or wish to use wood to achieve some degree of energy self-sufficiency. A larger number of people are likely to purchase wood stoves as insurance against oil or natural gas shortages or as a supplement to more conventional systems.

Gasification of Biomass

Gasification is the process of turning solid biomass into a gas suitable for use as a fuel or for chemical synthesis. There are several types of thermal gasification processes, or gasification induced by heat. Gases produced in blast furnaces or by the water gas process are low-Btu gases (80 to 180 Btu/stdft^3). Other gasifiers use pure oxygen and partial combustion of the feedstock to produce a medium-Btu gas (300 to 500 Btu/stdft^3) suitable for regional industrial pipelines or chemical synthesis. Still others (pyrolysis gasifiers) provide an external source of heat to produce a medium-Btu gas (e.g., dual fluidized bed gasifier described in the next section).

The gasifiers discussed in detail in this section are the airblown gasifiers. This type blows air through the feedstock to partially combust it. The heat generated is used to gasify the remaining material. The resultant gas from updraft and downdraft airblown gasifiers (termed intermediate-Btu gas) has a lower heat content (120 to 250 Btu/stdft^3) than with oxygen or pyrolysis gasification, due to the dilution effect of the nitrogen contained in the air. (Air is about 78 percent nitrogen and 21 percent oxygen.) This lower heat content makes the gas unsuitable for regional pipeline distribution, but it is not a disadvantage if the gasifier is attached directly to the boiler being fired (so-called close-coupled gasifier) or used directly for process or space heat. Gases with heat contents of 250 to 400 Btu/stdft^3, however, have been produced from an experimental fluidized-bed airblown gasifier, but the gas contains considerable tar and oil.[11]

Close-coupled airblown gasifier systems have the potential for higher efficiencies than direct combustion when a variety of feedstocks with different moisture contents are used (see "Generic Aspects of Biomass Thermochemistry"), and can be used for process heat. Moreover, they are likely to be more efficient and less expensive, in most applications, than oxygen-blown or pyrolysis gasifiers (due to the energy loss and cost associated with the added equipment needed to produce oxygen or the external supply of heat). Nevertheless, for methanol synthesis, these gasifiers would be necessary. Moreover, there may be some circumstances where regional industrial natural gas pipelines could be converted wholly or partially to gasified biomass. Consequently, cost calculations for two medium-Btu gasifiers are included in appendix B.

Airblown (and other) gasifiers have the flexibility of being able to be used together with or as a substitute for oil or natural gas in industrial boilers for crop drying, and for process heat. This means that even where biomass feedstocks are not available in large quantities, those that are available can be used to displace oil and natural gas to the extent of their availability; and (barring regulations prohibiting it) the users could return to oil or natural gas if the biomass is temporarily unavailable or in short supply. (It should be noted that the suspension burner retrofit mentioned in the last section also has this advantage but the types of feedstocks it will accept are more restricted than for gasifiers.) Furthermore, in properly designed close-coupled gasifiers, the fuel gas needs only minor cleanup (cyclone precipitator and perhaps fiberglass filter). This

---

[11]Steven R. Beck, Department of Chemical Engineering, Texas Tech University, Lubbock, Tex., private communication, 1979.

together with the fact that the volume of fuel gas that needs to be cleaned is less than the volume of flue gas (from direct combustion) means that the gas cleanup is likely to be less expensive for gasifiers.

Gasifiers, however, need further development to improve their reliability (particularly with respect to materials clogging), and, in some cases, to lower the tar and char produced. Furthermore, improvements in gasifier efficiency and throughput rates can lower the effective feedstock costs and capital investment, respectively. The types of airblown gasifiers, their efficiency, and the.costs are discussed below. Finally, gasifiers for internal combustion engines (ICEs) are considered briefly.

Airblown Gasifier Types

The types of reactors suitable for gasification include updraft, downdraft, fluidized-bed, and entrained-flow reactors. Each of these types is depicted schematically in figure 22. The entrained-flow reactor is the fastest of these four. It has the disadvantages, however, that it requires a finely ground feedstock and the fuel gas contains considerable ash. If the ash is cleaned from the gas by wet scrubbing, then the wastewater may contain toxic compounds (e.g., phenol).[12]

Fluidized-bed reactors can take a wide range of particle sizes. In addition the material throughput is more than three times as rapid as with the updraft and downdraft gasifiers[13] and the particle stays in the gasifier only minutes[14] or fractions of a second[15] rather than hours with the slower gasifiers. Fluidized-bed reactors release some ash into the gas stream, which must be cleaned from it. Tars in the fuel gas can also be a problem.

The updraft and downdraft gasifiers are the slowest, but they also are the simplest to construct. Updraft gasifiers tend to produce more ash and tar in the fuel gas than with downdraft reactors, but their construction is the simplest of all gasifiers. Both types require relatively large feedstock particles so that the gas can flow freely through the bed of biomass.

The ideal gasifier would be simple to construct and operate, produce no ash in the fuel gas, completely gasify the feedstock (producing no char or tar), accept a wide range of feedstock sizes and moisture contents, and gasify the feedstock rapidly. The downdraft and fluidized-bed gasifiers appear to be the most favorable types, but further development of all types is required before an unambiguous choice can be made. In the end it may well be found that different gasifier types are superior for different feedstocks and applications. A partial list of gasifiers currently under development is given in appendix C.

Efficiency of Airblown Gasifiers

The heat content of the fuel gas is an important consideration in determining the overall efficiency of using a gasifier. The Electric Power Research Institute has determined the efficiency of a boiler using gases with various heat values, as shown in figure 23. Both the sensible heat (gas temperature) and the fuel value of the gas can contribute to this heating value.[16] Typical gas values range from 120 to 200 Btu/stdft^3 from airblown updraft and downdraft gasifiers. Some researchers claim that the energy content of the gas is increased and its burning characteristics are improved by the presence of pyrolytic oils (incompletely decomposed biomass),[17] but these oils tend to condense in fuel lines, clog valves, and in some cases may cause excessive particulate formation when combusted (thereby requiring flue gas cleanup and reducing the combustion efficiency and applicability for process heat). Determining the optimum gas composition and

[12]Ralph Overend, "Gasification—An Overview," in *Retrofit '79, Proceedings of a Workshop on Air Gasification,* sponsored by the Solar Energy Research Institute, Seattle, Wash., Feb. 2, 1979.
[13]Ibid.
[14]Ibid.
[15]Beck, op. cit.

[16]T. B. Reed, et al., "Technology and Economics of Close-Coupled Gasifiers for Retrofitting Gas/Oil Combustion Units to Biomass Feedstocks," in *Retrofit '79, Proceedings of a Workshop on Air Gasification,* sponsored by the Solar Energy Research Institute, Seattle, Wash., Feb. 2, 1979.
[17]Ibid.

Figure 22.—Schematic Representation of Various Gasifier Types

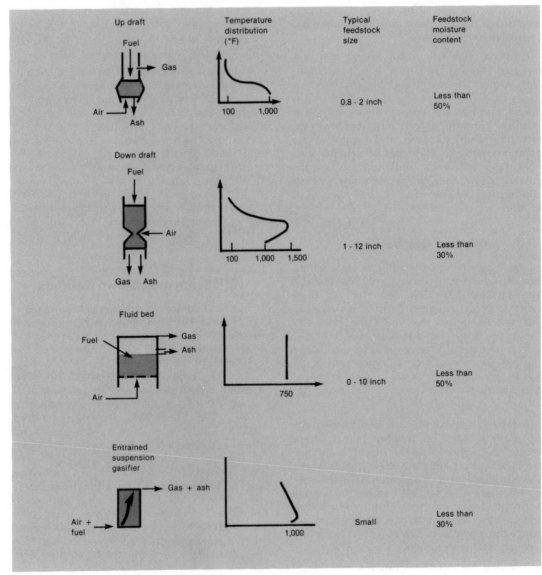

[a]Note that other schemes such as moving grate gasifier also exist.

SOURCE: From R. Overend, "Gasification - An Overview," Retrofit - 79, Proceedings of a Workshop on Air Gasification, Seattle, Wash., SERI/TP-49-183, Feb. 2, 1979.

Figure 23.—Boiler Efficiency as a Function of the Btu Content of the Fuel Gas

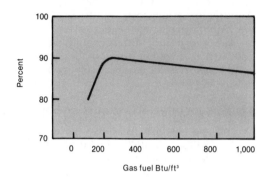

SOURCE: *Fuels From Municipal Refuse for Utilities: Technical Assessment* (Electric Power Research Institute, March 1975), EPRI report 261-1, prepared by Bechtel Corp.

how to obtain it requires further experimentation and a better understanding of biomass combustion chemistry. Nevertheless, some downdraft gasifiers have produced gases approaching 200 Btu/stdft³ from wood with little oil formation,[18] and there appears to be no fundamental reason why the optimum energy content (see figure 23) with low tars cannot be reached with additional gasifier development.

The other factor determining the overall efficiency of gasifier-boiler systems is the efficiency of the gasifier itself. Since both the sensible heat* and the chemical energy in the gas can be utilized with a close-coupled gasifier, the only gasification losses are the heat radiated from the gasifier, that lost during fuel gas cleanup, and the fuel value lost in condensed tars, oils, or char. Gasifiers have achieved efficiencies of 85 to 90 percent[19][20] and well-insulated gasifiers designed to minimize char, oil, and tar formation should be able to reach efficiencies of 90 percent or better. This would raise the overall efficiency of feedstock to steam to 85 percent or higher and provide high efficiencies for process heat needs.

[18]J. R. Goss, "The Downdraft Gasifier," *Retrofit '79, Proceedings of a Workshop on Air Gasification,* sponsored by the Solar Energy Research Institute, Seattle, Wash., Feb. 2, 1979.

*Sensible heat is the energy contained in the gas by virtue of its being hot; i.e., it is the heat that can be sensed or felt directly.

[19]Goss, op. cit.

[20]Reed, op. cit.

Airblown Gasifier Costs

It has been estimated that oil- or gas-fired boilers can be retrofitted with mass-produced airblown biomass gasifiers for $4,000 to $9,000/million Btu/hr ($5 to $12/lb of steam/hr), with gasifiers ranging from 14 million to 85 million Btu/hr.[21] Retrofit costs, however, can vary considerably depending on the difficulty of accessing the boiler and the possible need for an additional building, to house the gasifier. Voss, for example, has estimated the cost at $20,000/million Btu/hr when new buildings and foundations are needed.[22]

The favorable case cost estimates are compared with the costs of new oil/gas- and wood-fired package boilers in figure 24 (similar prob-

Figure 24.—Comparison of Oil/Gas Package Boiler With Airblown Gasifier Costs

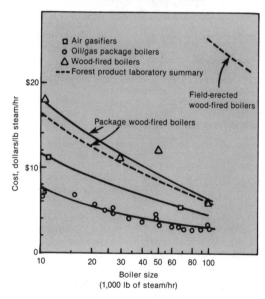

SOURCE: T. B. Reed, D. E. Jantzen, W. P. Corcoran, and R. Witholder, "Technology and Economics of Close-Coupled Gasifiers for Retrofitting Gas/Oil Combustion Units to Biomass Feedstocks," *Retrofit '79, Proceedings of a Workshop on Air Gasification,* sponsored by the Solar Energy Research Institute, Seattle, Wash., Feb. 2, 1979.

[21]Cited in Ibid.

[22]G. D. Voss, American Fyr-Feeder Engineers, Des Plaines, Ill., private communication, 1979.

lems with installation can occur with these boilers as well). It can be seen that the capital investment for a gasifier retrofit is roughly twice that for a new oil/gas-fired boiler, but only two-thirds of that for a new wood-fired boiler. From these preliminary estimates, it appears that a new gasifier-oil/gas boiler combination costs roughly the same as a new wood package boiler but more refined data on gasifiers are needed before accurate comparisons can be made.

With costs of $4,000 to $9,000/million Btu/hr and wood fuel at $30/dry ton ($21/air dry ton, 30-percent moisture) the resultant gas is estimated to cost about $2.70 to $2.90/million Btu (see table 46). In the unfavorable case of $20,000/million Btu/hr, the cost could be $3.35/ million Btu with this feedstock cost.

Table 46.—Cost Estimate for Fuel Gas From Wood Using a Mass-Produced Airblown Gasifier

Fixed investment	$4,000-$9,000 per 10^6 Btu/hr of capacity
	Dollars/10^6 Btu
Wood ($21/ton, 30% moisture, i.e., $30/dry ton)	$2.38
Labor, electricity	0.20
Capital charge (30% of fixed investment per year)	0.150 - 0.34
Total	$2.73 - $2.92
Estimated range ($20-$60/dry ton wood) . . .	$2 - $6

Input: 38 to 230 tons of air-dried wood (30% moisture) per day
Output: 14 to 85 10^6 Btu/hr of intermediate-Btu gas
Load: 330 operating days per year

SOURCE: Office of Technology Assessment.

Obviously from table 46, the dominant cost is the feedstock cost. If waste byproducts are used to fuel the gasifier, the gas could cost less than $1/million Btu. For the larger quantities of wood, grasses, and residues costing $20 to $60/ dry ton, the gas price is estimated to range from $2 to $6/million Btu. These costs are competitive with fuel oil at $6.50/million Btu ($0.90/gal), but less so with natural gas at about $3.50/million Btu. To achieve the full potential of gasifiers, however, units in the range of 0.1 million to 10 million Btu/hr should also be developed.

Field-erected gasifiers are considerably more expensive (see app. B). They may be economic, however, in cases where very large quantities of a low-cost feedstock are available. Alternatively, package gasifiers of several hundred million Btu/hr could be developed, which, together with smaller gasifiers, would cover most situations involving biomass feedstocks.

Gasifiers for Internal Combustion Engines

Wood and charcoal gasifiers were used during the 1930's and 1940's in Europe to fuel automobile and truck engines. After some development, the gasifiers operated satisfactorily, but even under favorable circumstances, operation and maintenance required an estimated 1 hour per day of operation.[23] Because of this and the 30-percent power loss associated with switching to the gas,[24] it is unlikely there would be a large market for gasifiers used in automobiles, except under cases of extreme shortages of gasoline. Gasifiers could, however, be used to fuel remote ICEs for irrigation water pumping or electric generation.

The principal difference between gasifying for close-coupled boiler operation and process heat and for ICEs is that the latter application requires that the gas be cooled before entering the engine and requires particularly low tar and ash content. The cooling is required to enable sufficient gas to be sucked into the cylinder to fuel the engine and to prevent misfiring. The careful gas cleanup is required to prevent fouling or excessive wear in the engine.

These problems were alleviated for charcoal and low-moisture wood by using downdraft gasifiers and various gas cooling and cleanup schemes in Europe before and during World War II.[25] (Charcoal tended to form more ash, while wood more tar, so somewhat different systems were required.) The applicability of these gasifiers to other feedstocks, however, is uncertain.

[23]Swedish Academy of Engineering, *Generator Gas—The Swedish Experience From 1939-1945*, Generalstabens Litografiska Anstalts Förlag, Stockholm, 1950, translated by the Solar Energy Research Institute, Golden, Colo., 1979.
[24]Ibid.
[25]Ibid.

Gasifiers could be used as the sole fuel for spark ignition (e.g., gasoline) engines or together with reduced quantities of diesel fuel in diesel engines (by fumigation, i.e., replacing the air intake with an air-fuel gas mixture). The energy lost in cooling the gas and removing the tar and the added cost of the cooling equipment are likely to more than double the gas costs over that for close-coupled gasifiers. (This is based on calculations by Reed,[26] in which it is estimated that about half of the close-coupled gas energy is sensible heat. The actual value, however, will vary with the gasifier.)

With waste byproducts having no value or giving a disposal credit, the gas would be com-

[26]Reed, op. cit.

petitive with electric irrigation, gasoline, diesel fuel, and, probably, natural gas. With crop residues costing $30/dry ton, the gas cost with conventional technology is likely to be over $7/ million Btu, which is competitive with electric irrigation and will soon be competitive with gasoline and diesel fuel, but is more expensive than natural gas at present.

Gasifiers suitable for ICEs could probably be manufactured immediately, but improvements in the gasifier efficiency and reliability could improve the applicability of gasifiers to ICEs for crop irrigation and other uses. The development could parallel the development of other gasifiers, and improved units could probably be available in 2 to 5 years.

Liquid Fuels From Thermal Processes

Numerous liquid fuels can be made from biomass through thermal processes and chemical synthesis. The liquid fuels considered here are methanol, pyrolytic oil, and ethanol. Cost estimates for the production of these fuels are shown in table 47, with further details given in appendix B. Each of the processes is discussed below.

Methanol

Methanol ("wood alcohol") was first produced from biomass as a minor byproduct of charcoal manufacturing. This process for methanol synthesis, however, is no longer economic. Most methanol today is produced from natural gas. The natural gas is reacted with steam and CO_2 to produce a CO-hydrogen mix-

ture. The gas composition is then adjusted to the correct ratio of these components and the resultant gas is pressurized in the presence of a catalyst to produce methanol. Finally, the crude methanol may be distilled to produce pure methanol.

Methanol can be produced from biomass by gasifying the biomass with oxygen or through pyrolytic gasification to produce the CO-hydrogen mixture, with the remainder of the process being identical to the processes which use natural gas. The oxygen-blown gasifier systems can be built today, whereas pyrolysis gasifiers require further development.

Cost estimates for an oxygen-blown gasifier used to produce methanol are given in table 48 and a flow diagram of the process is shown in figure 25. The cost is estimated at $0.75 to

Table 47.—Summary of Cost Estimates for Various Liquid Fuels From Wood via Thermochemical Processes

Fuel	$/bbl	$/gal	$/million Btu	Commercial facilities could be available by:
Methanol	$28 - $56	$0.67 - $1.33	$10.50 - $20.90	Now
Pyrolysis oil	30 - 50	0.70 - 1.20	7 - 12	Mid to late 1980's
Ethanol	23 - 68	0.55 - 1.62	6.50 - 19.10	1990's

SOURCE: Office of Technology Assessment.

Table 48.—1979 Cost of Methanol From Wood Using Oxygen Gasification

Fixed investment (field erected)	$80 million
Working capital (10% of fixed investment) . . .	8 million
Total investment	$88 million
	$/bbl
Wood ($15/green ton)	10.35
Labor, water, chemicals................	1.10
Electricity (3.8 kWh/gal, $0.04 kWh)	6.40
Capital charges (15-30% of total investment per year)................	$13.80-$27.60
Total.........................	$31.65-$45.45
	($0.75-$1.08/gal)
Estimated range:....................	$28-$56/bbl
($10-$30/green ton wood)	($0.67-$1.33/gal)
	($10.50-$20.90/10⁶ Btu)

Input: 2,000 green ton/d of wood (50% moisture)
Output: 2,900 bbl/d methanol (40 million gal/yr)
Load: 330 operating days per year

SOURCE: Office of Technology Assessment; and based on J. H. Rooker, Davy McKee, Inc., Cleveland, Ohio, private communication, May 1980; A. E. Hokanson and R. M. Rowell, "Methanol From Wood Waste: A Technical and Economic Study," Forest Products Laboratory, Forest Service, U.S. Department of Agriculture, general technical report FPL12, June 1977; and E. E. Bailey, manager, Coal and Biomass Conversion, Davy McKee Corp., Cleveland, Ohio, private communication, 1979.

$1.08/gal from $30/dry ton wood, and the capital investment is about $2.00 for each gallon per year of capacity, which is somewhat more expensive than grain ethanol distilleries.

Comparable cost calculations are given for a dual fluidized-bed pyrolysis gasifier in appendix B. In this gasifier, the fluidizing medium is heated in one fluidized-bed reactor which burns biomass and it is transferred to another fluidized bed where it gasifies biomass in the absence of air or oxygen. Although dual fluidized-bed gasifiers are not fully developed, the calculations in appendix B indicate that this method may produce methanol at somewhat lower costs than using oxygen-blown gasifiers, principally because it eliminates the equipment needed to produce oxygen. A more accurate comparison, however, must await development and demonstration of dual fluidized-bed and other pyrolysis gasifiers.

Figure 25.—Block Flow Diagram of Major Process Units

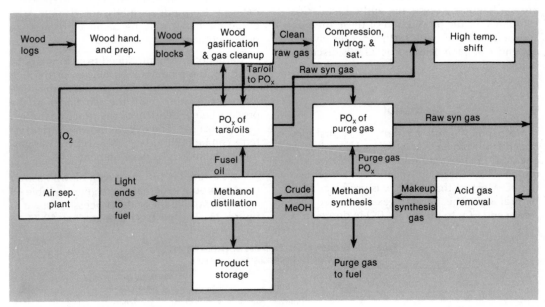

SOURCE: J. H. Rooker, *Methanol Via Wood Gasification* (Cleveland, Ohio: Davy Mckee, Inc., 1979).

The only part of methanol synthesis, for which there is any uncertainty is the operation of and yield from the gasifier. Oxygen-blown wood gasification can probably be accomplished with commercial fixed-bed gasifiers,[27] but a large part of the gasifier cost would be associated with cleaning tars, oils, and other compounds from the gas. Consequently, the costs would be reduced somewhat by developing advanced oxygen gasifiers that maximize the CO-hydrogen yields and reduce the tar and oil formation.

With plant herbage as the feedstock, additional problems may arise from the handling of this material and possible clogging of the gasifier. These problems probably can be solved with a relatively straightforward development of suitable gasifiers.

Methanol yields from wood would vary depending on the type of wood, but have been estimated at 120 gal/dry ton in a plant that purchases its electricity.[28] If the electricity is cogenerated onsite the yield would be about 100 gal/dry ton.[29] These yields correspond to conversion efficiencies of 48 and 40 percent, respectively. Yields from plant herbage are not available, but based on the above efficiencies, they may be 100 or 80 gal/ton depending on whether the electricity is purchased or generated onsite. In neither case would additional boiler fuel be needed. In theory, however, these yields can be increased significantly.

Accessing a large part of the potential biomass resource would be aided by the development of small, inexpensive package methanol plants. However, because small centrifugal compressors cannot achieve the pressures needed for methanol synthesis, plants smaller than about 3 million to 10 million gal/yr of methanol would require a different type of compressor, e.g., reciprocal compressor.[30][31]

This could increase the plant cost above that resulting from the normal diseconomy of scale, but engineering details and costs are uncertain at present.

There is little doubt that methanol can be synthesized from wood with existing technology. Since the only uncertainty is with the gasifier, the cost estimates are probably accurate to within 20 percent. This would put the cost per Btu of methanol from wood at about the same level as ethanol from grain. However, both alcohols are likely to be more expensive than methanol from coal, due primarily to the economy of scale that can be achieved by building very large coal conversion facilities.

Pyrolytic Oil

Pyrolytic oil can be produced by slowly heating biomass under pressure and in the presence of a catalyst. The pressure suppresses gas formation and the catalyst aids the formation of the oil. Other possibilities, however, such as rapid heating and cooling can also produce pyrolytic oils.

The process involving slow heating is currently under development and a pilot plant in Albany, Oreg., has produced a small quantity of oil, following earlier difficulties. The oil is about 30 percent lower in heat content (per gallon) than petroleum fuel oil and it may be corrosive but it contains negligible sulfur. The oil is said to be roughly equivalent to a low-grade fuel oil, but further testing is necessary to determine how well the oil stores and what modifications in boilers may be necessary to use this oil as a boiler fuel.

Since the pyrolytic oil is made from feedstocks that could be used in close-coupled, air gasifiers and would have some of the same uses as the gasifier fuel gas, pyrolytic oil production should be compared to close-coupled gasifiers. The pyrolytic oil is less expensive to transport than raw biomass and it is probably well suited to fully automatic boiler operation. It may also be possible to refine the oil to higher grade liquid fuels. At present, nevertheless, the costs appear to be high in relation

[27]J. H. Rooker, *Methanol Via Wood Gasification* (Cleveland, Ohio: Davy McKee, Inc., 1979).

[28]E. E. Bailey, Manager, Coal and Biomass Conversion, Davy McKee Corp., Cleveland, Ohio, private communication.

[29]A. E. Hokanson and R. M. Rowell, "Methanol From Wood Waste: A Technical and Economic Study," Forest Products Laboratory, Forest Service, U.S. Department of Agriculture, general technical report FPL 12, June 1977.

[30]Bailey, op. cit.

[31]J. H. Rooker, Davy McKee, Inc., Cleveland, Ohio, private communication, May 1980.

to air gasifiers and the efficiency of using the biomass feedstock in this way is considerably lower than with gasification, but the oil may be comparable in cost to some other synthetic fuels. Consequently, if gasifiers become widely available, markets for the pyrolytic oils may be limited to those users who are willing to pay for complete automation of their boilers.

Various other thermal processes are possible for the production of oils from biomass (see app. C), including processes which do not try to minimize oil production during gasification and collect the oil as one of the products. These latter types produce gas, oil, and char products.

The multiproduct systems, while being technically easier to develop, have decreased oil yields (since part of the biomass is not converted to oil) and the management and economics are more complicated due to the need to sell each of the various products. A technical solution to these problems being studied is to slurry the char with the oil. Although the char contains ash and the oil is corrosive and may deteriorate under storage,[32] the Department of Energy (DOE) is funding a feasibility study for burning this slurry of oil and char in gas turbines.[33] Since conventional turbines may not be able to tolerate gases with sodium and potassium the project proposes to use turbine combustion technology developed from military programs.[34]

It would seem, however, to be more technically and economically sound to develop conversion processes which produce little or no char and which produce only as much gas as can be utilized by the conversion facility.

Consequently, OTA has not analyzed the multiproduct liquefaction systems in detail.

Still another type of liquefaction process would subject medium-Btu gas to pressure in the presence of a catalyst (the biomass analog of the South African SASOL process for producing gasoline from coal). The capital investment, however, appears to be quite high,[35] and further development will be needed to lower these costs.

Ethanol

Conceptually, ethanol can be produced from biomass through rapid gasification to produce ethylene. The ethylene is then separated from the other gases and converted to ethanol using commercial technology.

The critical factor in determining the economics is the ethylene yield from rapid gasification. Present experimental yields have reached 6 percent (by weight) from biomass,[36] but some researchers[37] believe that yields as high as 30 percent (by weight) may be possible. If so, then this process could produce fuel ethanol at prices considerably below those for the fermentation of lignocellulosic materials and at costs (per million Btu) comparable to those projected for methanol from coal, or roughly $0.65/gal of ethanol.

The process, however, needs considerable research to determine if and how such ethylene yields can be achieved. Even under favorable circumstances, it is unlikely that commercial processes could be available before the 1990's.

[32]J. W. Birkeland and C. Bendersky, "Status of Biomass Waste and Residue Fuels for Use in Directly Fired Heat Engines," presented at the Conference on Advanced Materials for Alternate Fuel Capable Directly Fired Heat Engines, sponsored by the Electric Power Research Institute and the Department of Energy, Maine Maritime Academy, Castine, Maine, August 1979.

[33]Teledyne CAE, Toledo, Ohio, "Gas Turbine Demonstration of Pyrolysis-Derived Fuels," Department of Energy contract E778-C-03-1839.

[34]Birkeland and Bendersky, op. cit.

[35]Dow Chemical, U.S.A., Freeport, "Technical, Economic, and Environmental Feasibility Study of China Lake Pyrolysis System," report to the Environmental Protection Agency, 1978.

[36]S. Prahacs, H. G. Barclay, and S. P. Bhada, "A Study of the Possibilities of Producing Synthetic Tonnage Chemicals From Lignocellulosid Residues," Pulp and Paper Magazine of Canada, vol. 72, p. 69, 1971.

[37]See, e.g., M. J. Antal, Biomass Energy Enhancement—A Report to the President's Council on Environmental Quality (Princeton, N.J.: Princeton University, July 1978).

Environmental Impacts of Wood and Wood Waste Combustion

The major environmental impacts of wood combustion, aside from any impacts from growing and harvesting the wood fuel, arise from the generation of air pollution in the combustion units. A variety of other impacts, including safety problems with small units, water pollution from wood storage and ash disposal, and air pollution from wood fuel distribution may be of lesser importance, although wood appliance safety could easily become an important public concern. Because the magnitude of the impacts, even on a "per ton of wood burned" basis, is quite dependent on the size of the operation, this discussion treats residential and other small-scale use separately from utility and industrial wood boilers.

Small-Scale Burning

Residential use of wood as a heating fuel is usually a low combustion efficiency, low-temperature process compared to larger industrial fossil-fueled or wood boilers. The low combustion efficiency is reflected in relatively high emissions of CO and unburned hydrocarbons (see table 49). The low temperature, coupled with extremely low fuel-bound nitrogen in wood (about 0.1 percent compared to 1.5 per-

Table 49.—Emission Factors for Residential Wood Combustion Processes

Pollutant	g/kg[a]	lb/cord[a]
Particulate[c]	5-19	20-72
Carbon monoxide	60-130	240-520
Hydrocarbons	2-9	8-40
SO$_x$	0.2	0.8
NO$_x$	0.3	1.2
Formaldehyde	1.6	6.4
Acetaldehyde	0.7	3
Phenols	1	4
Acetic acid	6.4	26
Polycyclic organic matter	0.3-4.6% of total particulates	
Elemental metals	7	30

[a]Units are grams of species emitted per kilogram of wood burned. Wood moisture is not specified in the references cited.
[b]Alternate units are pounds of species emitted per cord of wood burned. One cord is assumed to equal 4,000 lb.
[c]Particulate includes inorganic ash, condensable organics, and carbon char. Note that other entries in the table, e.g., polycyclic organic matter and elemental metals, are somewhat redundant in that they are subcomponents of particulate matter and not separate species.

SOURCE: J. O. Miliken, "Airborne Emissions From Wood Combustion," Environmental Protection Agency/Research Triangle Park, N.C., Feb. 20, 1979, with revisions based on private communication with Milliken.

cent in coal,[38]) leads to levels of nitrogen oxide (NO$_x$) emissions well below those of fossil boilers. (Old Environmental Protection Agency (EPA) emission factors from "AP-42" showed NO$_x$ emissions to be as high as those from coal boilers, but these factors have been demonstrated to be inaccurate.) Wood sulfur levels are equal to or less than 0.05 percent,[39] and sulfur oxide (SO$_x$) emissions consequently are very low.

Particulates are an especially worrisome component of emissions from residential wood combustion. Areas with high concentrations of wood stoves are known to have particulate pollution problems, especially during winter inversion conditions. Rapid deployment of wood stoves could have significant effects on air quality in New England and the Northwest.[40]

Condensable organics make up about two-thirds of the particulate matter emitted by residential wood combustion units.[41] Polycyclic organic matter (POM), species of which are known animal carcinogens, makes up as much as 4 or 5 percent of these organics and may be the most dangerous component.[42] Based on available emission data, POM emissions from wood stoves are likely to be far greater (on a "per Btu" basis) than emissions from the systems they would replace—fossil-fueled powerplants and residential oil or gas furnaces.

POM is emitted by all combustion sources and is spread throughout the environment, although usually in low concentrations. Table 50 shows the major sources of benzo(a)pyrene (B(a)P), which is often used as an indicator species of POM. Aerosols containing B(a)P and other species of POM can survive long enough

[38]Comparison of Wood and Fossil Fuels (Washington, D.C.: Environmental Protection Agency, March 1976), EPA-600/2-76-056.
[39]R. H. Perry and C. H. Childton, eds., Chemical Engineer's Handbook, 5th Edition (McGraw Hill, 1973).
[40]M. D. Yokell, et al., Environmental Benefits and Costs of Solar Energy, vol. I (draft), Solar Energy Research Institute report SERI/TR-52-074, September 1979.
[41]J. O. Milliken, Environmental Protection Agency, Research Triangle Park, N.C., private communication, Oct. 26, 1979.
[42]Ibid.

Table 50.—Estimates of Total B(a)P Emissions
(metric tons/year)

Major sources	Minimum	Maximum
Burning coal refuse banks..............	280	310
Residential fireplaces.................	52	110
Forest fires........................	9.5	127
Coal-fired residential furnaces...........	0.85	740
Coke production....................	0.05	300

SOURCE: Energy and Environmental Analysis, Inc. ''Preliminary Assessment of the Sources, Control and Population Exposure to Airborne Polycyclic Organic Matter (POM) as indicated by benzo(a)pyrene [B(a)P],'' November 1978.

to travel 60 miles (100 km) or farther from their source.[43] However, sources that are far from population centers are less dangerous than urban sources both because of the dispersion that occurs with distance and because POMs eventually can be degraded to less harmful forms by photo-oxidative processes.[44]

POMs are dangerous for a number of reasons. First, because of their physical nature, they are more likely than most substances to reach vulnerable human tissues. They are formed in combustion as vapors and then condense onto particles in the flue gas. The smaller particles adsorb a proportionately high amount because they have large surface/weight ratios. These smaller particles are both less likely to be captured by particulate control equipment and more likely to penetrate deep into the lungs if breathed in. Second, several of the POM compounds produced by combustion are "the same compounds that, in pure form, are known to be potent animal carcinogens."[45] POM is suspected as a cofactor (contributor) to the added lung cancer risk apparently run by urban residents.[46] Finally, POM is suspected of causing or contributing to added incidence of chronic emphysema and asthma.[47]

[43]G. Lunde and A. Bjorjeth, "Polycyclic Aromatic Hydrocarbons in Long-Range Transported Aerosols," *Nature, 268*, 1977, pp. 518-519.

[44]M. J. Svess, "The Environmental Load and Cycle of Polycyclic Aromatic Hydrocarbons," *The Science of the Total Environment, 6*, 239, 1979.

[45]J. O. Milliken, "Airborne Emissions From Wood Combustion," presented at the *Wood Heating Seminar V*, Portland, Oreg., sponsored by the Wood Energy Institute, Mar. 22-24, 1979.

[46]J. O. Milliken and E. G. Bobaleck, *Polycyclic Organic Matter: Review and Analysis* (Research Triangle Park, N.C.: Special Studies Staff, Industrial Environmental Research Laboratory, Environmental Protection Agency, 1979).

[47]K. L. Stemmer, "Clinical Problems Induced by PAH," in *Carcinogenesis, Volume I.: Polynuclear Aromatic Hydrocarbons: Chemistry, Metabolism, and Carcinogenesis* (New York: Raven Press, 1976).

Because POM and other organic emissions, as well as CO, are the products of incomplete combustion, the new airtight stoves, which are beginning to take an increasing market share, will have to be evaluated carefully for their emission characteristics, especially under improper operation. Airtight stoves achieve a higher overall *heating* (but not necessarily combustion) efficiency by slowing down combustion, transferring more of the heat produced into the room rather than up the flue, and avoiding the establishment of an airflow from the room into the stove and up the flue. The reduction of excess air allowed into the combustion zone increases the emissions of CO and unburned hydrocarbons. Ideally, these pollutants will be burned in a secondary combustion zone fed with preheated air (air that is first routed through the primary combustion chamber). However, if the air fed into this zone is too cool, secondary combustion will not occur; under these circumstances, airtight stoves would be substantially more polluting than ordinary stoves. Also, the lower airflow and cooler exit gases of these stoves cause them to deposit more of their organic emissions—in the form of creosote—on the interior of their chimneys. Deposits of creosote from wood stoves and fireplaces have always been a fire hazard; this hazard will be increased by greater use of airtight stoves. An added safety problem associated with airtight stoves is the potential for "back-puffing"—surge back of flames—when the stove is opened. Both of these safety hazards are controllable by, respectively, having the flue cleaned regularly and increasing the intake airflow before opening the stove.

Utility and Industrial Boilers

Large wood-fired boilers should be more efficient energy converters than small units and therefore should have less problems with CO and unburned hydrocarbons. However, the potential exists to generate significant quantities of these pollutants, and some existing large boilers are fairly inefficient and thus fulfill this potential. (For example, emissions of CO from industrial boilers range from 1 to 30 g/kg of wood, compared to 60 to 130 g/kg from small wood stoves.)[48] Inefficient boilers will generate the same dangerous organic compounds—including species of POM—as do small residential stoves and fireplaces. These organics are mostly "low-molecular-weight hydrocarbons and alcohols, acetone, simple aromatic compounds, and several short-chain unsaturated compounds such as olefins.[49] Some of these emissions are photochemically reactive, although the amounts in question should not contribute significantly to smog problems. As the price of wood and wood waste increases, strong incentives for greater combustion efficiency should work to minimize the organic emission problem.

Sulfur dioxide (SO_2) emissions should be minimal because of wood's low sulfur content. An exception to this is the combustion of black liquor in the pulp and paper industry; some of these boilers should require SO_x scrubbing under Federal regulations.[50]

Although particulate emissions generated by wood-fired boilers can be high (6 g/kg, or about as high per unit of energy as a wood stove), efficient controls are available for the larger units. Available devices or combinations of devices include multicyclones coupled with low-energy wet scrubbers, dry scrubbers, electrostatic precipitators (ESPs), or baghouses (fabric filters). Although ESPs are the most widely used control mechanism for utility boilers, they have been said to be less practical for wood-fired boilers because of the very low resistivity of both the flyash and unburned carbon particles from wood combustion.[51] However, ESPs have been successfully used on some wood-fired boilers, and the problem of low resistivity apparently can be handled with appropriate precipitator design.

Current regulations for emission control from pollution sources do not distinguish particulates by their size. Most control devices in current use suffer from a severe drop in efficiency in controlling the finer, more dangerous particles. Baghouses appear to be the only feasible control devices currently available that are capable of collecting particles below a few microns in size with 99-percent efficiency or greater. It appears quite probable that emission standards for the finer particulates eventually will be promulgated; these standards would almost certainly lead to extensive use of baghouse controls.

Current EPA emission factors show NO_x emissions from wood combustion to be comparable to emissions from coal combustion.[52] If these factors were correct, large boilers subject to Federal new source performance standards would require NO_x reductions of 40 percent. This would pose a problem in the short term, because there is virtually no experience in reducing NO_x emissions from wood-fired boilers. Techniques used for fossil fuel boilers that may be applicable to wood are:

- low excess air firing,
- staged combustion, and
- flue gas recirculation.

Recent measurements conducted by Oregon State University[53] and TRW[54] show actual NO_x emissions from test boilers to be one-third or less than those predicted by using the current emission factors. These measurements are much more in line with the lower combustion temperatures in wood boilers and wood's low nitrogen content. EPA and DOE researchers are convinced that the current emission factors are in error[55][56] and it appears likely that the factors will soon be revised.

In the past, wood boilers have never attained the size normally associated with large coal-fired boilers. Whereas coal-fired utility boilers are typically a few hundred megawatts

[48]Milliken, "Airborne Emissions From Wood Gasification," op. cit.

[49]M. D. Yokell, op. cit.

[50]*Environmental Readiness Document, Wood Commercialization* (Department of Energy, 1979), draft.

[51]*Wood Combustion Systems: An Assessment of Environmental Concerns* (Mittelhauser Corp., July 1979), draft, contractor report to Argonne National Laboratory.

[52]*Compilation of Air Pollutant Emission Factors Revised* (Washington, D.C.: Office of Air Programs, Environmental Protection Agency, February 1972), publication No. AP-42.

[53]Memorandum from Paul A. Boys, Air Surveillance and Investigation Section to George Hofer, Chief, Support and Special Projects Section, U.S. Environmental Protection Agency, "Comparison of Emissions Between Oil Fired Boilers and Woodwaste Boilers," November 3, 1978.

[54]J. O. Milliken, Environmental Protection Agency, Research Triangle Park, N.C., private communication, June 6, 1979.

[55]Milliken, Oct. 26, 1979, op. cit.

[56]J. Harkness, Argonne National Laboratory, private communication, Oct. 26, 1979.

in generating capacity and range up to 1,000 MW, a 25- or 50-MW wood-fired boiler would be considered extremely large.

Higher capacities would require using wood suspension firing, analogous to firing with pulverized coal, or fluidized-bed combustion. Pulverizing wood to extreme fineness for suspension firing may be costly enough to offset other economic advantages of going to larger size plants, so future increases in wood boiler size may depend on further development of fluidized-bed combustion. The expense of transporting wood considerable distances has also been a constraint on boiler size in the past, but rising costs for alternative fuels may make longer distance transport of wood more attractive, increasing the effective radius of supply and the maximum practical size of the boiler.

The local impacts of utility or industrial wood-fired boilers will be moderated by their comparatively small size. However, the effects of low stacks (compared to the stacks on large coal-fired utility boilers) will be to allow less

diffusion of the emissions from the plants; a higher percentage of the pollution will fall out near the plants than would normally be expected for large generating facilities or industrial boilers. Also, the high water content of wood leads to higher concentrations of water vapor in the stack gases and greater visibility of the plumes. Although not harmful except in an esthetic sense, this increased visibility may lead to added local objections to wood-fired boilers.

In general, emissions from other portions of the fuel cycle are quite low compared to emissions from combustion. The single exception is CO, which is produced in substantial quantities by harvesting, chipping, and transport equipment. Table 51 presents a comparison of the emissions at all stages of the fuel cycle for coal, oil, and wood boilers. As noted above, CO and organic emissions from wood boilers are far higher than emissions from coal. Note that the emissions of SO_2 and particulates are dependent on the level of control, and can be reduced significantly if required.

Table 51.—''Source-to-Power'' Air Emissions for Coal, Oil, and Wood Fuel Systems

Fuel/energy system	Emissions ton/yr (basis 50-MW plant)			
	SO_2[a]	CO	Particulates	Total organic
Low-sulfur Western coal				
Surface mining	—	·—	113.1	—
Rail transport (1,800 miles)	20.2	218.1	21.8	22.2
Power generation	2,664.8	87.2	113.1	25.8
Total	2,685.0	305.3	248.0	28.0
Crude oil				
Domestic oil pipeline	25.8	0.0	3.2	0.5
New Jersey refined with desulfurization	193.8	4.8	3.2	40.4
Rail transport (300 miles)	1.5	16.3	1.6	1.7
Power generation	854.3	—	80.8	16.2
Total	1,075.4	21.1	88.8	58.8
Wood				
Wood recovery	6.5	48.5	3.2	8.1
Process chipping	14.5	116.3	6.5	19.4
Truck transport (60 miles)	4.4	36.3	2.1	6.0
Power generation	119.5	398.9	339.2	398.9
Total	144.9	600.2	351.0	432.4

NOTE: NO_x levels may be significant for wood fuel. There is inadequate data on NO_x emission levels. There are also production tradeoffs for various conversion systems.

[a]SO_2 emissions from coal-fired powerplant assume no scrubbers. 90% control required by new source performance standards would lower emissions from 2,664.8 tons to 266 tons.

SOURCE: E. H. Hall, et al., *Comparison of Fossil and Wood Fuels* (Washington, D.C.: Environmental Protection Agency, March 1976), EPA-600/2-76-056.

Environmental Impacts of Cofiring Agricultural and Forest Residues With Coal

Cofiring of coal and agricultural or forest residues has been proposed both as a means of expanding energy supply and as an economical way to lower sulfur emissions (from burning local high-sulfur coal) without importing low-sulfur coal. Since wood and most crop residues have very low sulfur contents (cotton gin trash is one exception), total SO_2 emissions can be significantly lowered if the residues can replace a large fraction of the coal normally burned in the boilers. Two situations where cofiring would appear to be attractive are:

- reducing SO_2 emissions from existing coal-fired powerplants that are marginally out of compliance with their State implementation plans, and
- allowing very high-sulfur coals to be used with scrubbers in new powerplants (achievement of the current 1.2 lb/million Btu SO_2 standard may be difficult with some very high-sulfur coals)

There are few examples of cofiring experiments in the literature and these examples generally do not examine emission changes caused by the addition of crop and wood wastes to the coal fuel. Because SO_2 is the only pollutant whose formation generally does not vary with combustion conditions (except that sulfur may be captured in the char from a pyrolytic reaction), it is probably the only pollutant that can be predicted reliably at this time. However, general emission *trends* for some pollutants can be predicted. For example, hydrocarbon and CO emissions may increase slightly, because combustion temperatures are lowered and complete combustion is more difficult to achieve when residues are added to the boiler fuel. The lower combustion temperature and low fuel-bound nitrogen in the residues should cause NO_x emissions to be lowered. If dryers are used for high-moisture-content residues, their emissions must be added to those of the boiler.

Particulate emissions are difficult to predict because they are affected by several site-specific factors. However, there appears to be some potential for increased particulate emissions under certain conditions. Although biomass residues generally have lower inorganic ash contents than the coal they would replace, they tend to generate more organics in particulate form. The ability of the boiler to maintain nearly complete combustion conditions will thus strongly affect particulate emissions. In large facilities with ESPs, the lower resistivity of the particles generated from combustion of the residues may allow a higher percentage to escape control. If the biomass is fed moist into the boiler, the steam generated during combustion will increase the flow of hot combustion gases and conceivably may lead to more entrainment of bottom ash and higher particulate emissions. On the other hand, if the biomass is first artificially dried, particulate emissions from the dryer could be high unless they are carefully controlled. The significance of any of these effects is uncertain at the present time.

The importance of these emission changes depends on the original quality of the coal, the nature of the residues added, the percentage fuel mixture, the type of pollution controls on the boiler, and its operating conditions. All of these factors vary considerably from site to site. However, it seems likely that emission increases will be small except in cases where the cofiring seriously degrades the operating characteristics of the boiler (it is unlikely that cofiring would continue under such conditions unless noneconomic pressures—such as the possibility of adverse publicity and/or embarrassment of company management—prevented cessation of operations). In addition, emissions changes will be limited by constraints on the amount of biomass that can be mixed with the coal. Logging residues and high-moisture crop residues have a considerably lower energy content per unit volume than coal. Because boiler systems are sized to allow a certain volumetric flow rate of fuel feed, a high percentage of biomass volume in the feed will limit boiler output capacity. An additional limit may be presented by the additional volume of combustion gases that would be generated if the biomass is fed moist into the boiler. These constraints do not apply when the energy output required is much lower than the boiler's rated capacity, or when the system is specifically designed for cofiring. Also, the high-moisture content of the biomass may cause condensation problems in the stack unless the biomass content is limited, stack temperatures are increased (by removing less energy from the gas and thus lowering system efficiency), or the biomass is first dried (which may also lower system efficiency).

Environmental Impacts of Gasification

Gasification technologies have a number of potential air and water impacts. Because few such gasifiers are in operation, quantification of these impacts is premature. The low concentrations of trace metals and sulfur in the biomass feedstocks and the lack of extreme temperature and pressure conditions imply that impacts should be substantially less than those associated with coal gasification. However, scientists working for DOE's Fuels from Biomass Branch profess to be unsure as to whether this supposed biomass "advantage" actually exists, especially in the water effluent stream; although the hydrocarbons present in biomass gasification wastewater should be more amenable to biological treatment than coal gasification hydrocarbons (they are more oxygenated), they may be produced in greater quantities and have a higher biological oxygen demand than those of a coal system.[57] Also, the potential for proliferation of small-scale biomass gasifiers may present monitoring and enforcement problems that would not exist with a few large coal gasifiers. Therefore, biomass gasification may require as much attention and concern as coal gasification.

The quantity and mix of air pollutants produced by biomass gasification plants will depend in large part on the combustion/gasification conditions maintained as well as the environmental controls and the chemical make-up of the feedstock. For example, the concentration of hydrogen in the reaction chamber and of sulfur and nitrogen in the feedstock will influence the formation of ammonia (NH_3),

hydrogen sulfide (H_2S), and hydrogen cyanide (HCN). Other products of the gasification process include carbonyl sulfide (COS) and carbon disulfide (CS_2) as well as phenols and polynuclear aromatic (PNA) compounds.[58] Gasification processes that are closer in their nature to pyrolysis and that produce considerable byproduct char will have lower nitrogen and sulfur-derived emissions; about half of the original sulfur and nitrogen in the biomass should remain in the char.[59]

The gas produced will either be burned onsite (producer gas) or cleaned and upgraded to pipeline gas. Either process should eliminate or reduce most of the more toxic pollutants, with the onsite burning oxidizing them to CO_2, SO_2, NO_2, and water. Recent tests of a close-coupled gasifier/boiler combination using woodchips for fuel showed emissions of CO, particulates, and hydrocarbons—which are of major concern in wood combustion—to be well below emissions expected from a direct-fired wood boiler, although a fuel oil boiler replaced by such a gasifier would have had considerably *lower* particulate and hydrocarbon emissions. NO_x emissions from the gasifier/boiler combination were lower than those expected from either oil- or wood-fired boilers.[60]

---

[57]Richard Doctor, Science Applications, Inc., private communication, November 1979.

[58]*Solar Program Assessment: Environmental Factors, Fuels From Biomass* (Washington, D.C.: Energy Research and Development Administration, March 1977), ERDA 77-47/7.

[59]Ibid.

[60]California Air Resources Board, "Source Test Report No. C-8-002-C, Source Test of Exhaust Gas From a Boiler Fired by Production Gas Generated From an Experimental Gasifier Unit Using Wood Chips for Fuel," Stationary Source Control Division, March 1978.

	Gasifier	Wood boiler	Oil boiler
Carbon monoxide (lb/hr)..	0	1.8-54	0.33
Particulates (lb/hr)	0.70	4.5-13.5	0.13
Hydrocarbons (lb/hr).....	0.90	63	0.07
Nitrogen oxides (lb/hr)	0.39	9	1.46

These results cannot be readily extrapolated to other situations, but they imply that the use of gasifiers *may* offer a less polluting alternative to direct combustion of biomass when a shift to renewables (from oil) is being contemplated.

Leaks of raw product gas represent a potential for significant impacts, especially on those in the immediate vicinity of the gasifier. The probability of such leakage is not known. Although impact analyses of high-pressure coal gasification technologies have identified fugitive hydrocarbon emissions as a likely problem, it is not clear that similar problems would occur with (lower pressure) biomass gasifiers.

The combustible char produced by the gasification process is another potential source of air pollution. It may be used as a fuel source elsewhere or else used to heat the bed in a fluidized-bed gasifier. In either case, its combustion will produce NO_x, flyash, and SO_x as well as trace metals either adsorbed on the flyash (potassium, magnesium, sodium, iron, boron, barium, cadmium, chromium, copper, lead, strontium, and zinc) or in gaseous form (beryllium, arsenic compounds, fluorides).[61][62]

Because most biomass feedstocks used in gasification processes have concentrations of

---
[61]*Solar Program Assessment,* op. cit.
[62]Doctor, op. cit.

trace elements, ash, and sulfur that are substantially lower than concentrations found in coal, combustion of the char should emit lower concentrations of related pollutants than would coal combustion. Depending on the farming and harvesting techniques, however, the feedstock may be somewhat contaminated with pesticides, fertilizers, and soil, which should add to combustion pollutants. Also, some forms of biomass—for example, cotton trash, with 1.7 percent—have sulfur levels comparable to levels in coal.

Aside from water impacts caused by construction activities and leaching from biomass storage piles, gasification facilities will have to control potential impacts from disposal and storage of process wastes and byproducts. Water initially present in the feedstock and that formed during the combustion accompanying gasification should provide significant amounts of effluent requiring disposal (although in close-coupled systems, the moist low-Btu gas may be fed directly into the boiler). Air pollutants identified above may appear also as water contaminants: NH_3 (as ammonium hydroxide), HCN and its ionized form, phenols, and trace elements found in the ash. Leaching from byproduct chars may be a problem if the char is (incompletely carbonized) brown char although (carbonized) black char should be similar to charcoal and far less likely to be polluting. Finally, the tars produced by gasification may well be carcinogenic; as yet no data confirm this potential. These water contaminants present a potential occupational as well as ecological and public health concern, because plant operators may be exposed unless stringent "housekeeping" is enforced.

Research, Development, and Demonstration Needs

Thermochemical conversion includes the least expensive, near-term processes for using the major biomass resources—wood and plant herbage. Moreover, R&D is likely to lead to interesting new possibilities for the production of fuels and chemicals from biomass. Some of the more important areas are:

- **Thermochemistry of biomass.**—Basic and applied research into the thermochemistry of biomass, including secondary gas phase reactions, is needed to better define the possibilities for fuel synthesis and to aid engineers in designing advanced reactors. The research should include studies of the effects of the various operating parameters on the nature and composition of products and ways to maximize the yields of various desirable products such as CO and hydrogen, ethylene, methane, and other light hydrocarbons.

- **Gasifier development and demonstration.** — Gasifiers should be developed further and demonstrated, so as to improve their reliability, efficiency, and flexibility with respect to feedstock type and moisture content. This should include airblown gasifiers for process heat, boiler retrofits, and ICEs, and oxygen-blown and pyrolysis gasifiers for methanol synthesis. It should also include the demonstration of gasifiers suitable for converting plant herbage to methanol and should investigate the tradeoff between densifying herbage before gasification versus gasification of herbage directly. Each of the uses for gasifiers will have unique requirements, which probably will dictate separate development and demonstration efforts.

- **Compressor development.** — One of the major costs of producing methanol in small plants is the relativley high price of small compressors. The cost of methanol synthesis from biomass would be lowered substantially if small, inexpensive compressors suited to the process are developed.

Each of the new biomass conversion technologies will require environmental assesment to ensure the development of appropriate control technologies and incorporation of environmental considerations in system design, siting, and operation. In general, the larger scale technologies are likely to be assessed as part of normal EPA and DOE environmental programs. The smaller technologies generally will not come under Federal new source performance standards (specifications of allowable emissions), but there is growing recognition in EPA and DOE of the potential environmental dangers of small-scale technologies such as wood stoves.

Key environmental R&D areas in thermochemical conversion are:

- development of wood stove designs (or controls) that achieve complete combustion and minimize emissions of unburned hydrocarbons;

- development of combustion controls that will allow efficient—and pollutant minimizing—thermochemical reactions regardless of feedstock characteristics;

- assessment of the potential health effects of emissions from wood stoves and other biomass conversion technologies, with a focus on particulates with a high unburned hydrocarbon component;

- evaluation of toxicity and carcinogenicity of biomass gasifier/pyrolysis tars and oils; and

- design of controls for gasifier/pyrolysis effluent streams.

Appendix A.—Optimum Size for a Wood-Fired Electric Powerplant

The annual cost, C, of producing electricity in a wood-fired electric powerplant can be expressed as:

$$C = C_c + C_f + C_t \qquad (1)$$

Where C_c represents the capital and other fixed charges, C_f represents the fuel and other variable costs, and C_t the cost of transporting the fuel.

Letting S represent a dimensionless scaling parameter

$$C_c = C_c{}^o S^{0.7} \qquad (2)$$

Here $C_c{}^o$ represents the fixed charges for a base case and it is assumed that these charges scale with a 0.7 scaling factor. Furthermore, the variable costs are:

$$C_f = C_f{}^o S \qquad (3)$$

Where $C_f{}^o$ represents the base case.

Assuming the fuel is collected from a circular area surrounding the powerplant, the transport costs can be expressed as:

$$C_t = C_t{}^o Q_o S \bar{r} \qquad (4)$$

Where $C_t{}^o$ is the transport cost per ton-mile, Q_o is the annual quantity of wood transported in the base case, and $\bar{r}$ is the average transport distance.

For a given scaling parameter S, the quantity of wood transported is:

$$Q = Q_o S = \varrho \pi r^2 \qquad (5)$$

where ϱ is the average availability of fuel wood collected in dry tons per square mile year and r is the radius of the circle from which wood is collected.

If one assumes that the actual transport distance from a harvest site to the powerplant is $\sqrt{2}$ times the direct line distance.

$$\bar{r} = \sqrt{2} \ \frac{\int_0^r a^2 da}{\int_0^r a \, da} = \frac{(2)^{1.5}}{3} r \qquad (6)$$

where the integrals represent the average straight line distance to the powerplant over the entire collection area. Substituting equation (6) and (5) into equation (4), one finds

$$C_t = \frac{C_t{}^o}{3} (\varrho \pi)^{-0.5} (2 Q_o S)^{1.5} \qquad (7)$$

$$= K S^{1.5}$$

where K is defined by this equation.

Combining equations (1) through (3) and (7) results in a cost per unit output (e.g., cost per kilowatthour), which is proportional to

$$\bar{C} = \frac{C}{S} = C_c{}^o S^{-0.3} + C_f{}^o + K S^{0.5} \qquad (8)$$

Taking the derivative of $\bar{C}$ with respect to S and setting it to zero (in order to find the minimum cost per kilowatthour) yields

$$S = \left(\frac{0.6 C_c{}^o}{K} \right)^{1.25} \qquad (9)$$

This represents the optimum size for the powerplant.

Evaluating the parameters for the base case given in appendix B results in:

$$S = 11.7 \ \varrho^{0.625} \qquad (10)$$

where ϱ is in dry tons per acre year, the base case corresponds to a 62-MW powerplant, and the transport costs are assumed to be \$0.20/dry ton-mile (\$0.10/green ton-mile).

As expected, the higher the density of biomass availability, ϱ, the larger is the optimum-sized powerplant. Ironically, however, as ϱ increases, the average transport distance decreases. In other words, it is more economic to keep the powerplant size smaller than to transport large quantities of wood for greater distances.

If one assumes that $\varrho = 0.5$ dry ton/acre-yr, then the optimum powerplant size is over 500 MW and the radius of the collection circle is about 50 miles. With $\varrho = 0.05$ dry ton/acre-yr, the optimum size is 110 MW and the circle radius is 75 miles. If the transport charges double, then for these values of ϱ, the optimum sizes are reduced to 200 and 50 MW, respectively, with collection radii of 30 and 50 miles, respectively.

In principle, then, large-scale biomass conversion facilities are not unrealistic. The values assumed for ϱ are probably less than what can be achieved in a region where the infrastructure for fuelwood harvests is fully developed. In practice, however, it is likely to be difficult to develop a mature harvest-supply infrastructure devoted to a single conversion facility. As the infrastructure is being developed, many small users are likely to compete for the fuelwood and the resultant availability to a single user may never reach the hundreds of thousands or millions of dry tons per year necessary for the larger facilities.

Clearly biomass farms dedicated to a single conversion facility would overcome these problems of obtaining a large feedstock source. It is unlikely, however, that these farms will be developed as described in the section on "Unconventional Biomass Production."

Appendix B.—Analysis of Break-Even Transport Distance for Pellitized Wood and Miscellaneous Cost Calculations

Two ways for producing 1 million Btu of steam from woodchips containing 50-percent moisture are presented in figure B-1. In the first case 389 lb of greenwood are transported to the boiler and burned directly; in the second 339 lb of greenwood are pelletized first and then transported to the boiler for burning.

In the first case the cost of the fuel needed to produce this steam is:

$$C = C_w W_w + C_t W_w d \qquad (1)$$

where C_w is the cost per ton of wood at a central yard, W_w is the weight of wood to be transported, C_t is the transport cost per ton-mile, and d is the distance from the yard to the boiler.

In the second case:

$$C = C_p W_p + C_t W_p d \qquad (2)$$

when C_p is the cost of the pellets at the pellet mill, W_p is the weight of pellets to be transported and the other symbols are as before (assuming the pellet mill is located at the wood yard).

Setting these two costs equal to one another and using the weights of wood and pellets as above, one finds that:

$$d = (0.65\,C_p - 1.65\,C_w)\,/\,C_t \qquad (3)$$

With transport costs of $0.10/ton mile and the wood and pellet costs given in the text, the break-even transport distance varies from 43 to 71 miles. If, however, the original wood is 40-percent moisture, only 324 lb are needed in the boiler (with the same efficiency) and the break-even transport distance becomes 123 to 134 miles. If the transport charges are lower, the break-even distance will increase. Conversely, where transport is more expensive, the break-even distance will be less. Numerous other local variables can also change the results.

Miscellaneous Cost Calculations

Following are estimates for the costs of various thermochemical conversion processes.

Figure B-1.—Two Ways to Produce 10⁶ Btu Steam From Wood

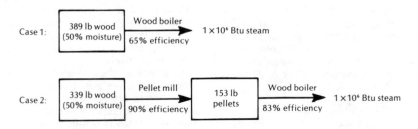

Table B-1.—Electricity From Wood by Direct Combustion

Input:	2,000 green ton/d of wood (50% moisture)		
Output:	62-MW electricity		
Load:	300 operating days per year		
Fixed investment (field erected). .	$50 million		
Working capital (10% of fixed investment)	5 million		
Total investment	$55 million		
		Mills/kWh	Million $/yr
Wood ($15/green ton).		20	9.0
Labor and water.		9	4.0
Capital charges (15% of total investment per year).		19	8.25
Total		48	21.3
Estimated range:		45-70 mills/kWh	

SOURCE: OTA from Steven R. Beck, Department of Chemical Engineering, Texas Tech University, Lubbock, Tex., private communication, 1979.

Table B-2.—Steam From Wood by Direct Combustion (package boiler)

Input:	270 green ton/d of wood (50% moisture)	
Output:	50,000 lb[a] of steam/hr	
Load:	330 operating days per year	
Fixed investment (package boiler)	$600,000	
	$1,000 lb steam	
Wood ($15/green ton).	3.38	
Labor ($75,000/yr).	0.19	
Capital charges (15-30% of fixed investment per year).	0.23-0.46	
Total	3.80-4.03	
Estimated range:	$3.50-$6.00/ 1,000 lb steam ($2.80-$4.80/10⁶ Btu)	

[a]1,000 lb of steam = 1.25 million Btu of steam.

SOURCE: OTA from A Survey of Biomass Gasification (Golden, Colo.: Solar Energy Research Institute, July 1979).

Table B-3.—Electricity and Steam From Wood by Direct Combustion

Input:	2,000 green ton/d of wood (50% moisture)	
Output:	21.4-MW electricity and 390,000 lb steam[a]/hr	
Load:	300 operating days per year	
Fixed investment (field erected)........	$40 million	
Working capital (10% of fixed investment).....................	4 million	
Total investment	$44 million	
	Million $/yr	
Wood ($15/green ton)...............	10.9	
Labor and water...................	4	
Capital charges (15% of total investment per year).............	6.6-13.2	
Total	21.5-28.1	

Product costs:	Steam (assumed cost) $/1,000 lb	Electricity (derived) mills/kWh
	4	67-109
	5	48-91
	6	30-73

[a]1,000 lb of steam = 1.25 million Btu of steam.

SOURCE: OTA from Steven R. Beck, Department of Chemical Engineering, Texas Tech University, Lubbock, Tex., private communication, 1979.

Table B-4.—Medium-Btu Gas From Wood in a Dual Fluidized-Bed Field Erected Gasifier

Input:	2,000 green ton/d of wood (50% moisture)
Output:	460 10⁶ Btu/hr medium-Btu gas
Load:	330 operating days per year
Fixed investment (field erected)........	$43 million
Working capital (10% of fixed investment).....................	4.3 million
Total investment	$47.3 million
	$/10⁶ Btu gas
Wood ($15/green ton)...............	3.26
Labor and water...................	0.82
Capital charges (15% of total investment per year).............	$1.95-$3.90
Total	$6.03-$7.98
Estimated range:	$5.50-$9.00/10⁶ Btu

SOURCE: OTA from Steven R. Beck, Department of Chemical Engineering, Texas Tech University, Lubbock, Tex., private communication, 1979.

Table B-5.—Medium-Btu Gas From Manure in a Fluidized-Bed Gasifier

Input:	1,000 dry ton/d of manure
Output:	400 10⁶ Btu/hr medium-Btu gas
Load:	330 operating days per year
Fixed investment (field erected)........	$36 million
Working capital (10% of fixed investment).....................	3.6 million
Total investment	$39.6 million
	$/10⁶Btu of gas
Manure ($3/dry ton)	0.31
Labor, water, chemicals, ash disposal, electricity	1.64
Capital charges (15% of total investment per year).............	$1.88-$3.75
Total	$3.83-5.70
Estimated range:	$3.50-$7.00/10⁶ Btu

SOURCE: OTA from Steven R. Beck, Department of Chemical Engineering, Texas Tech University, Lubbock, Tex., private communication, 1979.

Table B-6.—Methanol From Wood Through Gasification in a Dual Fluidized-Bed Gasifier

Input:	2,000 green ton/d of wood (50% moisture)
Output:	3,150 bbl methanol/d (44 million gal/yr)
Load:	330 operating days per year
Fixed investment (field erected)........	$64 million
Working capital (10% of fixed investment).....................	6.4 million
Total investment	$70.4 million
	$/bbl
Wood ($15/green ton)...............	9.52
Labor, water, and chemical	4.91
Capital charges (15-30% of total investment per year).............	10.16-20.32
Total	$24.59-$34.75 ($0.58-$0.83/gal)
Estimated range:	$22-$40/bbl ($0.52-$0.95/gal) ($8.20-14.96/10⁶ Btu)

SOURCE: OTA from Steven R. Beck, Department of Chemical Engineering, Texas Tech University, Lubbock, Tex., private communication, 1979.

Table B-7.—Pyrolysis Oil From Wood by Catalytic Direct Liquefaction

Input:	2,000 green ton/d of wood (50% moisture)
Output:	2,500 bbl/d of pyrolytic oil (4.2 10⁶ Btu/bbl)
Load:	330 operating days per year
Fixed investment (field erected)........	$50 million
Working capital (10% of fixed investment).....................	5 million
Total investment	$55 million
	$/bbl
Wood ($15/green ton)...............	12.00
Labor, water, and chemicals	7.27
Capital charges (15-30% of total investment per year).............	10.00-20.00
Total	$29.27-$39.37
Estimated range:	$30-$50/bbl ($7-$12/10⁶ Btu)

SOURCE: OTA from Steven R. Beck, Department of Chemical Engineering, Texas Tech University, Lubbock, Tex., private communication, 1979.

Table B-8.—Ethanol From Wood via Gasification in a Dual Fluidized-Bed Gasifier

Input:	2,000 green ton/d of wood (50% moisture)
Output:	1,620 bbl/d of ethanol (assuming 14 wt. % yield of ethylene from dry wood)
Load:	330 operating days per year
Fixed investment (field erected)........	$60 million
Working capital (10% of fixed investment).....................	6 million
Total investment	$66 million
	$/bbl
Wood ($15/green ton)...............	18.27
Labor, water, chemicals, and electricity	12.53
Capital charges (15-30% of total investment per year).............	18.52-37.04
Total	$49.32-$67.84 ($1.17-$1.62/gal) ($13.90-$19.20/10⁶ Btu)
With ethylene yield of 30 wt. %:	$23-$32/bbl ($0.55-$0.76/gal) ($6.50-$9.00/10⁶ Btu)

SOURCE: OTA from Steven R. Beck, Department of Chemical Engineering, Texas Tech University, Lubbock, Tex., private communication, 1979.

Appendix C.—Survey of Gasifier Research, Development, and Manufacture[a]

	Gasifier type				Size
Organization	Input	Contact mode	Fuel products	Operating units	Btu/hr
Air gasification of biomass					
Alberta Industrial Dev., Edmonton, Alb., Can.	A	Fl	LEG	1	30 M
Applied Engineering Co., Orangeburge, S.C. 29115	A	U	LEG	1	5 M
Battelle-Northwest, Richland, Wash. 99352.	A	U	LEG	1-D	—
Century Research, Inc., Gardena, Calif. 90247.	A	U	LEG	1	80 M
Davy Powergas, Inc., Houston, Tex. 77036.	A	U	LEG-Syngas	20	—
Deere & Co., Moline, Ill. 61265.	A	D	LEG	1	100 kW
Eco-Research Ltd., Willodale, Ont. N2N 558	A	Fl	LEG	1	16 M
Forest Fuels, Inc., Keene, N.H. 03431	A	U	LEG	4	1.5-3.0 M
Foster Wheeler Energy Corp., Livingston, N.H. 07309	A	U	LEG	1	—
Fuel Conversion Project, Yuba City, Calif. 95991	A	D	LEG	1	2 M
Halcyon Assoc. Inc., East Andover, N.Y. 03231	A	U	LEG	4	6-50 M
Industrial Development & Procurement, Inc., Carie Place, N.Y. 11514	A	D	LEG	Many	100-750 kW
Pulp & Paper Research Inst.,[b] Pointe Claire, Quebec H9R 3J9	A	D	LEG	—	—
Agricultural Engr. Dept., Purdue University, W. Lafayette, Ind. 47907	A	D	LEG	1	0.25 M
Dept. of Chem. Engr., Texas Tech University, Lubbock, Tex. 79409	A	Fl	LEG	1	0.4 M
Dept. of Chem. Engr., Texas Tech University, Lubbock, Tex. 79409	A	U	LEG	1	—
Vermont Wood Energy Corp., Stowe, Vt. 05672	A	D	LEG	1	0.08 M
Dept. of Ag. Engr., Univ. of Calif., Davis, Calif. 95616	A	D	LEG	1	64,000
Dept. of Ag. Engr., Univ. of Calif., Davis, Calif. 95616	A	D	LEG	1	6 M
Westwood Polygas (Moore)	A	U	LEG	1	
Bio-Solar Researc & Development Corp., Eugene, Oreg. 97401	A	U	LEG	1	—
Oxygen gasification of biomass					
Environmental En. Eng., Morgantown, W.V.	O	D	MEG	1P	0.5
IGT-Renugas	O, S	Fl	MEG		
Pyrolysis gasification of biomass					
Wright-Malta, Ballston Spa, N.Y.[c]	PG	O	MEG (C)	1R, 1P	4
Coors/U. of Mo.	P	Fl		1P	
U. of Arkansas	P	O	MEG (C)	1R	
A & G Corp., Jonesboro, Ark.	P	O	MEG (C)	1C	
ERCO, Cambridge, Mass.	P	Fl	PO, C	1P, (1C)	16, (20)
ENERCO, Langham, Pa.	P		MEG, PO, C	1P, 1C	
Garrett Energy Research	MH		MEG	1P	
Tech Air Corp., Atlanta, Ga. 30341.	P	U	MEG, PO, C	4P, 1C	33
M. Antal, Princeton Univ., NS	PG	O	MEG, C	1R	—
M. Rensfeit, Sweden	PG	O	MEG, C	1R	
Texas Tech, Lubbock, Tex.	PG	Fl	MEG	1P	
Battelle-Columbus, Columbus, Ohio					
Air gasification solid muncipal waste (SMW)					
Andco-Torrax,[a] Buffalo, N.Y.	A	U	LEG	4C	100 M
Battelle-Northwest, Richland, Wash. 99352.					

Table Notation: (by columns)
Input: A = air gasifier; O = oxygen gasifier; P = pyrolysis process; PG = pyrolysis gasifier; S = steam; C = char combustion.
Contact mode: U = updraft; D = downdraft; O = other (sloping bed, moving grate); Fl = fluidized bed; S = suspended flow; MS = molten salt; MH = multiple hearth.
Fuel products: LEG = low energy gas (~150-200 Btu/SCF) produced in air gasification; MEG = medium energy gas produced in oxygen and pyrolysis gasification (350-500 Btu/SCF; PO = pyrolysis oil, typically 12,000 Btu/lb; C = char, typically 12,000 Btu/lb.
Operating units: R = research; P = pilot; C = commercial size; CI = commercial installation; D = demonstration.
Size: Gasifiers are rated in a variety of units. Listed here are Btu/h derived from feedstock throughput on the basis of biomass containing 16 MBtu/ton or 8,000 Btu/lb, SMW with 9 MBtu/ton. ( ) indicate planned or under construction.
[a]Unless noted otherwise, the gasifiers listed here produce dry ash (T > 1,100° C) and operate at 1 atm pressure. (Coal gasifiers and future biomass gasifiers may operate at much higher pressures.)
[b]Operates at 1-3 atm pressure.
[c]Operates at 10 atm pressure.
[d]These gasifiers produce slagging (T > 1,300° C) instead of dry ash.

SOURCE: A Survey of Biomass Gasification (Golden, Colo.: Solar Energy Research Institute, July 1979).

| Organization | Gasifier type | | | Operating units | Size |
	Input	Contact mode	Fuel products		Btu/hr
Oxygen gasification of SMW					
Union Carbide (Linde), Tonowanda, N.Y.[d]	O	U	MEG	1	100 M
Catorican, Murray Hills, NS	O	U			9 M
Pyrolysis gasification of SMW					
Monsanto, Landgard, Enviro-chem	P, C	K	LEG, O, C	1D	20 (375)
Envirotech, Concord, Calif.	P	MH	LEG	1P	
Occidental Res. Corp., El Cajon, Calif.	P	Fl	PO, C, MEG	1C	
Garrett En. Res. & Eng., Hanford, Calif.	P	MH	MEG	1P	
Michigan Tech., Houghton, Mich.	P	ML	MEG		
U. of W. Va.-Wheelebrator, Morgantown, W. Va.	P, G, C	Fl	MEG	1P	
Pyrox, Japan.	P, G, C	Fl	MEG	1C	
Nichols Engineering	P		MEG, C		
ERCO, Cambridge, Mass.	P	Fl	MEG	1P	16
Rockwell International, Canoga Park, Calif.	P	MS	MEG, C	1P	16
M. J. Antal, Princeton, NS	P	O	MEG, C	2R	—

Table Notation: (by columns)
Input: A =air gasifier; O = oxygen gasifier; P = pyrolysis process; PG = pyrolysis gasifier; S = steam; C = char combustion.
Contact mode: U = updraft; D =downdraft; O = other (sloping bed, moving grate); Fl = fluidized bed; S = suspended flow; MS =molten salt; MH = multiple hearth.
Fuel products: LEG = low energy gas (~150-200 Btu/SCF) produced in air gasification; MEG = medium energy gas produced in oxygen and pyrolysis gasification (350-500 Btu/SCF; PO = pyrolysis oil, typically 12,000 Btu/lb; C = char, typically 12,000 Btu/lb.
Operating units: R = research; P = pilot; C = commercial size; CI =commercial installation; D =demonstration.
Size: Gasifiers are rated in a variety of units. Listed here are Btu/h derived from feedstock throughput on the basis of biomass containing 16 MBtu/ton or 8,000 Btu/lb, SMW with 9 MBtu/ton. ( ) indicate planned or under construction.
[a]Unless noted otherwise, the gasifiers listed here produce dry ash (T   1,100° C) and operate at 1 atm pressure. (Coal gasifiers and future biomass gasifiers may operate at much

Chapter 8
FERMENTATION

Chapter 8.—FERMENTATION

FERMENTATION

Introduction

Ethanol, or "grain alcohol," is a versatile and commercially important liquid which has been used for a variety of purposes for centuries. Ethanol is the intoxicant in alcoholic beverages and, prior to the industrial age, society's most common contact with ethanol was as an ingredient of beer, wine, or liquor.

Beverage alcohol is a major item of commerce and a source of substantial tax revenues. In addition, ethanol is also a key industrial chemical and is used as a solvent or reactant in the manufacture of organic chemicals, plastics, and fibers. Ethanol has a long history as a combustible fuel for transportation vehicles and space heating. Except under unusual circumstances (e.g., wartime Europe), ethanol has been little used for these purposes in the 20th century, having been largely displaced by petroleum-based motor and boiler fuels.

Beverage alcohol is usually produced by fermentation processes, but the processes are designed to achieve various qualities of taste and aroma which are irrelevant to fuel alcohol production. Most industrial ethanol is produced from ethylene, a gas derived from petroleum or natural gas liquids. Rising oil prices have made biomass-derived ethanol competitive with ethanol derived from petroleum but it is unclear whether the chemical industry will turn to biomass or coal for its supply of ethanol.

All processes for the production of ethanol through fermentation consist of four basic steps: 1) first the feedstock is treated to produce a sugar solution; 2) the sugar is then converted to ethanol and carbon dioxide (CO_2) by yeast or bacteria in a process called fermentation; 3) the ethanol is removed from the fermented solution by a distillation* process which yields a solution of ethanol and water that cannot exceed 95.6 percent ethanol (at normal pressures) due to the physical properties of the ethanol-water mixture; and 4) in the final step, the water is removed to produce dry ethanol. This is accomplished by distilling once again in the presence of another chemical.

The main distinctions among the processes using different feedstocks are the differences in the pretreatment steps. Sugar crops such as sugarcane, sweet sorghum, and sugar beets yield sugar directly, but the sugar often must be concentrated to a syrup or otherwise treated for storage or the sugar will be destroyed by bacteria. Starch feedstocks such as corn and other grains require a rather mild treatment with enzymes (biological catalysts) or acid to reduce the starch to sugar. And cellulosic (cellulose containing) feedstocks such as crop residues, grasses, wood, and municipal wastepaper require more extensive treatment to reduce the more inert cellulose to sugar.

Processes utilizing each of the ethanol feedstock types are considered below. In addition, the environmental effects of ethanol distilleries are discussed as are various process changes that could lower costs. Although ethanol is emphasized in this chapter, it should be remembered that other alcohols (e.g., butanol) and chemicals could be produced from the sugar solutions, but technical and economic uncertainties are too great to include a detailed consideration of these alternatives at present.

---

*Distillation consists of heating the ethanol-water solution and passing the vapor through a column in which the vapor condensed and revaporized numerous times, a process that successively concentrates the ethanol and removes the water.

Ethanol From Starch and Sugar Feedstocks

Ethanol can be produced from starch and sugar feedstocks with commercially available technology. Starch feedstocks are primarily grain crops such as corn, wheat, grain sorghum, oats, etc., but also include various root plants such as potatoes. The sugar feedstocks are plants such as sugarcane, sweet sorghum, sugar beets, and Jerusalem artichokes. Since these feedstocks are all crops grown on agricultural lands under intensive cultivation and can be converted with commercial technology, they are considered together.

The processes for producing ethanol from starch and sugar feedstocks are shown schematically in figures 26 and 27. The energy consumption of these processes is discussed next, followed by a description of process byproducts, cost calculations, and onfarm processes.

Figure 26.—Process Diagram for the Production of Fuel Ethanol From Grain

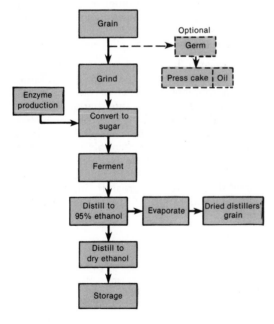

SOURCE: Office of Technology Assessment.

Figure 27.—Process Diagram for the Production of Fuel Ethanol From Sugarcane or Sweet Sorghum

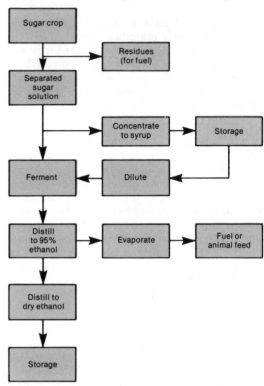

SOURCE: Office of Technology Assessment.

Energy Consumption

Most ethanol ditilleries in the United States today were designed for beverage alcohol production, with little emphasis on energy usage. A fuel ethanol distillery can take advantage of newer technology and the low purity requirements of fuel ethanol to reduce its energy consumption. Nevertheless, both the type of fuel used and the amount of energy consumed at the distillery will continue to be important determinants of the efficacy of fuel ethanol production in displacing imported fuels.

In the plant currently producing most of the fuel ethanol today, the germ (protein) in the corn feedstock is removed in a separate feed processing plant. Consequently, the distillery receives a more or less pure starch from the grain processing plant and the waste stillage (material left in the fermentation broth after the ethanol has been removed) is fed into a municipal sewage system, so that the energy needed to pretreat the corn and to process the waste stream is not included in the distillery energy usage. Nevertheless, the distillery consumes 30,000 Btu/gal of ethanol and 96.5 percent of this is in the form of natural gas.[1] If all processing energy inputs are included, fuel consumption is about 65,000 to 75,000 Btu/gal of ethanol (exclusive of the energy needed for waste stream treatment).[2] Furthermore, the economics of this process are predicated on income from process byproducts, such as corn oil, for which the markets are uncertain if large volumes are produced.

OTA's analysis indicates that the fuel used at the distillery cannot soon be reduced to an insignificant fraction of the energy contained in the ethanol. Thus, if the displacement of imported fuels (oil and natural gas) is to be maximized, fuel ethanol distilleries should be required to use abundant or renewable domestic fuels such as coal or solar energy (including biomass).

A distillery that might be more common in a large-scale ethanol program has been designed by Raphael Katzen Associates.[3] This distillery would produce a dry animal feed byproduct, known as distillers' grain (DG) (see next section on byproducts). Although the distillery uses some equipment to dry the DG which is not in common use in ethanol distilleries, all of the equipment is commercially available. The design reduces the number of distillation columns to the minimum using conventional technology (two columns: one to produce 95 percent ethanol and one to produce dry ethanol)

and uses "vapor recompression" evaporation for drying the DG. The distillery is coal-fired and consumes 42,000 Btu of coal and 13,000 Btu of purchased electricity* to produce 1 gal of ethanol which has a lower heating value of 76,000 Btu.** The energy breakdown for the Katzen design is shown in table 52.

Table 52.—Energy Consumption in a Distillery Producing Fuel Ethanol From Corn

Process step	Thousand Btu of coal/gal of ethanol[a]
Receiving, storage, and milling	0.8
Conversion to sugar (including enzyme production)	16.0
Fermentation	0.6
Distillation	24.8
Distillers' grain recovery	6.2
Miscellaneous	6.6
Total	55.0

[a]Assumes 10,000 Btu of coal per kilowatthour of electricity.

SOURCE: Raphael Katzen Associates, *Grain Motor Fuel Alcohol, Technical and Economic Assessment Study* (Washington, D.C.: Assistant Secretary for Policy Evaluation, Department of Energy, June 1979), GPO stock No. 061-000-00308-8.

At first thought, one might expect the energy demand of a distillery using sugar plant feedstocks to be less than that for starch feedstocks, since the energy needed to reduce the starch to sugar is no longer required. The situation is, in fact, quite the opposite. The processes for extracting the sugar from the feedstock and concentrating it to a syrup (highly concentrated sugar solution) are quite energy intensive. The average energy usage for a sugar feedstock (based on sugarcane) would be about 85,000 Btu of coal per gallon of ethanol produced on the average,[4] or slightly more than the energy content in the ethanol. If the bagasse, i.e., plant matter left over after the sugar is extracted, is used to fuel the boiler, then 110,000 Btu of bagasse would be needed to produce 1 gal of ethanol. (This assumes a 70-percent boiler efficiency for bagasse, as opposed to 90 percent for coal.)

For both the grain and sugar feedstocks, crop residues could be used to fuel the distilleries. In both cases there is sufficient resi-

[1]K. Strasma, "Domestic Crude Oil Entitlements, Application for Petroleum Substitutes, ERA-03" submitted to the Department of Energy by Archer Daniels Midland, Co., Decatur, Ill., May 17, 1979 update.

[2]Ibid.

[3]Raphael Katzen Associates, *Grain Motor Fuel Alcohol, Technical and Economic Assessment Study* (Washington, D.C.: Assistant Secretary for Policy Evaluation, Department of Energy, June 1979), GPO stock No. 061-000-00308-9.

*10,000 Btu/kilowatthour.

**Lower heating value is measured when water vapor is the product of combustion. The higher heat value, when liquid water is the product, is 84,000 Btu/gal.

[4]Ibid.

due produced together with the starch or sugar to fuel an energy-efficient distillery, although the quantity may be only marginally adequate for sugar feedstocks.[5] If one requires that sufficient residues be left on the land to provide adequate soil erosion protection, then the available residues are not adequate in most cases.[6]* Crop residues gathered from adjacent croplands where the crops are not used for ethanol production could easily supplement the shortfall, however.

Since the sugar feedstocks are generally delivered to the distillery with much of the residue, which subsequently arises as a waste byproduct of the sugar extraction step, it is more likely that residues will be used to fuel these distilleries, although it is technically feasible in both cases.

If crop residues are used to fuel distilleries, then the fossil fuel usage at the distillery will be negligible. The fossil energy used to collect and transport residues and replace their nutrient value to the soil would have to be included. OTA estimates this energy to be about 10,000 Btu/gal of ethanol for grain feedstocks and about 3,000 Btu/gal for sugar feedstocks. (These estimates assume that no grain residues are normally harvested with the grain and that the entire sugar plant is harvested and transported to the distillery. Therefore, the grain-fed distillery needs 10.3 lb of residue per gallon of ethanol and the sugar-fed distillery needs a supplement of 3 lb of residue per gallon of ethanol.)

[5]R. A. Nathan, "Fuels From Sugar Crops," published by Technical Information Center, Department of Energy, TID-22781, July 1978.
[6]Ibid.
*As an example, the national average available crop residues for corn are about 7.3 lb/gal of ethanol (see ch. 3). With a 70-percent boiler efficiency, this would provide 70 percent of the energy needed at the distillery (assuming 6,500 Btu/lb).
, For sugarcane and sweet sorghum (syrup variety), the total crop residues are about 11 lb of combustible matter per gallon of ethanol. The residues required to protect against soil erosion vary greatly. If all of the residue is used, one gets about 80 to 85 percent of the distillery energy requirement (assuming 30-percent leaves with 6,500 Btu/lb and 70-percent cane with 9,000 Btu/lb and 70-percent boiler efficiency). And in areas where residues are needed to protect the soil from erosion, the available residues might be only the cane, which would be about 60 percent of the distillery energy requirement.

Process Byproducts

All of the material in the feedstock, except for the sugar or starch (most of which is converted to alcohol), become byproducts of distillation. In addition, the excess yeast or bacteria grown in the fermentation step can also serve as a byproduct. The grain feedstocks are high in protein and, consequently, the byproduct credits will be larger than with sugar feedstocks.

The grain protein can be removed as "gluten" before distillation and oil, such as corn oil, can be extracted. As mentioned above, however, the oil market is uncertain and the required selling price for such oil is too high for it to be considered as a fuel.

The grain processes considered most likely for large-scale fuel ethanol production would ferment a mash (crushed, cooked, and treated grain plus water) that still contains all the non-starch components of the grain. The material left after the ethanol has been removed, called "stillage," has in it protein, dead yeast, and bacteria as well as various other materials contained in the grain. This stillage can be fed to animals directly or can be dried (to produce DG) for transport and, again, used as an animal feed. The wet stillage, however, spoils in 1 to 2 days, so care must be exercised when feeding the stillage wet.[7]

The high protein content makes DG a suitable protein supplement to animal feed, although its high fiber content limits the quantity that can be fed and the types of animals that can consume it. Although DG contains about half the protein per pound of material as does soybean meal, a common protein supplement, the types of protein in DG are such that the cattle use it more effectively and experiments indicate that 1.5 lb of DG can substitute for 1

[7]E. W. Kienholz, et al., "Grain Alcohol Fermentation Byproducts for Feeding in Colorado," Department of Animal Sciences, Colorado State University, Fort Collins, Colo., 1979.

lb of soybean meal.[*][8][9] Consequently, the by-product of distilling 1 bu of corn can displace the meal from about 0.25 bu of soybeans.[**]

Other experiments have indicated that DG causes the cattle to digest more of the starch in their feed than would be digested without DG[10] thereby giving DG an enhanced feed value, since less total corn could be fed to animals if part of the corn were converted to ethanol and the resulting DG fed in place of the corn. These results, however, occur only when the animal is fed a starch-rich and protein-poor diet. Feed rations commonly used today have a more nearly optimum protein-starch balance, so this effect would not occur,[11] and the feed value of DG is only as a replacement for other protein concentrates used in animal rations.

The quantity of DG that can be fed to cattle has been estimated to correspond to an ethanol production level of 2 billion to 3 billion gal/yr.[12] As mentioned above, the protein in the grains could be removed before fermentation, and this protein feed ("gluten") would be suitable for a larger variety of animals. Theoretically, if the byproduct replaces all domestic consumption of crushed soybeans used for animal feed,[13] a production level of 7 billion gal/yr could be achieved before all crushed soybeans had been replaced with distillery by-product (assuming the byproduct of fermenting 1 bu of corn displaces the soybean meal from 0.25 bu of soybeans). The byproduct, however, is not a perfect substitute for soybean meal and the actual level at which the animal feed market becomes saturated is probably considerably lower than this.

Other uses for DG are possible. Brewers' yeast is used as a B vitamin source by some people and the protein could possibly be used as a human protein source. It is not clear, however, whether this source of protein will gain consumer acceptance. The distiller byproduct could also be exported as an animal feed supplement, but if it competes with indigenous soybean meal producers (such as in Europe), import tariffs or quotas may be imposed.

While there are numerous possibilities, most proposals are vague and involve some obvious problems. Consequently, byproduct credits could drop or disappear in a large-scale ethanol program based largely on grain feedstocks.

If the protein in grains is removed in the pretreatment or sugar feedstocks are used, the stillage consists primarily of yeast or bacteria, and has smaller feed value than DG. (The distillery producing most of the fuel ethanol used today removes the protein in the pretreatment and returns the stillage to sewage treatment.) Although there is a limited market for this stillage, it is likely that it will either be dried and used as a fuel or subjected to anaerobic digestion with the resulting biogas used as a fuel. Drying and burning the byproduct result in slightly more energy—an estimated 8,000 Btu/gal of ethanol.[*]

Other possible byproducts of fermentation include oils, vitamins, other alcohols, various organic acids (e.g., vinegar), fuel oil (a mixture of alcohols), and other chemicals. The processes, however, are generally controlled so that the major chemical byproduct is fuel oil. This would probably be combined with the

*Cattle break down some proteins in the rumen and later use the resultant ammonia in the intestines to synthesize new proteins. Other proteins pass through the rumen and are absorbed directly in the intestine. Depending on the relative proportions of the two classes of proteins, the effective quantity of usable protein will vary.

[8]T. Klopfenstein, Department of Animal Sciences, University of Nebraska, Lincoln, Nebr., private communication, 1979.

[9]M. I. Poos and T. Klopfenstein, "Nutritional Value of Byproducts of Alcohol Production for Livestock Feeds," Cooperative Extension Service, University of Nebraska, Lincoln, Nebr., Animal Science Publication No. 79-4, 1979.

**One bushel of distilled corn yields about 18 lb of DG. One bushel of soybeans produces about 48 lb of soybean meal.

[10]W. P. Garrigus, University of Kentucky, Proceedings of 10th Distillers' Feed Conference, Cincinatti, Ohio, Mar. 3, 1955.

[11]Klopfenstein, op. cit.

[12]R. L. Meekhof, W. E. Tyner, and F. D. Holland, "Agricultural Policy and Gasohol," Purdue University, West Lafayette, Ind., May 1979, contractor report to OTA. These authors assume a 2:1 substitution of DG for soybean meal and 3 billion gal of ethanol per year as the saturation point. Using 1.5:1 as the ratio, however, reduces this to 2.25 billion gal/yr.

[13]Agricultural Statistics, 1979 (Washington, D.C.: U.S. Department of Agriculture, 1979), GPO stock No. 001-000-04069-3.

*If the material is dried, 11,000 Btu (2 lb) of material result per gallon of ethanol. The drying however, requires an estimated 3,000 Btu additional input energy. Anaerobic digestion would produce about 5,000 Btu of biogas (assuming 4 ft³ biogas/lb solids) with the process requiring about 1,000 Btu.

fuel alcohol, resulting in a 0.7-percent increase in the quantity of fuel produced.

CO_2 is also a byproduct of fermentation which is used in carbonated beverages, dry ice, and, to a small extent, in chemical processes. Moreover, CO_2 has many interesting properties that are currently being researched and recovery may eventually become more widespread and profitable.

Ethanol Production Costs

Raphael Katzen Associates has performed a detailed cost calculation on a 50-million-gal/yr coal-fired distillery that purchases its electricity from an electric utility.[14] Including coal-handling and pollution control equipment and allowing the production of dried DG, the total distillery would cost an estimated $53 million.

Inflating this to early 1980 dollars (20 percent) results in a distillery investment of $64 million. (These figures do not include engineering fees which could be small if a large number of distilleries are built, but which are estimated at $6 million, in 1978 for a single distillery.)

A distillery designed solely for sugar crop feedstocks would cost considerably more. As mentioned above, the sugar has to be concentrated to a syrup for storage, since the feedstock is available for only part of the year, during and somewhat after the harvesting season. Hence, the pretreatment equipment has to be able to handle a larger capacity than the distillery for part of the year, while standing idle for part of the year. In addition storage tanks are needed for the syrup. If the bagasse and crop residues are used as fuel, however, then some of the pollution control equipment needed to remove sulfur emissions can be eliminated, due to the very low sulfur content of the biomass. In all, a 50-million-gal/yr distillery for sugarcane or sweet sorgham would cost an estimated $100 million in 1978[15][16] or $120

million in 1980, assuming the feedstock is available for half of the year and half year's syrup storage is required. These assumptions about the length of time that the feedstock will be available may be somewhat optimistic for Midwestern grown sweet sorghum, however, and the cost could be higher. If the raw feedstock is available for only 3 months per year, OTA estimates the distillery would cost about $140 million in 1978 dollars.

Although it might be possible to avoid concentrating the extracted sugar solution to a syrup by using antibiotics or various chemicals, a major cost of the pretreatment is the equipment needed to remove the sugar solution from the raw plant material. Furthermore, storage of large quantities of dilute sugar solution would be expensive. Consequently, improvements in the economics of using sugar feedstock will require methods for storing the raw sugar feedstocks inexpensively and in a way that the sugar need not be removed and concentrated. Possibilities include pretreatment with chlorine gas, ammonia, or sulfur dioxide (to change the acidity and provide a toxic environment for bacteria). OTA is unaware, however, of any work in this area that would serve as a basis for cost calculation.

An alternate approach is to build a distillery capable of handling either starch or sugar feedstocks. Katzen has calculated that this 50-million-gal/yr distillery would cost $93 million in 1978 dollars.[17]

The ethanol costs are influenced by the capital investment in and financing of the distillery, the distillery operating costs, and the byproduct credits. For a coal-fired 50-million-gal/yr distillery using starch feedstock, the capital charges are about $0.21 to $0.42/gal of ethanol, depending on the financing arrangements. These charges, however, can vary significantly with interest rates, depreciation allowances, tax credits, and other economic incentives.

The major operating expense is the feedstock cost less the byproduct credit. For corn at $2.50/bu, the feedstock costs $0.96/gal of

[14]Raphael Katzen Associates, op. cit.
[15]Ibid.
[16]F. C. Schaffer, Inc., in E. S. Lipinsky, et al., *Sugar Crops as a Source of Fuels; Vol. II: Processing and Conversion Research*, final report to Department of Energy, Aug. 31, 1978.

[17]Raphael Katzen Associates, op. cit.

ethanol and the byproduct credit is about $0.38/gal ($110/ton of DG), resulting in a net feedstock cost of $0.58/gal. Because farm commodity prices are extremely volatile, the net feedstock and resultant ethanol cost could be quite variable. A $0.50/bu increase in corn grain prices (and a proportionate increase in the byproduct credit), for example, would raise the ethanol cost by $0.12/gal.

Tables 53 and 54 show the cost of ethanol produced from various feedstocks. Although the costs will vary depending on the size of the distillery, ethanol can be produced from corn ($2.50/bu) in a coal-fired 50-million-gal/yr distillery for $0.95 to $1.20/gal. About $0.10 to $0.30/gal should be added to these costs for deliveries of up to 1,000 miles from the distillery. (Most ethanol is currently delivered in

Table 53.—Early 1980 Production Costs for Ethanol From Grain and Sugar Crops
(in a 50-million-gal/yr distillery)

	Grain[a]	Sugar[b]
Fixed capital	$64 million	$120 million
Working capital (10% of fixed capital)	6.4 million	12 million
Total investment	$70.4 million	$132 million

	$ per gallon of 99.6% ethanol	
Operating costs		
Labor	$0.08	$0.09
Chemicals	0.01	0.01
Water	0.01	0.01
Fuel (coal at $30/ton for grain feedstock and crop residues at $30/ton for sugar feedstock)	0.08	0.04[c]
Subtotal	$0.18	$0.15
Capital charges		
15 to 30% of total investment per year[d]	0.21 - 0.42	0.40 - 0.79
Total	$0.37 - $0.60	$0.55 - $0.94

[a]Includes drying of distillers' grain.
[b]Includes equipment for extracting the sugar from the feedstock concentrating it to a syrup for storage.
[c]Bagasse-fueled distillery appropriate for sweet sorghum and sugarcane, supplemental fuel requirement is 3 lb of residue per gallon of ethanol.
[d]There are many often complex formulae to compute actual capital costs. Economic factors considered include debt/equity ratio, depreciation schedule, income tax credit, rate of inflation, terms of debt repayment, operating capital requirements, and investment lifetime. However, a realistic range of possibilities for annual capital costs would lie between 15 and 30% of total capital investment.

The upper extreme of 30% may be obtained assuming 100% equity finance and a 13% aftertax rate of return on investment. The lower extreme of 15% may be obtained assuming 100% debt financing at a 9% rate of interest. Both calculations assume constant dollars, a 20-year project lifetime, and include a charge for local taxes and insurance equal to 3% of fixed capital costs. For a more detailed treatment of capital costs see OTA, *Application of Solar Technology to Today's Energy Needs*, vol. II, ch. 1.

SOURCE: Office of Technology Assessment; and Raphael Katzen Associates, *Grain Motor Fuel Alcohol, Technical and Economic Assessment Study* (Washington, D.C.: Assistant Secretary for Policy Evaluation, Department of Energy, June 1979), GPO stock No. 061-000-00308-9.

Table 54.—Cost of Ethanol From Various Sources

Feedstock	Price[a]	Net feedstock cost[b] ($/gal ethanol)	Ethanol cost ($/gal)	Yields[c] (gallons of ethanol per acre)
Corn	$2.44/bu	$0.57	$0.94-$1.17	220
Wheat	3.07-4.04/bu[d]	0.73-1.08[d]	1.10-1.68	86
Grain sorghum	2.23/bu	0.49	0.86-1.09	130
Oats	1.42/bu	0.59	0.96-1.19	75
Sweet sorghum	15.00/ton[e]	0.79	1.34-1.73	380[e]
Sugarcane	17.03/ton[f]	1.26	1.81-2.20[f]	520

[a]Average of 1974-77 seasonal average prices.
[b]The feedstock cost less the byproduct credit. The difference in feedstock costs might not hold over the longer term due to equilibration of prices through large-scale ethanol production.
[c]Average of 1974-77 national average yields.
[d]Range due to different prices for different types of wheat.
[e]Assuming 20 fresh weight tons/acre yield, $300/acre production costs.
[f]Excludes 1974 data due to the anomalously high sugar prices that year.

SOURCE: *Agricultural Statistics, 1978* (Washington, D.C.: U.S. Department of Agriculture); and Office of Technology Assessment.

tank trucks, but as production volume grows other forms of transportation, such as barge shipments, rail tank cars, and petroleum product pipelines,* could decrease the transportation cost to as low as $0.03 to $0.05/gal under favorable circumstances.)

As shown in tables 53 and 54 the major tradeoff between starch and sugar feedstocks is that the starch-fed distilleries require considerably less investment than the sugar-fed ones, but the ethanol yield per acre cultivated may be larger with the sugar feedstocks. As noted in chapter 3, however, these yield figures are highly unreliable for sweet sorghum, and sugarcane cannot be grown on most cropland potentially available for energy crop production. If comparative studies of potential ethanol feedstocks grown under comparable conditions show that certain sugar crops produce more ethanol per acre than the starch crops, then there may be a tendency to turn to sugar feedstocks as farmland prices rise. Moreover, if the grain byproducts are difficult to sell, then economics could favor sugar crop feedstocks. For now, however, the lower capital investment required for grain-fed distilleries gives them an advantge over sugar-fed distilleries.

Onfarm Distillation

Apart from commercial distilleries, considerable interest has been expressed in individual farmers or farm coops producing ethanol. A number of factors, however, could limit the prospects of such production.

Technology for producing 90 to 95 percent ethanol (5 to 10 percent water) is relatively simple. Several farmers are or have constructed their own distilleries for this purpose. In addition prefabricated distilleries for producing 90 to 95 percent ethanol are available both at the farm size (15,000 gal/yr)[19] and coop size (several hundred thousand gallons per year)[20] for a cost of about $1 for each gallon per year of capacity, but there is insufficient onfarm operating experience to establish the reliability or expected operating life of these distilleries. OTA is not aware of smaller distilleries, but there is no fundamental reason why they cannot be built. There will, however, be a tradeoff between the cost of small distilleries and the amount of labor required to operate them.

A farmer must consider a number of site-specific factors before deciding to invest in an onfarm skill. Some of the more important of these are:

- **Investment.**—How much does the still and related equipment cost?
- **Use of the ethanol.**—Will the ethanol be used onfarm or sold? What equipment modifications are necessary? Will the farmer be dependent on a single buyer, such as a large distillery that will upgrade 95 percent ethanol to dry ethanol?
- **Labor.**—Does the farmer have access to cheap, qualified labor, or is it better to make a larger investment for an automatic distillery?
- **Skill.**—Although ethanol can be produced easily, the process yield—and thus the cost—as well as the safety of the operation can depend critically on the skill of the operator.
- **Equipment lifetime.**—Less expensive distilleries may be constructed of materials that are destroyed by rust after a few years' operation.
- **Fuel.**—Does the farmer have access to wood, grass, or crop residues and combustion equipment that can use these fuels? Can reliable, inexpensive solar stills be constructed for the distillation step?

If oil or natural gas is used in the distillery, would it be less expensive to use this

---

*Various strategies can be used to eliminate potential problems with the water sometimes found in petroleum pipelines. If ethanol is being transported, the total volume of ethanol in the batch can be kept large enough so that the percentage of water in the delivered ethanol is within tolerable limits. If gasohol is transported, it can be preceded by a few hundred barrels of ethanol which will absorb any water found in the pipeline, thereby keeping the gasohol dry. Other strategies also exist or can be developed.[18]

[18]L. J. Barbe, Jr., Manager of Oil Movements, EXXON Pipeline Co., Houston, Tex., private communication, August 1979.

[19]Paul Harback, United International, Buena Vista, Ga., private communication, October 1979.
[20]Robert Chambers, President, ACR Process Corp., Urbana, Ill., private communication, September 1979.

fuel directly as a diesel fuel supplement in a retrofitted diesel engine?

- **Byproduct.**—Can the farmer use the wet byproduct on his/her farm? Will this unduly complicate the feeding operations or make the animal operation dependent on an unreliable still? What will drying equipment cost and how much energy will it consume?
- **Water.**—Does the farmer have access to sufficient water for the distillery?

Under favorable circumstances, it might be possible to produce 95 percent ethanol for as little as $1/gal* plus labor with a labor-intensive distillery. If the ethanol is used in a diesel tractor, the ethanol would be equivalent to diesel fuel costing $1.70/gal, or about twice the current diesel fuel prices. Under unfavorable circumstances, the cost could be several times as great. Due to a lack of experience with onfarm distilleries, however, these cost estimates may be low.

Onfarm or coop production of dry ethanol could become competitive with commercially distilled ethanol, however, if relatively automatic, mass-produced distilleries capable of using fuels found onfarm and producing dry ethanol and dry DG could be sold for about $1 for each gallon per year of capacity and if farmers charge little for their labor. OTA is not aware of any package distilleries for producing dry ethanol that are available at this price.

*Assuming equipment costs of $1 for each gallon per year of capacity, the costs per gallon of ethanol are: $0.58 for net feedstock cost, $0.20 for equipment costs (operated at 75 percent of capacity), $0.20 for fuel (assuming $3/million Btu and 67,000 Btu/gallon), and $0.05 for enzymes and chemicals, resulting in $1.03/gal of ethanol or $0.98/gal of 95 percent ethanol.

Meeting this price goal for automatic, onfarm, dry ethanol production facilities will probably require process innovations, particularly in the ethanol-drying step, and could well involve the use of small, inexpensive computers (microprocessors) for monitoring the process. A major constraint, however, could be the cost of sensors, automatic valves, etc. that would be required.

For some farmers, however, the cost or labor required to produce ethanol may be of secondary importance. The value of some degree of liquid fuel self-sufficiency and the ability to divert limited amounts of corn and other grains when the market price is low may outweigh the inconvenience and/or costs. In other words, farmers may consider the technology to be an insurance against diesel shortages and hope that it will raise grain prices. **Although insurance against diesel shortages certainly can be achieved by purchasing large diesel storage tanks at a cost below an ethanol distillation and storage system, increased grain prices for the entire crop would make the economics considerably more favorable to farmers but would be a very expensive way for the nonfarm sector to provide fuel to farmers.** As evidence of the interest, the Bureau of Alcohol, Tobacco, and Firearms had received over 2,800 applications for onfarm distillation permits by mid-1979 and they expected 5,000 by the end of the year.[21] As a profitable venture in the absence of large subsidies or grain price increases, however, onfarm production of ethanol is, at best, marginal with current technology.

[21]William Davis, Bureau of Alcohol, Tobacco, and Firearms, U.S. Treasury, Washington, D.C., private communication, July 1979.

Cellulosic Feedstocks

The feedstocks with the largest potential for ethanol production—both in terms of the absolute quantity of ethanol and in terms of the quantity of ethanol per acre of cultivated land—are the cellulosic, or cellulose containing, feedstocks. These include wood, crop residues, and grasses, as well as the paper fraction of municipal solid waste.

Wood, grasses, and crop residues contain cellulose, hemicellulose, and lignin. The cellulose can be reduced, or hydrolyzed, to sugars that can be fermented to alcohol. The hemicellulose can also be reduced to sugars capable of being converted to ethanol with other types of bacteria. The lignin, however, does not convert to alcohol and can be used as a source of chemicals or dried and used as a fuel. Generally, paper is primarily cellulose with varying amounts of partially broken lignin.

The removal of hemicellulose from wood, grass, or crop residues and its reduction to sugar are relatively straightforward. In fact, hemicellulose from biomass is the prinicpal source of the chemical feedstock furfural. Although hemicellulose is not now used as a source of ethanol, the fermentation step can probably be developed without excessive difficulty.

The cellulose, on the other hand, is embedded with lignin, which protects it from biological, but to a much lesser extent chemical, attack. Thus, the reduction of cellulose involves treating the lignocellulose material with acid or pretreating the material either chemically or mechanically to make it susceptible to biological reduction with enzymes.

What was apparently the first acid hydrolysis of wood was described in a German patent issued in 1880.[22] Modifications of this process were used to produce animal fodder in several countries (mostly for the sugar) during World War I. At the end of the war, the economic basis became obsolete. Between World Wars I and II, however, other acid hydrolysis processes were used mostly in Germany to produce sugar and alcohol, partly because of materials shortages but partly in an attempt at self-sufficiency.[23] Other plants were also built in Switzerland and Korea.

During World War I, pilot plants were built in the United States for producing ethanol from wood wastes. Acid hydrolysis processes underwent a series of modifications during World War II. Following World War II, however, virtually all of the wood-ethanol plants were closed for economic reasons.* Today commercial wood sugar plants are in operation only in the U.S.S.R. and in Japan but sever-

al other countries have expressed interest in developing the technology, and one plant in Switzerland is again being used for pilot studies.[24]

Clearly it is technically possible to produce ethanol from lignocellulosic feedstocks today. The failure of these processes to remain economically viable except under special circumstances has been due, in large part, to the relatively low costs of petrochemicals and ethylene-derived ethanol. With oil prices rising, the primary competitor is likely to be grain- and sugar-derived ethanol. There are, however, improvements and developments in the lignocellulose processes which can make them competitive with the current costs of ethanol from these other feedstocks. Alternatively, large rises in farm commodity prices could make the cellulosic processes competitive without technical developments.

While there are processes whose economics rely on large byproduct credits or special financing that could be in commercial operation before 1985, the key to achieving economic competitiveness without these conditions is to develop processes which:

- produce high yields of ethanol per ton of biomass,
- do not require expensive equipment,
- allow nearly complete recovery of any expensive process chemicals, and
- do not produce toxic wastes.

No processes currently in existence fully satisfy all of these criteria, although there are processes that satisfy two and sometimes three of the criteria. Nevertheless, R&D currently underway could yield significant results in 3 to 5 years. With a normal scaleup of 5 years, one or more processes satisfying these criteria could become commercial by the late 1980's.

The generic aspects and historical problems with producing sugars from lignocellulosic feedstocks are now discussed, followed by a slightly more detailed description of various processes currently under investigation. Final-

---

[22]H. F. J. Wenzl, *The Chemical Technology of Wood*, translated by F. E. Brauns (New York: Academic Press, 1970).

[23]Ibid.

*One ethanol plant that uses the sugar-containing waste stream of a sulfite paper-pulping plant is still in operation. It is, however, primarily a waste treatment plant and less than 10 percent of the paper-pulping processes used in the United States produce a suitable waste stream.

[24]J. L. Zerbe, Program Manager, Forest Service Energy Research, U.S. Department of Agriculture, Forest Products Laboratory, Madison, Wis., private communication, 1980.

ly, a generic economic analysis is presented for a hypothetical advanced distillery for producing ethanol from lignocellulosic feedstocks.

Generic Aspects and Historical Problems With Pretreatment

As mentioned above, lignocellulosic materials consist of cellulose, hemicellulose, and lignin. Typically, such material would first be treated with dilute acid to remove the hemicellulose, which then would be fermented in a separate step to ethanol. The remaining lignin-cellulose combination would be treated with concentrated acid at low temperatures (perhaps 100° to 110° F) or dilute acid at high temperatures (300° to 400° F) to either dissolve the cellulose from the lignin or to cause the material to swell, thereby exposing the cellulose for hydrolysis. Alternatively, the material can be exposed to a number of different chemical or mechanical pretreatments which render the cellulose susceptible to hydrolysis. The hydrolysis is then accomplished by further exposure to acid or by the action of enzymes (biological catalysts).

The relative amounts of cellulose, hemicellulose, and lignin can vary considerably among the various lignocellulosic materials. If pure cellulose is converted completely to ethanol, however, the theoretical maximum yield is 170 gal of ethanol per ton of cellulose. The yields per ton of hemicellulose are similar. Consequently for a lignocellulosic material that is 50 percent cellulose, 20 percent hemicellulose, and 25 percent lignin, the theoretical yield is about 120 gal/dry ton of biomass fermented. A yield of 85 to 90 percent of this is a reasonable practical goal, which would result in yields of 100 to 110 gal of ethanol per ton of biomass fermented. The expected yield, however, will vary with the exact composition of the feedstock. For municipal solid waste (29 percent paper and 21 percent yard wastes and wood packaging[25]), the average yield could be about 60 gal of ethanol per ton assuming a 90-percent overall conversion efficiency.

[25]*Materials and Energy From Municipal Waste* (vol. 1, Washington, D.C.: Office of Technology Assessment, 1979), GPO stock No. 052-003-00692-8.

The historical processes have generally used acid hydrolysis. The dilute acid methods (Modified Rheinau, Scholler-Tornesch, Madison, Tennessee Valley Authority, and Russian Modification of Percolation processes) all suffer from a similar ailment.[26] The high temperatures and acidic conditions needed in the processes cause the resultant sugars to decompose, thereby lowering the overall ethanol yield. The concentrated acid processes (Rheinau-Bergius and Hokkaido), on the other hand, have resulted in good product yields. The economics, however, have historically suffered due to the loss of large quantities of acid in the processes. Nevertheless, one of the oldest concentrated acid processes (Rheinau-Bergius) is currently being reexamined to see if this economic conclusion necessarily pertains today (see below).

Publications over the past 20 years in the Soviet Union have reported good experimental results with impregnating wood with acid followed by mechanical grinding. The details for an assessment of the commercial viability of this process, however, are not available. On the other hand, a mechanical pretreatment is also involved in the Emert (formerly Gulf Oil Chemicals) process discussed below. Historically, the mechanical pretreatments needed have been quite expensive, but the researchers indicate that this is not a problem with the Emert process.[27] Finally, a variety of other processes or combinations of processes aimed at exposing the cellulose to hydrolysis are currently being researched. The most important of these are considered below.

Processes Currently Under Development

Emert Process

The development of this process started in 1971 under Gulf Oil Chemicals Corp., but was transferred to the University of Arkansas Foun-

[26]I. Goldstein, Department of Wood and Paper Science, North Carolina State University, Raleigh, N.C., private communication, 1979.
[27]G. H. Emert and R. Katzen, "Chemicals From Biomass by Improved Enzyme Technology," presented in the symposium *Biomass as a Non-Fuel Source*, sponsored by the ACS/CSJ Joint Chemical Congress, Honolulu, Hawaii, Apr. 1-6, 1979.

dation for scaleup (the transfer reportedly occurred because Gulf had made a management decision to concentrate its efforts on fossil fuels). This process is the most advanced of the enzymatic hydrolysis methods and, with proper financing, can probably be brought to commercial-scale operation by 1983-85.

The method consists of a pretreatment developed for this process which involves grinding and heating the feedstock followed by hydrolysis with a mutant bacterium also developed for this purpose. A unique feature is that the hydrolysis and fermentation are performed simultaneously in the same vessel, thereby reducing the time requirements for a separate hydrolysis step, reducing the costs and increasing the yield (since a sugar buildup during hydrolysis could slow the hydrolysis and decrease the overall yield). Also the process does not use acids, which would increase equipment costs. The sugar yields from the cellulose are about 80 percent of what is theoretically achievable,[28] but the small amount of hemicellulose in the sawdust is not being converted.

The process has been brought to the pilot plant stage and funds are currently being sought for a demonstration (1 million gal/yr) facility as part of the scaleup process. Based on the pilot plant experience, Emert estimates the selling price for the ethanol to be $1.49/gal (1983 dollars, 100-percent private equity financing, and 10-year amortization).[29] With 80-percent municipal bond financing, he estimates the selling price to be $1.01/gal (1983 dollars, 20-year amortization).

These cost estimates are based on a feedstock of 50-percent "air classified" municipal solid waste (i.e., the paper and plastic fraction) at $14/ton, 25-percent saw mill waste at $21/ton, and 25-percent pulp mill waste at $14/ton. These costs are all on the low end of estimates for 1978-79 prices and consequently represent optimistic estimates. Furthermore, by 1983, inflation would increase these costs. More realistic 1983 feedstock costs (50 to 100 percent higher than those cited) would raise the ethanol cost by about $0.10 to $0.20/gal.

The cost estimates also assume a large by-product credit for dried fermentation yeast and hydrolysis bacteria ($0.40/gal ethanol). Most of this comes from the hydrolysis bacteria and an animal feed value for this material has not been established. In addition, large-scale production could lead to a saturated animal feed market similar to that with grain distillation and subsequent loss of the byproduct credit.

Furthermore, problems encountered with scaling up a process virtually always lead to cost increases above those estimated. Consequently, these cost estimates could be too low by $0.20 to $0.70 or more per gallon of ethanol. Nevertheless, with municipal bond financing, this process could well be competitive with ethanol produced from corn in a privately financed distillery by 1983. (Assuming 7-percent annual inflation as apparently was done in Emert's calculations, $1.10/gal ethanol in 1979 would sell for about $1.45/gal in 1983).

While no cost estimates are available for this process using woodchips, grasses, or crop residues as feedstocks, Emert reports that experiments have shown that modifications in the thermal-mechanical pretreatment enables ethanol yields of 70 to 75 gal/ton of feedstock.[30] The increased costs for these feedstocks ($40 to $50/ton in 1983 up from $30 to $40/ton in 1979) would add $0.30 to $0.45/gal to the ethanol price. Consequently, it is less likely that this process using these feedstocks would be competitive with corn-derived ethanol, unless corn and other grain prices rise more rapidly than general inflation.

In sum, it appears that this process could be competitive with grain-derived ethanol if municipal wastepaper is used as a feedstock and the distillery receives special financing. A reliable determination of the competitive position of other feedstocks and financing arrangements are less certain and probably cannot be determined until a full-scale plant has been built.

[28] Ibid.
[29] Ibid.
[30] G. H. Emert, private communication, October 1979.

Reexamination of Rheinau-Bergius Process

Much of the detailed information on the Rheinau-Bergius process has been lost. Since the acid hydrolysis of wood involves subtle chemical processes which can change dramatically with small changes in the process conditions, the detailed process chemistry of hydrolysis with concentrated hydrochloric acid is being reexamined at North Carolina State University. The research should provide a basis for reevaluating the process as a source of ethanol and chemicals and determining whether sufficient quantities of the acid can be recovered to make the process economic at today's prices.

Tsao Process

This process is being developed at Purdue University with the major emphasis on crop residues as a feedstock and is currently progressing to the pilot plant stage. Although there have been numerous changes in the process as the research has proceeded, in the currently preferred process hemicellulose is removed first with dilute acid and then, the cellulose and lignin are dissolved in concentrated (70 percent) sulfuric acid. The acid is recovered by precipitating the cellulose-lignin from the acid through the addition of methanol, then the methanol is removed from the acid by distillation. Following this pretreatment, enzymes hydrolyze the cellulose.

The use of methanol to aid in recovering the acid is a novel aspect of this process. As the recovery has been proposed, however, the methanol is likely to react to form toxic byproducts such as dimethyl sulfate, dimethyl ether, dimethyl sulfoxide, and other compounds. The loss of process methanol as well as the disposal of these toxic wastes would increase the costs. In addition, there are several places in the process where more expensive equipment will be needed than has been included in most cost calculations due primarily to the corrosive effects of the acid.[31][32] Although novel acid recovery processes of this

[31]Raphael Katzen Associates, op. cit.
[32]I. Goldstein, op. cit.

type should be thoroughly investigated, it has not yet been satisfactorily demonstrated that the process proposed would be economically competitive as a source of fuel ethanol.

University of Pennsylvania—General Electric Process

In this process, woodchips are heated in an alkaline solution containing water, sodium carbonate, and butanol (a higher alcohol). Since butanol is only partly soluble in water, the solution consists of two phases (similar to oil floating on water). The hemicellulose goes to the water phase, the lignin dissolves in the butanol, and the cellulose remains undissolved. Following removal of the cellulose, and cleaning to remove traces of butanol, it can be hydrolyzed either with acid or enzymes and the hemicellulose can be converted to ethanol without removing it from solution. The butanol is then cooled, which causes the lignin to precipitate from solution, the solution is filtered, and the butanol recycled to the process.

Clearly the process economics will depend heavily on the cost of producing the process butanol and the quantity of butanol lost to the waste stream. On the other hand, the butanol-water sodium carbonate solution is considerably less corrosive than other chemicals used to remove lignin and therefore could result in lower equipment costs. At this stage, however, the processes are not well enough defined to provide a meaningful cost calculation.

U.S. Army—Natick Laboratories

Work done at this laboratory has contributed substantially to the basic knowledge about the enzyme system that converts cellulose to sugar.[33] These researchers first identified the three-enzyme system involved in the hydrolysis and have developed fungus mutants with improved enzyme productivities. Not only is this research applicable to ethanol production, but it also provides information for those interested in retarding cellulose degradation such as that which occurs with jungle rot.

[33]E. T. Reese, "History of the Cellulose Program at the U.S. Army Natick Development Center," *Biotechnology and Bioenergy Symposium,* No. 6, p. 9, 1976.

The system developed at Natick, however, requires relatively pure cellulose (such as in paper); it has not been effective on lignin-containing materials such as grasses, crop residues, and wood. Recently, attention has been directed at a mechanical process (ball milling) for reducing raw materials to extremely fine particles in order to use the Natick fungus, but this pretreatment is expensive and would probably make the process uneconomic, although detailed economic analyses are not available from the current pilot plant operation.

University of California at Berkeley (Wilke)

Wilke has concentrated on changing the pretreatment step of the Natick process by using acid and hammer milling of the wastepaper and field residues feedstocks. Nevertheless, a critical step involving the recycling of enzymes has not yet been demonstrated.

Iotech Process

This process is proprietary and the subject of patent applications in the name of the Canadian Research and Development Corp. Apparently, the novel aspect of the process is the pretreatment of the material before hydrolysis. In this process woodchips are exposed to high-pressure steam for several seconds, followed by explosive decompression. The product is said to be highly susceptible to hydrolysis.

Generic Economics of Lignocellulosic Materials to Ethanol

The processes described above represent a sampling of the possible approaches to ethanol production from lignocellulosic materials. The descriptions were necessarily brief and could not include all of the ramifications or aspects of the various research groups' efforts.

The chemistry and physics of lignocellulosic materials are complex, and there are few predictive theories that enable one to evaluate unambiguously the various approaches. Furthermore, the competition between research groups is enormous and details are often proprietary.

Nevertheless, the process at the most advanced stage of development (of those being developed) appears to be the Emert process. But as this process now stands and with a successful scaleup, the ethanol could sell for $0.30 to $0.60/gal more than corn-derived ethanol and the price difference could be greater if woodchips rather than sawdust are used as a feedstock. As mentioned above, however, special financing of the distillery (and an inexpensive feedstock source) could lower the selling price to a level competitive with the corn-derived ethanol from distilleries not specially financed. (Because of the larger investment, special financing lowers the price more than it would for corn distilleries.)

Alternatively, distilleries based on the older acid hydrolysis methods can be built to produce ethanol and chemical feedstocks. Katzen Associates, for example, has reevaluated the Madison process* on this basis and found that the ethanol could be sold at about $1.50/gal without byproduct credits (1978 dollars).[34] The economics, however, depend on the byproduct credits for the chemical feedstocks, but the chemical industry is unlikely to make the commitment necessary to support a large fuel ethanol industry until more information is available on the relative merits of biomass- and coal-derived chemical feedstocks.

As suggested earlier, the key to producing ethanol from lignocellulosic materials at a price competitive with corn-derived ethanol without relying on special financing or large byproduct credits is the R&D currently aimed at reducing equipment costs, increasing overall yields, and ensuring a good recovery of process chemicals without the production of toxic wastes.

---

*Dilute acid hydrolysis process. Products are ethanol, furfural, and phenol.

[34]Raphael Katzen Associates in The Feasibility of Utilizing Forest Residues for Energy and Chemicals (Madison, Wis.: Forest Products Laboratories, March 1976), report No. PB-258-630.

R&D currently underway could fulfill these criteria. If so, the production costs might look something like those in table 55. These costs represent plausible cost goals for the production of ethanol from lignocellulosic materials.

Distilleries can and may be built before these criteria are fulfilled, but the economics will depend on favorable financing and atypically low feedstock costs or in securing a market for chemical byproducts. Some distilleries based on these circumstances are likely to be built before the late 1980's. It is unlikely, however, that such circumstances will sustain a large fuel ethanol industry.

Table 55.—Plausible Cost Calculation for Future Production of Ethanol From Wood, Grasses, or Crop Residues
(in a 50-million-gal/yr distillery, early 1980 dollars)

	Dollars
Fixed investment	$120 million
Working capital	12 million
Total investment	$132 million

	$/gallon
Labor, chemicals, fuel	$0.30
Feedstock ($30/ton, 110 gal/ton)	0.27
Capital charges (15 to 30% of total investment)	0.36 - 0.72
Total	$0.93 - $1.29

SOURCE: Office of Technology Assessment.

Environmental Impact of Ethanol Production

The major potential causes of environmental impacts from ethanol production are the emissions associated with its substantial energy requirements, wastes from the distillation process, and hazards associated with the use of toxic chemicals (especially in small plants). A variety of controls and design alternatives are available to reduce or eliminate adverse effects, however, so actual impacts will depend more on design and operation of the plants than on any inevitable problems with the production process.

New large energy-efficient ethanol plants probably will require at least 50,000 Btu/gal of ethanol produced to power corn milling, distilling, stillage drying, and other operations (see "Energy Consumption" discussion). Small plants will be less efficient. Individual distilleries of 50-million-gal/yr capacity will use slightly more fuel than a 30-MW powerplant;* a 10-billion-gal/yr ethanol industry (the approximate requirement for a 10 percent alcohol blend in all autos) will use about the same amount of fuel needed to supply 6,000 to 7,000 MW of electric power capacity.

New source performance standards have not been formulated for industrial combustion facilities, and the degree of control and subsequent emissions are not predictable. The most

*Assuming 10,000 Btu/kilowatthour.

likely fuels for these plants will be coal or biomass (crop residues and wood), however, and thus the most likely source of problems will be their particulate emissions. Coal and biomass combustion sources of the size required for distilleries—especially distilleries designed to serve small local markets—must be carefully designed and operated to avoid high emission levels of unburned particulate hydrocarbons (including polycyclic organic matter). Fortunately, most distilleries will be located in rural areas; this will reduce total population exposure to any harmful pollutants. Particulate control equipment with efficiencies of 99 percent and greater are available, especially for the larger plants. If all energy requirements are provided by a single boiler, high efficiency control would be easier to provide. This is also true for any sulfur oxide (SO_x) controls (scrubbers) that may be required if the facility is fueled with high-sulfur coal.

Other air emissions associated with ethanol production include fugitive dust from raw material and product handling; emissions of organic vapors from the distillation process (as much as 1 percent of the ethanol, as well as other volatile organics, may be lost in the process); and odors from the fermentation tanks. These emissions may be tightly controlled by water scrubbing (for odors and organics) and cyclones (for dust).

The "stillage"—the waste product from the first still (or "beer still")—will be extremely high in organic material with high biological and chemical oxygen demand, and will also contain inorganic salts, and possibly heavy metals and other pollutants. When corn is the biomass feedstock, the stillage is the source of dried DG, which is a valuable cattle feed whose byproduct value is essential to the economics of the process. Thus, it will be recovered as an integral part of the plant operation and does not represent an environmental hazard. On the other hand, sugarcane stillage has far lower economic potential as a byproduct; its recovery is unlikely except as a response to regulation.

The stillage and other wastes from all ethanol plants have severe potential for damaging aquatic ecosystems if they are mishandled. The high biological and chemical oxygen demand levels in the stillage, which would result in oxygen depletion in any receiving waters, will be the major problem.[35] Control techniques are available for reducing impacts from these wastes. Biological treatment methods (activated sludge, biological filters, anaerobic digestion, etc.) and land disposal techniques used in the brewing industry are suitable for ethanol production, but controls for stillages from some crop materials will require further development and demonstration.

Because fermentation and distillation technologies are available in a wide range of sizes, small-scale onfarm alcohol production may play a role in a national gasohol program. The scale of such operations may simplify water effluent control by allowing land disposal of wastes. On the other hand, environmental control may in some cases be more expensive because of the loss of scale advantages. Current experience with combustion sources indicates that high emissions of unburned particulate hydrocarbons, including polycyclic organic matter, are a more common problem with

smaller units. Because smaller units are unlikely to have highly efficient particulate controls, this problem will be aggravated. Also, SO_x scrubbers are impractical for small boilers, and effective SO_x control may be achieved only with clean fuels or else forgone. Because local coals in the Midwest tend to have high sulfur contents (5 percent sulfur content is not unusual), small distilleries in this region may have objectionably high SO_x emission rates. Finally, small plants will be less efficient than large plants and will use more fuel to produce each gallon of alcohol.

The decentralization of energy processing and conversion facilities as a rule has been viewed favorably by consumer and environmental interests. Unfortunately, a proliferation of many small ethanol plants may not provide a favorable setting for careful monitoring of environmental conditions and enforcement of environmental protection requirements. Regulatory authorities may expect to have problems with these facilities similar to those they run into with other small pollution sources. For example, the attempts of the owners of late-model automobiles to circumvent pollution control systems conceivably may provide an analog to the kinds of problems that might be expected from small distilleries if their controls prove expensive and/or inconvenient to operate.

The same may be true for considerations of occupational safety. The current technology for the final distillation step, to produce anhydrous (water-free) alcohol, uses reagents such as cyclohexane and/or ether that could pose severe occupational danger (these chemicals are toxic and highly flammable) at inadequately operated or maintained distilleries. Similar problems may exist because of the use of pressurized steam in the distillation process. Although alternative (and safer) dehydrating technologies may be developed and automatic pressure/leak controls may eventually be made available (at an attractive cost) for small plants, in the meantime special care will have to be taken to ensure proper design, operation, and maintenance of these smaller plants.

[35]Caribbean Rum Study: Effects of Distillery Wastes on the Marine Environment (Washington, D.C.: Office of Research and Development, Environmental Protection Agency, April 1979).

Process Innovations

The processes for producing ethanol from sugar and grains are well established, but the traditional concern of the industries who operate them has been the flavor (or, in some cases, chemical purity) of the product. With the production of fuel ethanol, on the other hand, the principal concerns are cost and energy efficiency. There are several possible process improvements—at various stages of development—which can result in modest reductions in the processing cost and energy usage. Except for improvements in grain and sugar processing, the R&D could also be applicable to the production of ethanol from cellulosic materials. Some possible improvements in grain processing, fermentation, and alcohol recovery are mentioned below.

Grain and Sugar Processing

Developments in the last 20 years have led to more or less continuous grain preprocessing techniques which have lowered the costs over the traditional batch processes. Novel methods have been proposed, however, such as heating the mash with electrical current rather than process steam. This allows production of a more concentrated sugar solution, thereby reducing the load on evaporators at later stages in the operation. While this is a more energy-intensive pretreatment, it could lower the overall processing energy.[36]

The principal problem with sugar feedstocks, as noted, is the necessity of processing large quantities of feedstock to a syrup for storage. At least one research group is studying ways to store the sugar crops without reduction to syrup,[37] but the details are proprietary.

Fermentation

The key to cost reductions in fermentation is the use of methods for maintaining a high yeast or bacteria concentration in the mash, so that the fermentation proceeds rapidly—thereby reducing the size and number of fermentation vessels required. The two ways of doing this are through continuous fermentation or through recycling of the yeast.

Continuous fermentation processes have been tested in full-scale operation. Due to the possibility of infection of the mash (resulting in the production of products other than ethanol), the processes have two complete fermentation systems to allow periodic switchover and sterilization. The added cost for this equipment effectively nullifies the cost advantage of continuous fermentation.[38] Improved handling techniques, which can assure sterile operation, may obviate the necessity for this redundancy in equipment.

One type of continuous fermentation that is under R&D uses a vacuum over the fermentation mash. The ethanol is drawn off by the vacuum as it is produced, with the necessary heat for the evaporation of the alcohol being supplied by the fermentation process itself. This would reduce the need for cooling water as well as accelerate the fermentation (which is slowed by high ethanol concentrations). While added equipment costs might reduce or nullify the potential savings, the question of whether this will be the case has not been resolved.

Another way of maintaining a high yeast concentration is by recycling the yeast (after it is separated from any grain solids that are to be sold as a byproduct). A hybrid of yeast recycling and continuous fermentation involves a device called a countercurrent flow fermentation tower,[39] in which the yeast flows one way (counter to the current) while the sugars to be fermented flow in the opposite direction The high yeast concentrations require additional cooling of the mash, which increases the cooling equipment costs somewhat, but research in this area can probably result in some overall cost savings.

---

[36]Raphael Katzen Associates, *Grain Motor Fuel Alcohol, Technical and Economic Assessment Study,* op. cit.

[37]E. Lipinsky, Battelle Columbus Laboratories, Columbus, Ohio, private communication, 1979.

[38]Raphael Katzen Associates, *Grain Motor Fuel Alcohol, Technical and Economic Assessement Study,* op. cit.

[39]Ibid.

Distillation

The distillation process in the corn-to-fuel-ethanol distillery considered above consumes nearly half of the energy used at the distillery. Lowering the energy requirements for separating the ethanol from the mash is desirable for a fuel ethanol facility in any case, but the increased equipment costs for advanced ethanol separation techniques could counter part or all of the potential cost savings from lower fuel use and smaller boiler and fuel-handling requirements. Consequently, R&D into this area must address both the energy use and the equipment cost.

One way to lower the energy requirements of distillation is to produce a mash with an ethanol concentration higher than the usual 10 percent. This would require development of yeast or bacteria that are tolerant of the high alcohol concentrations. Since it would be expected that any yeast or bacteria producing ethanol would produce it more slowly at the higher ethanol concentrations, this might require longer fermentation times with a consequent increase in the cost of fermentation equipment. It may be possible, however, to combine this with advanced fermentation methods to provide an overall savings.

Several methods have been suggested for removing the ethanol from the water. These include:

- membranes using reverse osmosis (something like a super filter that allows the water or ethanol to pass through the membrane while preventing the other component from doing so);
- absorption agents (solids which selectively absorb the ethanol are then separated from the solution, with the ethanol finally being removed from the solid); and
- liquid-liquid extraction (extracting the ethanol into a liquid that is not soluble in water, physically separating the liquids, and removing the ethanol from the other liquid).

All of these processes, however, are likely to require that the yeast and grain solids be re-moved from the mash first, so that they do not interfere with the ethanol concentration step (e.g., by clogging the membrane). Little research has been done in producing a clarified solution from the mash, hence, the costs for these methods are highly uncertain.

Numerous other suggestions exist, and research in these areas may eventually produce usable results. One example is the use of supercritical CO_2. When gases are subject to high pressures at suitable temperatures, they form a fluid which is neither gas nor liquid, but is called a supercritical fluid. The properties of supercritical fluids are largely unresearched, but there are proprietary claims that supercritical CO_2 could be suitable for extracting ethanol from the mash. The pressure would then be lowered, the CO_2 would become a gas, and the ethanol would liquefy.

Another possibility is the use of phase separating salts. Salts, when dissolved in a liquid change the liquid's structure and properties. It has been suggested that there may be salts which would attract the water (or ethanol) so vigorously and selectively that the ethanol-water mixture would separate into two phases, with one being predominantly water and the other predominantly ethanol.

These novel approaches should be investigated, but it is not possible to predict when or if results applicable to commercial fuel ethanol production will emerge.

Producing Dry Ethanol

In a large, commercial distillery, the production of dry ethanol only costs $0.01 to $0.03/gal (of ethanol) more than the production of 95 percent ethanol.[40] (The difference in the selling price per gallon of 99.5 percent ethanol and 95 percent ethanol is due primarily to the fact that the latter contains 4.5 percent less ethanol per gallon of product.) Furthermore, with modern heat recovery systems, the production of dry ethanol requires very little additional energy. Consequently, little economic or energy savings are available here.

[40]Ibid.

On the other hand, the additional cost of equipment for producing dry ethanol automatically onfarm with conventional technology may be prohibitive. If the distillery is of the labor-intensive type, however, the additional equipment cost would be small since the same still could be used to produce 95 percent ethanol and then later used to distill to dry ethanol.

Drying agents or desiccants, however, may be a suitable substitute for the conventional process. These materials would selectively remove the water from 95 percent ethanol. Various chemicals are known to do this and recent research indicates that corn stover or corn grain may even be suitable.[41] It is not known, however, how much ethanol would be lost in the process or, if grain is used, whether the absorbed ethanol would inhibit the production of sugar from the starch. While the processes are undoubtedly technically possible, the economics are still highly uncertain.

[41]M. R. Ladisch and K. Dyck, *Science,* vol. 205, p. 898, 1979.

Chapter 9
ANAEROBIC DIGESTION

Chapter 9.—ANAEROBIC DIGESTION

ANAEROBIC DIGESTION

Introduction

Anaerobic ("without air") digestion is the process that occurs when various kinds of bacteria consume plant or animal material in an airtight container called a digester. Temperatures between 95° and 140° F favor bacteria that release biogas (50 to 70 percent methane —essentially natural gas—with most of the remainder as carbon dioxide—CO_2). The bacteria may be present in the original material when charged (as is the case with cattle manure) or may be placed in the digester when it is initially charged. The gas has the heat value of its methane component, 500 to 700 Btu/stdft^3, and can be used directly as a heat fuel or in internal combustion engines. In some cases there is enough hydrogen sulfide (H_2S) present to cause corrosion problems, particularly in engines. H_2S can be removed by a sim-

ple, inexpensive, existing technology. CO_2 can be removed by a somewhat more complex and expensive technology, which would need to be employed if the gas is to be fed into a natural gas pipeline.

The anaerobic digestion process is especially well adapted to slurry-type wastes and has environmental benefits in the form of treating wastes to reduce pollution hazards and to reduce odor nuisances. Furthermore, the residual from the process can be returned to land, either directly or through animal refeeding technologies, and thus retain nitrogen and organic levels of soil. Most other biomass energy conversion processes more nearly totally destroy the input material.

Generic Aspects of Anaerobic Digestion

The anaerobic digestion process involves a number of different bacteria and a digester's performance depends on a large number of variables. The basic process is considered first and then the feedstocks and byproducts of the process.

Basic Process

Not all of the bacteria involved in anaerobic digestion have been identified and the exact biochemical processes are not fully understood. Basically, however, the process consists of three steps:[1][2] 1) decomposition (hydrolysis) of the plant or animal matter to break it down to usable-sized molecules such as sugar, 2) conversion of the decomposed matter to or-

ganic acids, and 3) conversion of the acids to methane. Accomplishing these steps involves at least two different types of bacteria.

The rate at which the biogas forms will depend on the temperature (higher temperature usually gives a faster rate) and the nature of the substrate to be digested. Cellulosic materials, such as crop residues and municipal solid waste, produce biogas more slowly than sewage sludge and animal manure. Disturbances of the digester system, changes in temperature, feedstock composition, toxins, etc., can lead to a buildup of acids that inhibit the methane-producing bacteria. Generally, anaerobic digestion systems work best when a constant temperature and a uniform feedstock are maintained.

When a digester is started, the bacterial composition is seldom at the optimum. But if the feedstock and operating conditions are held constant a process of natural selection

[1] J. J. Wolis, *American Journal of Clinical Nutrition,* 27 (11), p. 1320, 1974.

[2] E. C. Clausen and J. L. Gaddy, "Stagewise Fermentation of Biomass to Methane," Department of Chemical Engineering, University of Missouri, Rolla, Mo., 1977.

takes place until the bacteria best able to metabolize the feedstock (and thus grow) dominate. Biogas production begins within a day or so, but complete stabilization sometimes takes months.

Numerous sources for good anaerobic bacteria have been tried, though the process is basically one of hit and miss. The potential for improvement cannot be assessed at this time. Future developments could produce superior genetic strains of bacteria, but too little is known about the process to judge if or when this can be accomplished. It is quite possible that if such strains are to be effective, the input material may first require pasteurization.

Biogas yields vary considerably with feedstock and operating conditions. Operating a digester at high temperatures usually increases the rate at which the biogas is formed, but raising the temperature can actually decrease the net fuel yield as more energy is required to heat the digester. The optimum conditions for biogas yields have to be determined separately for each feedstock or combination of feedstocks.

Feedstocks

A wide range of plant and animal matter can be anaerobically digested. Both the gas yields and rates of digestion vary. Generally materials that are higher in lignin (e.g., wood and crop residues[3]) are poor feedstocks because the lignin protects the cellulose from bacterial attack. Pretreatment could increase their susceptibility to digestion.[4] However, even then digestion energy efficiencies generally do not exceed 50 to 75 percent. Thus, more usable energy can generally be obtained through combustion or thermal gasification of these feedstocks (see ch. 5).

The best feedstocks for anaerobic digestion usually are wet biomass such as fresh animal manure, various aquatic plants, and wet food-processing wastes such as those that occur in the cheese, potato, tomato, and fruit-processing industries. See table 56 for a summary of the suitability of various feedstocks for digestion.

Byproducts

The digester effluent contains bacteria as well as most of the undigested material in the feedstock (mostly lignocellulose) and the solubilized nutrients. The process has the potential for killing most disease-causing bacteria, but volatile losses of ammonia may increase with anaerobic digestion.[5]

The most generally accepted technology for disposal of the effluent is to use it as a soil conditioner (low-grade fertilizer). Animal manure is already used widely for this but there is some controversy over whether the digester effluent is a better source of nitrogen than the undigested manure.[*] The actual added value (if any) as a fertilizer, however, will have to be determined experimentally and is likely to be highly feedstock specific. The effluent may also be used as fertilizer for aquatic plant systems. In one case the effluent is dewatered and used as animal bedding in place of sawdust.[6]

Another potential use of the effluent is as an animal feed. It has been claimed that the protein mix in the cake obtained from dewatering the effluent is superior to that of undigested manure.[7] Biogas of Colorado has concluded a successful animal feeding trial of digester cake and Hamilton Standard has also done feeding trials.[8]

---

[3]See also, J. T. Pfeffer, "Biological Conversion of Crop Residues to Methane," in Proceedings of the Second Annual Symposium on Fuels for Biomass, Troy, N.Y., June 20-22, 1978.

[4]P. L. McCarty, et al., "Heat Treatment of Biomass for Increasing Biodegradability," in Proceedings of the Third Annual Biomass Energy Systems Conference, sponsored by the Solar Energy Research Institute, Golden, Colo., June 1979, TP-33-285.

[5]J. A.,Moore, et al., "Ammonia Volatilization From Animal Manures," in Biomass Utilization in Minnesota, Perry Blackshear, ed., National Technical Information Service.

[*]The nitrogen is more concentrated in the effluent, but it is also more volatile.

[6]John Martin, Scheaffer and Roland, Inc., Chicago, Ill, private communication, 1980.

[7]B. G. Hashimoto, et al., "Thermophillic Anaerobic Fermentation of Beef Cattle Residues," in Symposium on Energy From Biomass and Wastes, Institute of Gas Technology, Washington, D.C., Aug. 14-18, 1978.

[8]D. J. Lizdas, et al., "Methane Generation From Cattle Residues at a Dirt Feedlot," DOE report COO-2952-20, September 1979.

Table 56.—Suitability of Various Substrates for Anaerobic Digestion

Feedstock	Availability	Suitability for digestion	Special problems
Animal wastes			
Dairy	Small- to medium-sized farms, 30-150 head	Excellent	No major problems, some systems operating.
Beef cattle	Feedlots, up to 1,000-100,000 cattle	Excellent	Rocks and grit in the feed require degritting, some systems operating.
Swine	100-1,000 per farm	Excellent	Lincomycin in the swine feed will inhibit digestion—full-scale systems on university farms.
Chicken	10,000-1,000,000 per farm	Excellent	Degritting necessary, broiler operations need special design due to aged manure, tendency to sour.
Turkey	30,000-500,000 per farm	Excellent	Bedding can be a problem, manure is generally aged, no commercial systems operating.
Municipal wastes			
Sewage sludge	All towns and cities	Excellent	Vast experience.
Solid wastes	All towns and cities	Better suited to direct combustion	Designed landfill best option.
Crop residues			
Wheat straw	Same cropland	Poor, better suited to direct combustion	Particle size reduction necessary, low digestibility, no commercial systems.
Corn stover	Same cropland	Poor, better suited to direct combustion	No commercial systems, no data available, particle size reduction necessary.
Grasses			
Kentucky blue	Individual home lawns	Very good	Distribution of feedstock disperse, no commercial systems.
Orchard grass	Midwest	Fair	No commercial systems, no data on sustainability of yields.
Aquatic plants			
Water hyacinth	Southern climates, very high reproduction rates	Very good	No commercial operations, needs pregrinding.
Algae	Warm or controlled climates	Good	Longer reaction time than for animal wastes.
Ocean kelp	West coast, Pacific Ocean, large-scale kelp farms	Very good	Full-scale operations not proven, no present value for effluent.
Various woods	Total United States	Poor, better for direct combustion or pyrolysis	Will not digest.
Kraft paper	Limited	Excellent, need to evaluate recycle potential and other conversion processes	Premixing watering necessary.

SOURCE: Tom Abeles and David Ellsworth, ''Biological Production of Gas,'' contractor report to OTA by I. E. Associates, Inc., Minneapolis, Minn., 1979.

Although most of the disease-causing bacteria are killed by digestion of the manure, several questions about refeeding of digester effluents need to be resolved. Buildup of toxic materials, development of resistance to antibiotics by organisms in the cake, permissible quantities of cake in the diet, storage, and product quality are all issues that have been raised. There is no firm evidence that these will present significant problems, however.

To avoid some of these problems, the Food and Drug Administration has generally favored cross-species feeding, but has not sanctioned its use as a feed or feed ingredient.[9] The use of digester effluents as an animal feed, however, would greatly improve the economics of manure digestion. Consequently, the value, use, and restrictions on using digester effluents as animal feeds should be thoroughly investigated. Moreover, the animal feed value of effluents from the digestion of feedstocks other than manure should also be investigated.

---

[9]T. P. Abeles, ''Design and Engineering Considerations in Plug Flow Farm Digesters,'' in *Symposium on Clean Fuels From Biomass,* Institute of Gas Technology, 1977.

Reactor Types

There are numerous possible designs for anaerobic digesters, depending on the feedstock, the availability of cheap labor, and the purpose of the digestion. The most complex and expensive systems are for municipal sewage sludge digestion, but the primary purpose of these has been to stabilize the sludge and not to produce biogas.

Digester processes have been classified into three types, depending on the operating temperature: 1) psychophilic (under 68° F), 2) mesophilic (68° to 113° F), and 3) thermophilic (113° to 150° F). The cost, complexity, and energy use of the systems increase with the temperature, as does the rate of gas production. The amount of gas produced per pound of feedstock, however, can either increase or decrease with temperature. Retention time is also an important consideration, wherein maximum gas production per pound of feedstock is sacrificed for reduced size and cost of the digester. Anaerobic digesters in the mesophilic and thermophilic ranges have used agricultural wastes, residues, and grasses, to produce biogas. The optimum temperature appears to be both site and feedstock specific. There are still unresolved technical questions about the tradeoffs between mesophilic and thermophilic digesters, but most onfarm systems have been mesophilic.

Other design parameters include continuous versus batch processes, mixed versus unmixed reactors, and other features. Some of the major types are summarized in table 57 and discussed briefly.

Single-Tank Plug Flow

This system is the simplest adaptation of Asian anaerobic digester technology (figures 28-30). The feedstock is pumped or allowed to flow into one end of a digester tank and removed at the other. Biogas is drawn off from the top of the digester tank. The feed rate is chosen to maintain the proper residence time*

---

*The time the feedstock remains in the digester.

in the digester and the feed or digester contents can be heated as needed. Depending on the placement of the heating pipes, some convective mixing can also occur.

Multitank Batch System

This system consists of a series of tanks or chambers which are filled sequentially with biomass and sealed. As each unit completes the digestion process, it is emptied and recharged. This type of reactor is best suited to operations where the feedstock arrives in batches, for example, grass or crop residues that are collected only at certain times of the year or turkey or broiler operations that are cleaned only when the flocks are changed. This digester system, however, is relatively labor intensive.

Single-Tank Complete Mix

The single-tank complete mix system (figure 31) has a single rigid digester tank which is heated and mixed several times a day. It has been argued that mixing enhances the contact of bacteria with the feedstock and inhibits scum formation, which can interfere with digester operation. Theoretical calculations,[10] however, indicate that the mixing does not improve bacterial contact, and these calculations have been confirmed experimentally in one case.[11] Single-tank complete-mixed digesters are used to treat municipal sewage sludge and have been used in the larger anaerobic digester systems (exclusive of landfills).

Anaerobic Contact

The single-tank complete-mix system effluent can be transferred to a second unmixed

---

[10]P. C. Augenstein, "Technical Principles of Anaerobic Digestion," Dynatech R&D Co., presented at course *Biotechnology for Utilization of Organic Wastes,* Universidad Autonoma Metropolitana, Iztapalapa, Mexico, 1978.

[11]K. D. Smith, et al., "Design and First Year Operation of a 50,000 Gallon Anaerobic Digester at the State Honor Farm Dairy, Monroe, Washington," Department of Energy contract EG-77-C-06-1016, ECOTOPE Group, Seattle, Wash., 1978.

Table 57.—Anaerobic Digester Systems

Type of system	Application and inputs	Scale[a]	Stage of development	Advantages	Disadvantages
Landfill	Existing and planned landfills municipal solid wastes, sewage sludge, warm climates	2×10^6 tons of waste and up (28-acre landfill)	Commercial for "as is" landfills, controlled land filling in pilot stages	Low cost, tanks not required, high loading rates possible, no moving parts	Gas generation may last only 10 years in "as is" landfills, gas usage onsite may present problems
Single-tank plug flow	All types of organics, farm and feedlot operations	Small to large	Commercial	Low cost, simple design can run high solids wastes, can have gravity feed and discharge	Low solids wastes may stratify
Multitank batch system	Can accept all types of wastes, limited application crop residues, grasses, chicken broilers, turkeys	Small to large	Commercial, in Asia	Simple, low maintenance, low cost, complete digestion of materials	Gas generation not continuous, labor-intensive feed and discharge, low gas production per day
Single-tank complete mix	All types of organics sewage treatment, farm and feedlot, municipal solid wastes	Small to large	Commercial	Proven reliability, works well on all types of wastes	Greater input energy to run mixers, higher cost than plug flow
Anaerobic contact	Sewage sludge and other organics, limited application (see variable feed)	Medium to large	Commercial, for sewage treatment	Smaller tank sizes, operation not overly critical	Two tanks necessary
Two or three phase	Cellulosic feedstocks	Medium to large	Pilot scale	Allows more complete decomposition, greater gas yields, greater loading rates, lower retention times	Feed rates vary with feedstocks, have not been attempted full scale, require tight controls and management of the operation
Packed bed	Dilute organics—sewage, food-processing wastes, very dilute animal wastes—industrial and commercial	Medium to large	Commercial, as waste treatment technology	High loading rates possible, short retention times	Tends to clog with organic particles, limited to dilute wastes
Expanded bed	Dilute organics—sewage, food-processing wastes, very dilute animal wastes	Undetermined	Laboratory	High loading rates, low temperature digestion, high quality gas, short retention times	Not developed, high energy input to operate pumps, no operating data
Mixed bed	Sewage sludge, animal wastes, food-processing wastes—fairly dilute mixtures	Small to large	Pilot scale	Fast throughputs, high loading rates, higher solids input than packed beds	Tends to clog, high pumping energy input no operating data
Variable feed	All types of organics, farms and feedlots	Small to medium	Conceptual—combines plug flow with anaerobic contact	Allows seasonal peaking of gas production, preserves nutrient value of material, low cost	Feed-discharge may require extra pump

[a]Scale defined as: small—0 to 30,000 gal; medium—30,000 to 80,000 gal; andlarge—over 80,000 gal.

SOURCE: Tom Abeles and David Ellsworth, "Biological Production of Gas," contractor report to OTA by I. E. Associates, Inc., Minneapolis, Minn., 1979.

Figure 28.—Chinese Design of a Biogas Plant

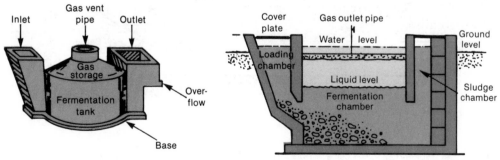

SOURCE: K. C. Khandeleval, "Dome-Shaped Biogas Plan," *Compost Science*, March/April 1978.

Figure 29.—Diagram of a Gobar Gas Plant (Indian)

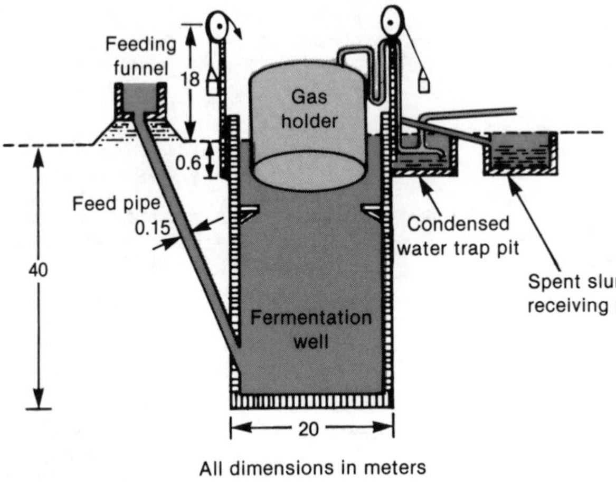

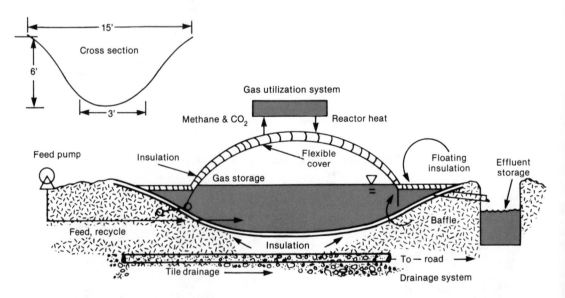

All dimensions in meters

SOURCE: R. B. Singh, "Biogas Plant," Gobar Gas Research Station, Ajitmal, Etaweh (V.P.), India, 1971.

Figure 30.—Plug Flow Digestion System

SOURCE: W. J. Jewell, et al., "Low Cost Methane Generation on Small Farms," presented at Third Annual Symposium on Biomass Energy Systems, Solar Energy Research Institute, Golden, Colo., June 1979.

Figure 31.—Single-Tank Complete Mixed Digester

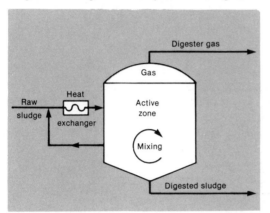

SOURCE: Environmental Protection Agency, "Process Design Manual, Sludge Treatment and Disposal," EPA 625/1-29-001, September 1979.

and unheated storage tank. Here the biomass undergoes further digestion and solids settle out (figure 32). In other words, by adding a second, inexpensive digester tank gas yields can be improved. These systems have been used extensively in sewage treatment and may receive wide application where preservation of the effluents nutrient value requires covered lagoons or in short throughput systems located in warm climates.

Two or Three Phase

As mentioned previously under "Generic Aspects of Anaerobic Digestion," the basic process consists of a series of biochemical steps involving different bacteria. The idea behind the multitank systems is to have a series of digester tanks (figure 33) each of which is separately optimized for one of the successive digestion steps. The rationale behind such system is the hypothesis that they: 1) can accept higher feedstock concentrations without inhibiting the reactions in successive stages, 2) have greater process stability, 3) produce higher methane concentrations in the biogas, and 4) require lower retention times in the digester than with most single-phase digesters. The majority of the work on this approach has been on municipal sewage sludge, although the Institute for Gas Technology hopes to eventually transfer the technique to kelp digestion. The need for uniform feed rates and controls may limit the use of two or multiphased systems to larger or extremely well-managed operations; but this type of reactor should be carefully examined for other anaerobic digestion applications because of its potentially high efficiencies.

Figure 32.—Two-Stage Digester

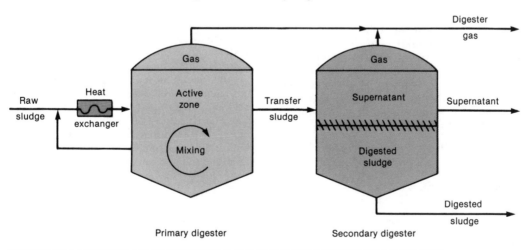

SOURCE: Environmental Protection Agency, "Process Design Manual, Sludge Treatment and Disposal," EPA 625/29-001, September 1979.

Figure 33.—Two-Phase Digestion of Cellulosic Feed

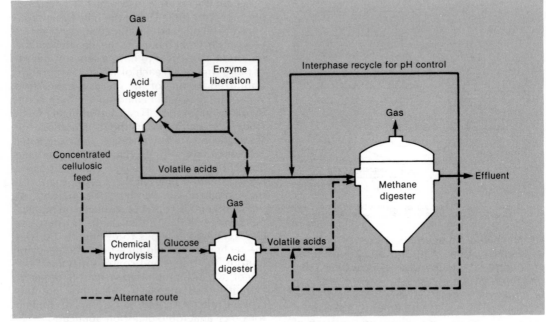

SOURCE: S. Ghos and D. L. Klass, "Two Phase Anaerobic Digestion," *Symposium on Clean Fuels From Biomass* (Institute of Gas Technology, 1977).

Packed Bed (Anaerobic Filter)

In this system a dilute stream of feedstock is fed up through a verticle column packed with small stones, plastic balls, ceramic chips, or other inert materials (figure 34). Because the bacteria attach themselves to the inert material, it is possible to pass large quantities of feedstock through it while maintaining a high bacterial concentration in the digester. The system is best suited to municipal sewage (and other dilute feedstocks). More concentrated feedstocks tend to clog the column.

Analyses of bench-scale laboratory results on the ANFLOW system indicate that the system could produce enough energy to make this sewage treatment step energy self-sufficient.[12]

As it now exists, however, it is not well suited to energy production.

Like the packed and expanded bed, the mixed-bed systems are intended to provide an inert substance to which the bacteria can attach, thereby preventing them from being flushed out with the effluent. The digester maintains a higher bacteria population. Various designs include netting,[13] strips of plastic, and rough porous digester walls. In all cases, the inert substance increases the resistance to flow and thus the energy needed for pumping increases too, but it decreases the necessary reactor size. Sufficient data are not yet available for a detailed analysis of this tradeoff.

[12]R. K. Genung and C. D. Scott, "An Anerobic Bioreactor (ANFLOW) for Wastewater Treatment and Process Applications," briefing presented to the Subcommittee on Energy and Power, House Interstate and Foreign Commerce Committee, Nov. 1, 1979.

[13]S. A. Serfling and C. Alsten, "An Integrated, Controlled Environment Aquaculture Lagoon Process for Secondary or Advanced Wastewater Treatment," Solar Aquasystems, Inc., Encinitas, Calif., 1978.

Figure 34.—Packed-Bed Digester

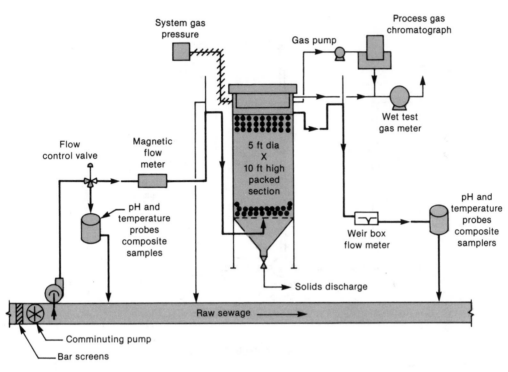

SOURCE: R. K. Genung, W. W. Pitt, Jr., G. M. Davis, and J. H. Koon, "Energy Conservation and Scale-Up Studies for a Wastewater Treatment System Based on a Fixed-Film, Anaerobic Bioreactor," presented at 2nd Symposium on Biotechnology in Energy Production, Gatlinburg, Tenn. Oct. 2-5, 1979.

Expanded Bed

A variation on the packed-bed concept is the expanded-bed reactor (figure 35). In this case the column packing is sand or other very small particles. The feedstock slurry is fed up through the column and the bed of inert material expands to allow the material to pass through. A semifluidized state results, reducing the potential for clogging when relatively concentrated material is fed into the reactor. The process has been found to be quite stable with high organic inputs, short residence times in the digester, and relatively low temperatures (50° to 70° F).[14] The study did indicate, how-ever, that the process would not be a net energy producer due to the energy required to expand the bed.

Variable Feed

The idea behind variable feed systems (figure 36) is to store undigested manure in times of low gas demand for use during periods of high demands. The key is to be able to store the manure for long periods (e.g., 6 months) without excessive deterioration. The effect of long-term storage is being investigated,[15] but the systems may be limited to areas with cool summers or to operations in which the gas is used to generate electricity for export during the peak electric demand periods in summer.

[14]M. S. Switzenbaum and W. J. Jewell, "Anaerobic Attached Film Expanded Bed Reactor Treatment of Dilute Organics," presented at *51st Annual Water Pollution Control Federation Conference,* Anaheim, Calif., 1978.

[15]W. J. Jewell, Cornell University, Ithaca, N.Y., private communication, 1978.

Figure 35.—Expanded Bed Reactor

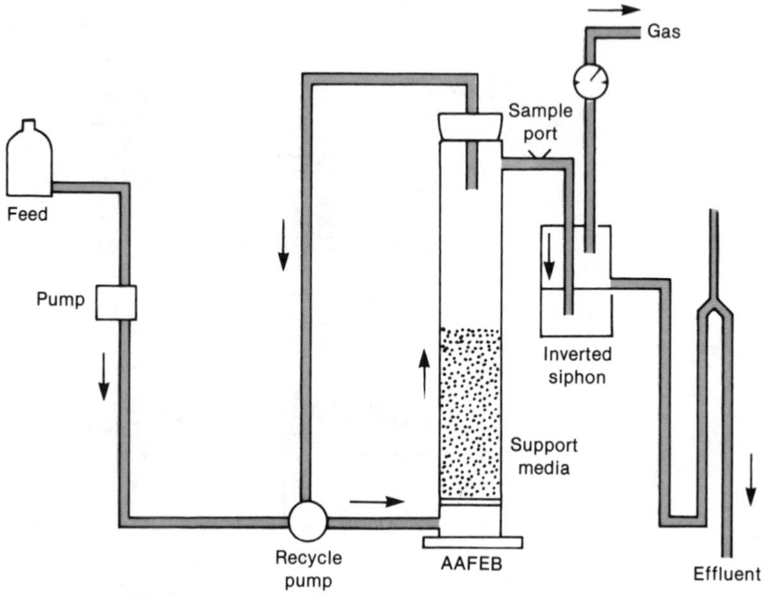

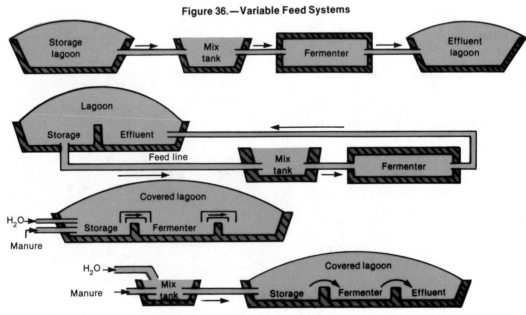

SOURCE: M. S. Switzenbaum and W. J. Jewell, "Anaerobic Attached Film Expanded Bed Reactor Treatment of Dilute Organics," presented at 51st Annual Water Pollution Control Federation Conference, Anaheim, Calif., 1978.

Figure 36.—Variable Feed Systems

SOURCE: Tom Abeles and David Ellsworth, "Biological Production of Gas," contractor report to OTA by I. E. Associates, Inc., Minneapolis, Minn., 1979.

Existing Digester Systems

Fourteen experimental and prototype digesters of animal manure were identified as operational in 1978.[16] The capacity of these plants varies from less than 1,500 gal to 4 million gal. Two of the prototypes are owned by individual farmers and are sized for farm use. The 12 others are owned by private firms, universities, or the Federal Government.

Since then, however, the field of anaerobic digestion has been advancing rapidly, and any list of existing operations would be quickly outdated. Several companies currently design and sell digesters and the support equipment. Most systems are currently designed for cattle manure. One example of an apparently successful digester system is on a dairy farm in Pennsylvania. The digester is fed by 700 head of cattle and has been functioning since late fall 1979. The biogas is fed into a dual-fuel diesel engine and supplies about 90 percent of the engine's fuel needs. The engine drives a 125-kW generator (for peak electric demands) and the generator has an average output of 45 kW. The system supplies essentially all of the operation's direct energy needs.

Other systems are operational or are likely to become operational soon. Nevertheless, operating experience is limited and suitable digesters for all types of manures and combined animal operations are currently not available. Consequently, commercialization of the technology could be helped by demonstrating a wide range of digester systems in a variety of confined animal operations so as to provide operating experience and increase the number of operations for which suitable digester systems exist.

[16]D. L. Klass, "Energy From Biomass and Wastes," in *Symposium on Energy From Biomass and Waste,* Institute of Gas Technology, Washington, D.C., Aug. 14-18, 1978.

Economic Analysis

Aside from the paper and other digestable matter in municipal solid waste (which is not included in this report), the best feedstocks for anaerobic digestion are animal manure, some types of grasses, aquatic plants, and various processing wastes. The supply of aquatic plants is likely to be small in the next 10 years and little information is available on the digester requirements for grasses. Furthermore, with grass at $30/dry ton, the feedstock cost alone would be $4.50/million Btu. More energy at a lower cost can usually be produced from grass by thermal gasification or combustion. Hence, animal manure and some processing wastes are the most promising near-term sources of biogas by far. The larger of these two sources is animal manure.

More than 75 percent of the animal manure resource is located on confined animal operations that have less than 800 dairy cows or the equivalent weight of other animals (e.g., 250,000 chickens). Large feedlots account for less than 15 percent of the resource or less than 0.04 Quad/yr (see ch. 5 in pt. I). Since it is relatively expensive to transport manure long distances, this economic analysis concentrates on digesters appropriate for onfarm use.

The system analyzed consists of a plug flow digester operating at 70° to 90° F, with a feedstock pond and effluent residue storage pit. (See top schematic, figure 36.) After removal of the hydrogen sulfide (H_2S), the biogas is burned in an internal combustion engine to generate electricity. The electricity is used on the farm (replacing retail electricity) and the excess is sold wholesale to the electric utility. The waste heat from the generator engine is also used onfarm, but any excess heat goes to waste. On the average, 15 percent of the energy produced is used to heat the manure entering the digester and for the other energy needs of the digester (e.g., pumping). (Other systems vary from 10 to 40 percent, depending on the type of digester and the operating conditions.)

There is sufficient gas storage capacity to limit electric generation to those times of the day when the utility or the farmer has peak electric demands; and the feedstock storage allows for seasonal variations in the average daily energy production. Therefore, this system can be used either as a peakload or baseload electric generating system. In a mature system, the electric utility would be able to call for more or less electric generation from onfarm units by sending coded signals along the electric powerlines.

Other systems are possible, including one in which the water in the digester effluent is largely removed (dewatered) and the resultant material sold as a fertilizer or animal feed. Table 58 shows the cost of various systems; the basic digester cost represents the sum of the costs for digester, pumps, pipes, hot water boilers, H_2S scrubber, low-pressure gas compressor, heat exchangers, and housing. The cost of the manure premixing equipment is also included with tanks larger than 40,000 gal. However, these costs should be viewed as preliminary and approximate.

Removal of the CO_2 and sales of the gas to natural gas pipelines were assumed not to be feasible in small operations because: 1) the gas pipelines often are not readily accessible, 2) the cost of CO_2 removal equipment is high, and 3) revenues from the gas sales would probably be relatively low. In very large systems, though, production of pipeline quality gas may be feasible.

Table 59 gives the energy that could be produced with onfarm digesters. It also shows the quantities that could be used onfarm and exported for various animal operations in some of the major producing States if farm energy use stays at 1974-75 levels or if it decreases 25 percent due to energy conservation. In most cases, the digester energy output is sufficient to meet the energy needs of the livestock operation and in more than half of the cases considered, it also fills the farmer's home energy needs and enables a net export of electricity. With conservation, the situation is even more favorable with respect to energy exports. The lower revenue that the farmer receives for surplus energy as opposed to the replacement of retail energy, however, makes conservation less economically attractive unless the farmer is not energy self-sufficient without conservation. In other words, it is more economically attractive to replace retail electricity than to generate surplus electricity for sales at wholesale rates.

The digester size, capital investment, and operating costs for anaerobic digester-electric generation systems for these various operations are shown in table 60. Assuming the farmers can displace retail electricity costing 50 mill/kWh, sell wholesale electricity for 25 mill/kWh, and displace heating oil used onfarm costing $6/million Btu, the returns from the digester system are shown in table 61. Also shown are the farmer's costs for two assumed capital charges: 1) where the annual capital charges are 10.8 percent of the investment, i.e.,

Table 58.—Investment Cost for Various Anaerobic Digester System Options

Tank size (gallons)	Median capital costs ($1,000)				
		Options			
	Basic digester	Dewatering	Electric generator[a]	Feedstock lagoon[b]	Residue pit[c]
10,000	$ 19.6	$34.0	$ 4.0	$ 6.7	$ 0.4
20,000	33.7	34.0	5.0	13.4	0.8
40,000	48.1	34.0	6.0	26.9	1.6
80,000	65.6	45.0	12.0	53.7	3.2
200,000	98.8	60.0	45.0	133.0	7.5
400,000	143.6	90.0	70.0	268.3	15.2

[a]Generator is in operation 12 hours each day.
[b]Storage for 6 months.
[c]Storage for 9 months.

SOURCE: Tom Abeles and David Ellsworth, ''Biological Production of Gas,'' contractor report to OTA by I. E. Associates, Inc., Minneapolis, Minn., 1979.

Table 59.—Anaerobic Digestion: Energy Balance, or the Energy Production Potential and Energy Consumption Potential of Individual Farms in Major Producing States

1974 livestock average number sold (inventory)/farm	Methane energy options — Direct use only	Methane — Electricity kWh	Methane — Waste heat	Demands of livestock operation — 1974 levels		Demands — 25% conserv.		Household use (1975 levels)		Excess energy (1974 levels) — Direct use only	Excess — Electricity kWh	Excess — Waste heat	Excess energy (25% conservation) — Direct use only	Excess — Electricity kWh	Excess — Waste heat
	Btu 10⁶	kWh	Btu 10⁶	Btu 10⁶	kWh	Btu 10⁶	kWh	Btu 10⁶	kWh	Btu 10⁶	kWh	Btu 10⁶	Btu 10⁶	kWh	Btu 10⁶
Turkeys															
Minnesota (124,000)	14,914	877	9,545	8,754	248	6,559	186	326	16	5,835	613	465	8,029	675	2,660
California (98,000)	11,787	693	7,544	4,714	147	3,536	110	180	16	6,893	530	2,650	8,071	567	3,828
North Carolina (89,000)	10,705	630	6,851	1,931	160	1,448	120	128	22	8,646	448	4,792	9,129	488	5,275
Broilers															
Minnesota (52,000)	522	31	334	357	14	268	11	163	8	2	9	−186	91	12	−97
(198,000)	1,986	117	1,271	1,366	55	1,024	41	163	8	457	54	−258	799	68	84
California (56,000)	562	33	360	350	10	262	8	90	8	122	15	−80	210	17	8
(1,377,000)	13,812	812	8,840	8,606	244	6,454	183	180	16	5,026	552	54	7,178	613	2,206
Arkansas (63,200)	634	37	406	314	9	236	7	107	9	213	19	−15	291	21	63
(186,000)	1,866	110	1,194	937	27	703	20	107	9	822	74	150	1,056	81	384
Swine															
Iowa	325ᵃ	19	208	147	28	110	21	155	8	23	−17	−94	60	−10	−57
Missouri	325ᵃ	19	208	69	22	52	16	136	8	120	−11	3	137	−5	20
North Carolina (500)	325ᵃ	19	208	29	13	22	10	64	11	232	−5	115	239	−2	123
Dairy cows															
Wisconsin (36)	261	15	167	11	16	8	12	161	8	89	−9	−5	92	−5	−2
New York (71)	521	31	333	25	33	19	25	126	8	370	−10	182	376	−2	188
California (337)	2,459	145	1,574	370	133	278	100	270	24	1,819	−12	934	1,911	21	1,026
Laying hens															
Minnesota (13,000)	846	50	541	213	34	160	26	163	8	470	8	165	523	16	218
(41,000)	2,670	157	1,709	672	106	504	80	163	8	1,835	43	874	2,003	69	1,042
Georgia (14,000)	912	54	584	140	34	105	26	72	10	700	10	372	735	18	407
(41,000)	2,670	157	1,709	410	98	308	74	72	10	2,188	49	1,227	2,290	73	1,329
California (14,000)	912	54	584	224	55	168	41	90	8	598	−9	270	654	5	326
(105,000)	6,838	402	4,376	1,680	414	1,260	310	180	16	4,978	−28	2,516	5,398	76	2,936
Beef (500)	1,527	90	977	—	—	—	—	160	10	1,427	80	817	—	—	—

ᵃIncludes breeding stock.

Assumptions: 15% of biogas used to run digester system, electric generation at 20% efficiency, and 80% of the engine waste heat can be recaptured.

SOURCE: Tom Abeles and David Ellsworth, ''Biological Production of Gas,'' contractor report to OTA by I. E. Associates, Inc., Minneapolis, Minn., 1979; *Energy and U.S. Agriculture, 1974 Data Base* (Washington, D.C.: Energy Research and Development Administration, 1974).

9-percent interest loan with 20-year amoritization, and 2) where the annual capital charges are 15 percent of the investment.

The principal cost factor in anaerobic digestion is the capital charge, or the cost of the digester itself—thus, favorable financing is the most effective way of reducing the cost to the farmer.

Financing aside, the anaerobic digestion operations that are most economically attractive are relatively large poultry, dairy, beef, or swine operations (enabling an economy of scale) which are also relatively energy intensive (enabling the displacement of relatively large quantities of energy at retail prices). For example, anaerobic digestion on a broiler farm with 198,000 birds in Minnesota is more economically attractive than an equivalent operation with only 52,000 birds (see table 61). On the other hand, the 52,000-bird broiler farm (equivalent to 250 head of cattle in terms of the quantity and quality of manure) is more economically attractive than a 500-head cattle feedlot, because the poultry operation consumes considerably more energy, and thus could better utilize the digester output within its own enterprise.

Based on 1978 fuel prices, the feasibility of anaerobic digestion was assessed for various types of farm animal operations in the various regions of the country. It was found that it would be feasible to digest 50 percent of the

Table 60.—Cost of Various Digesters With Electric Generating Capabilities

1974 livestock average number sold (inventory)/farm	Digester size (1,000 gal)	Capital investment ($1,000)	Annual operating costs ($1,000)
Turkeys			
Minnesota (124,000). . . .	300	220	4.6
California (98,000)	250	195	3.9
North Carolina (89,000). .	225	182	3.7
Broilers			
Minnesota (52,000)	12	25	0.9
(198,000). . . .	45	62	2.6
California (56,000)	13	27	0.9
(1,377,000). . .	300	220	4.4
Arkansas (63,200)	14	28	1.0
(186,000)	40	59	2.6
Swine			
Iowa (500)	10	27	0.8
Missouri (500)	10	27	0.8
North Carolina (500)	10	27	0.8
Dairy cows			
Wisconsin (32)	10	24	0.6
New York (64)	10	30	0.8
California (337)	80	92	1.6
Laying hens			
Minnesota (13,000)	34	50	1.6
(41,000)	100	103	1.7
Georgia (14,000)	34	50	1.6
(41,000)	100	103	1.7
California (14,000)	35	51	1.6
(105,000)	250	169	3.0
Beef (500)	20	43	0.7

SOURCE: SOURCE: Tom Abeles and David Ellsworth, ''Biological Production of Gas,'' contractor report to OTA by I. E. Associates, Inc., Minneapolis, Minn., 1979.

Table 61.—Annual Costs and Returns From Digester Energy Only

1974 livestock (inventory)/farm	Return from energy displacement and sales of electricity ($1,000)	Digester costs (operating + capital) ($1,000)	
		10.8% annual capital charge	15% annual capital charge
Turkeys			
Minnesota (124,000). . . .	81	28	38
California (98,000)	51	25	33
North Carolina (89,000). .	36	12	31
Broilers			
Minnesota (52,000)	3.3	3.6	4.7
(198,000). . . .	12	9.3	12
California (56,000)	3.4	3.8	5.0
(1,377,000). . .	80	28	37
Arkansas (63,200)	3.8	4.0	5.2
(186,000)	9.9	9.0	11
Swine			
Iowa (500)	2.2	3.7	4.9
Missouri (500)	2.2	3.7	4.9
North Carolina (500)	1.5	3.7	4.9
Dairy cows			
Wisconsin (32)	1.8	3.2	4.2
New York (64)	2.5	4.0	5.3
California (337)	11	12	15
Laying hens			
Minnesota (13,000)	4.6	7	9.1
(41,000)	12	13	17
Georgia (14,000)	3.7	7	9.1
(41,000)	9.5	13	17
California (14,000)	4.6	7.1	9.3
(105,000)	31	21	28
Beef (500)	3.2	5.3	7.2

SOURCE: SOURCE: Tom Abeles and David Ellsworth, ''Biological Production of Gas,'' contractor report to OTA by I. E. Associates, Inc., Minneapolis, Minn., 1979.

animal manure to produce electricity and on-site heat if the effective annual capital charges were 6.6 percent of the investment. (Digestion was deemed feasible if the returns from displacing onsite energy use and wholesaling excess electricity were greater than the capital and operating costs for the anaerobic digestion energy system.) This effective capital charge could be achieved by a 9-percent interest, 20-year loan with 4.2-percent annual tax writeoff. Other possible credits could be available through combinations of Agricultural Stabilization and Conservation Service pollution abatement cost sharing, soil conservation district cost sharing, energy credits, and other incentives, although these are not included in the feasibility calculations. In table 62, the percent of the manure resource that would be feasible for energy production with the 6.6-percent capital charge is shown for various manure types and regions. Also shown

are the quantities of manure that would be feasible if the digester effluent were dewatered and sold as a fertilizer at $10/dry ton over the revenues available from sales of the raw manure as a fertilizer. Furthermore, if the dewatered effluent could be sold for feed, higher credits may be possible based on the protein content of the effluent.[17] Although the feasibility of these higher credits is unproven as yet, the selling of digester effluent as feed or fertilizer would substantially expand the quantity of manure that could be digested economically.

Turkey farms tend to be the most economic because of their rather large average size and

---

[17]Biogas of Colorado, "Energy Potential Through Bioconversion of Agricultural Wastes," *Phase II, Final Report to the Four Corners Regional Commission,* demonstration project FCRC No. 672-366,002, Arvada, Colo., 1977.

Table 62.—Economic Feasibility of Anaerobic Digestion (percent of total manure resource that can be utilized economically[a])

Region	Total usable manure problem	Layers	Broilers	Turkeys	Cattle on feed	Dairy	Swine
Northeast	52%	68%	82%	98%	6%	43%	—
Southeast	65	54	75	90	45	80	—
Appalachia	47	51	85	89	8	26	—
Corn Belt	17	38	67	89	9	19	8%
Lake States	30	50	90	98	6	23	8
North Plains	39	37	46	96	55	12	—
Delta	69	68	82	86	21	37	—
South Plains	82	76	87	94	90	49	—
Mountain	75	82	44	95	83	49	—
Pacific	88	78	97	98	90	89	—
Alaska	69	0	0	0	0	82	—
Hawaii	70	35	82	0	89	99	—
National totals	49 + 20[b]	59	81	94	61	41	5
With fertilizer enhancement assumption of $10/dry ton	69 + 30[b]	85	95	99	72	60	35

[a]Assuming feedstock lagoon (6-month storage), residue pit (9-month storage), generation of heat and electricity for onsite use and electricity for wholesale sales, and an annual capital charge of 6.6% of the investment. Also assumes 1978 energy costs as follows: home heating: $3.80/million Btu; farming heat: $5.40/million Btu; retail electricity: according to DOE, Typical Electric Bills, January 1978, October 1978; and wholesale elecrricity: 25 mill/kWh.

[b]Estimated uncertainty. These correspond to weighted average percentages.

SOURCE: Tom Abeles and David Ellsworth, "Biological Production of Gas," contractor report to OTA by I. E. Associates, Inc., Minneapolis, Minn., 1979.

the relatively large amount of thermal energy consumed by them. Swine operations, however, are usually too small to be economically attractive, for the energy alone, but because of odor problems these may also be attractive.

If 50 percent of the animal manure on confined animal operations in the United States is converted to electricity and heat, about 7 billion kWh of electricity per year (equivalent to about 1,200 MW of electric generating capacity) and about 0.08 Quad/yr of heat would be produced by 0.12 Quad/yr of biogas. At 70-percent utilization, electricity equivalent to about 1,600 MW of electric generating capacity would be produced along with 0.11 Quad/yr of heat.

In either case some of the heat would be wasted in the systems described above. There is, however, the possibility of expanding the operation to use the excess heat, for example, by building greenhouses. This could improve the economics, but it would require major adjustments in the farmer's operation. In the end, site-specific economics and the inclination of the individual farmer will determine whether such options are adopted.

Care should be exercised when using these data. They are based on a number of approximations and they cannot be taken too literally. They do, though, indicate the general trends as to economic feasibility and they show that a substantial quantity of the manure produced on livestock operations could be used economically to produce energy, if the effective capital charges are reduced through various economic incentives.

Environmental Impacts of Biogas Production: Anaerobic Digestion of Manure

Anaerobic digestion of feedlot manure is considered to be an environmentally beneficial technology because it is an adaptation of a pollution control process. The energy product—biogas—is basically a byproduct of the control process, which converts the raw manure (which often represents a substantial disposal problem) into a more benign sludge waste. Where the manure was used as a fertilizer and soil amendment, the digestion wastes can substitute for the manure while eliminating some of its drawbacks.

The environmental benefits associated with reducing feedlot pollution are extremely important. The runoff from cattle feedlots is a source of high concentrations of bacteria, suspended and dissolved solids, and chemical and biological oxygen demand (COD/BOD). This type of runoff has been associated with: large and extensive fish kills because of oxygen depletion of receiving waters; high nitrogen concentrations in ground and surface waters, which can contribute to the aging of streams and to nitrate poisoning of infants and livestock; transmission of infectious disease organisms (including *salmonella, leptospirosis,* and *coliform* and *enterococci* bacteria) to man, livestock, and wildlife; and coloring of streams.[18] Other problems associated with feedlots include attraction of flies and obnoxious odors.

Because anaerobic digestion is a relatively simple process not requiring extreme operating conditions or exotic controls, biogas facilities may be designed for very small (10 cow) operations as well as large feedlots. The environmental impacts will vary accordingly. For example, recycling of wastewater may be possible for the larger operations; it is not likely to be possible for the small onfarm digesters because of high water treatment costs. The product gas from the smaller units is likely to be used onsite and, depending on its use, may or may not be scrubbed of its H_2S and ammonia (NH_3) content; the product from very large units may be upgraded to pipeline quality by removing these pollutants as well as the 30 to 40 percent of the CO_2 fraction in the biogas.

The major problem associated with the digestion process is waste disposal and the associated water pollution impacts that could result. As noted above, anaerobic digestion is basically a waste treatment technology, but although it reduces the organic pollution content of manure it does not eliminate it. The combination of liquid and solid effluent from the digester contains organic solids, fairly high

concentrations of inorganic salts, some concentrations of H_2S and NH_3, and variable amounts of potentially toxic metals such as boron, copper, and iron. For feedlot operations where the manure is collected only intermittently, small concentrations of pesticides used for fly control may be contained in the manure and passed through to the waste stream.

A variety of disposal options for the liquid and sludge wastes exist. Generally, wastes will be ponded to allow settling to occur. The liquid, which is high in organic content, can be pumped into tank trucks (or, for very large operations, piped directly to fields) to be used for irrigation and fertilization. The high salt content and the small concentrations of metals in the fluid make it necessary to rotate land used for this type of disposal. Large operations may conceivably treat the water and recycle it, but treatment cost may prove to be prohibitive. Other disposal methods include evaporation (in arid climates), discharge into waterways (although larger operations are likely to be subject to zero discharge requirements by the Environmental Protection Agency), and discharge into public sewage treatment plants.[19] In all cases, infiltration of wastewater into the ground water system is a possibility where soils are porous and unable to purify the effluent through natural processes. As with virtually all disposal problems of this nature, this is a design and enforcement problem rather than a technological one; if necessary, ponds can be lined with clay or other substances for ground water protection.

The organic content of the effluent, which varies according to the efficiency of the digester, will represent a BOD problem if allowed to enter surface waters that cannot dilute the effluent sufficiently. Similar problems can occur with organics leached from manure storage piles. However, this problem exists in more severe form in the original feedlot operation.

The sludge product can be disposed of in a landfill, but it appears that the sludge has

[18]*Environmental Implications of Trends in Agriculture and Silviculture, Volume II: Environmental Effects of Trends* (Washington, D.C.: Environmental Protection Agency, December 1978), EPA-600/3-78-102.

[19]*Solar Program Assessment: Environmental Factors, Fuels From Biomass* (Washington, D.C.: Energy Research and Development Administration, March 1977), ERDA 77-47/7.

value either as a fertilizer or cattle feed if the heavy metals content is not too great. Successful experience with anaerobically digested municipal sludges, which clearly have higher concentrations of heavy metals, indicates that use of the feedlot-derived sludge as fertilizer should present no metals problem.[20] In numerous applications overseas, the sludge is considered a substantial improvement over the previously used manure fertilizer. In areas where chemical fertilizers are not available or are too expensive, the retention of the manure's fertilizer value is a particularly critical benefit of the biogas process.

Although the H_2S (and related compounds) content of the effluent may present some odor problems, this problem, as well as that of the very small pesticide content, should be negligible.[21]

The gas produced by the digester will contain small (less than 1-percent each)[22] concentrations of H_2S and NH_3. If the gas is burned onsite without scrubbing out these pollutants, combustion will oxidize these contaminants to sulfur and nitrogen oxides. Because the H_2S will form mostly sulfurous and sulfuric acids, which are extremely corrosive to metal, the biogas has limited use if it is not scrubbed. For example, scrubbing is a requirement if the gas is to be used in an internal combustion engine. Simple and inexpensive scrubbing methods are available, using an "iron sponge" of ferric oxide and wood shavings to react the gas with the iron to form ferric sulfide.[23] However, even if the gas were not scrubbed, the pollutant concentrations caused by biogas combustion should be of little consequence to public health as long as the combustion did not take place in a confined area. For example, combustion of biogas produced from fresh cow manure might generate sulfur oxides on the order

of 0.1 lb/million Btu,[24] compared to the Federal requirement of 1.2 lb/million Btu for coal combustion in large utility boilers.

The major air pollution problem of anaerobic digestion, therefore, is not from combustion of the product gas, but from leaks of raw gas from the system. For a manure sulfur content of 0.2 percent and digester pH of 7.2, the raw biogas can contain H_2S in concentrations of nearly 2,000 parts per million (ppm).[25] Although exposure to this full concentration seems extremely unlikely, concentrations of 500 ppm can lead to unconsciousness and death within 30 minutes to 1 hour, and concentrations of 100 ppm to respiratory problems of gradually increasing severity over the course of a few hours; the Occupational Safety and Health Administration's standard is a maximum permissible exposure level of 20 ppm.[26]

Because of rapid diffusion of the gas, health problems associated with H_2S exposures are likely to be confined to these occupational exposures. However, venting of raw gas can cause severe odor problems to the general public. In this case, odor problems associated with gas venting should be compared to the similar (but more certain) odor problems associated with the sometimes haphazard treatment of manure that the biogas operation replaces.

Because methane is explosive when mixed with air, strong precautions must be taken to avoid biogas leakage into confined areas and to prevent any possibility of the gas coming into contact with sparks or flames. Although this will be a universal problem with biogas facilities, it is particularly worrisome if small units proliferate.

If normal operating conditions hold biogas leakage to near-zero levels, the powerful odor of the H_2S contaminant would serve as an early warning of a leak. Because low concentrations of H_2S will deactivate the sense of smell, the acceptance of small leaks as "stand-

[20]*Methane Generation From Human, Animal, and Agricultural Wastes* (Washington, D.C.: National Research Council, National Academy of Sciences, 1977), Library of Congress catalog No. 77-92794.
[21]M. C. T. Kuo and J. L. Jones, "Environmental and Energy Output Analysis for the Conversion of Agricultural Residues to Methane," *Symposium on Energy From Biomass and Waste,* Institute of Gas Technology, Washington, D.C., Aug. 14-18, 1978.
[22]*Solar Program Assessment,* op. cit.
[23]Kuo and Jones, op. cit.

[24]Ibid.
[25]*Solar Program Assessment,* op. cit.
[26]Ibid.

ard operating practice" would eliminate this safety factor.

The institutional problems associated with assuring that there is adequate control of digester impacts are very similar to those of ethanol plants: there is an attraction towards smaller size ("onfarm") plants which may have some advantages (mainly ease of locating sites for waste disposal and smaller scale local impacts) but which cannot afford sophisticated waste treatment, are unlikely to be closely monitored, and may be operated and maintained by untrained (and/or part-time) personnel. Some of the potential safety and health problems probably will respond to improved system designs if small onfarm systems become popular and the size of the market justifies increased design efforts on the part of the manufacturers. The ease of building homemade systems, however, coupled with farmers' traditional independence should provide potent competition to the sale of manufactured (and presumably safer) systems.

Research, Development, and Demonstration Needs

Below, the more important research, development, and demonstration needs for anaerobic digestion are divided into the general areas of microbiology, engineering, and agriculture.

Microbiology

The whole range of studies related to how anaerobic digestion works should be addressed. This includes identifying the bacteria and enzymes involved, studying the bacteria's nutrient requirements (including trace elements), identifying optimum conditions for the various conversion stages, and investigating why some feedstocks are superior to others. Much of this is in the realm of basic research needed to understand the processes involved so that the yields, rates, control, and flexibility of anaerobic digestion can be improved.

Engineering

A large number of different digester types need to be demonstrated to aid in optimizing the safety and reliability of digester systems while reducing the cost for onfarm use or for large-scale systems. The unique problems and opportunities of various types of animal operations should also be addressed by the various digester systems. There are also numerous design alternatives that could lower the digester costs, and these should be thoroughly examined.

Electric power generation and the related feedstock pumping is a weak area in digester systems, particularly with respect to reliability, maintenance, and efficiency of the engine-generator units. Development work for small engines intended to use biogas and the related pumps could lead to improvements in these areas, and fuel cells capable of using biogas should be developed. Development work is also needed into the best ways the farm generator can supply the electric utility grid during periods of high demand without undue inconvenience for the livestock operation.

Agriculture

More needs to be known about the difference between digested and undigested manure. The digested manure should be investigated in order to determine its value as a fertilizer, animal feed, and nutrient source for aquatic plants. High-value uses for the digester effluent, proved through thorough testing, could significantly improve the economics of anaerobic digestion.

USE OF ALCOHOL FUELS

Chapter 10.—USE OF ALCOHOL FUELS

USE OF ALCOHOL FUELS

Introduction

Liquid fuels have some unique advantages over solids and gases that make them important fuels in some applications. They contain a large amount of energy per unit volume (as compared to gases) and their combustion can easily be controlled (as compared to solids). However, there are substantial differences among liquid fuels. At one end of the spectrum is residual fuel oil, which can produce considerable emissions when burned and is generally best suited as a boiler fuel (an application also open to solid fuels). At the other end are light distillate oils, gasolines, and alcohol fuels. The oils and gasolines are superior to alcohols with respect to their energy content per unit weight (ethanol has two-thirds and methanol one-half of the energy per gallon of gasoline), which makes them better suited for aviation. The alcohols are superior to oils with respect to their lower particulate emissions and higher octane values. These properties make them particularly useful for marine and ground transportation where energy density is not critical and for gas turbines used for peakload electric generation, the applications considered here.

While both ethanol and methanol are alcohols, they have different physical and chemical properties. Of the two, methanol is less soluble in gasoline, separates easier, and can more easily damage certain plastics, rubbers, and metals used in current automobiles. Furthermore, methanol requires more heat to vaporize it. Both alcohols, as contrasted to gasoline, contain oxygen and conduct electricity. These properties are important when considering the use of alcohol fuels.

While oil and hydrocarbon (HC) crops may some day produce fuels for transportation, their costs and yields are highly uncertain at this time. For the near to mid-term, the most likely biomass substitutes for gasoline, diesel, and light fuel oil are the alcohol fuels.

Biomass is the only solar technology for producing liquid fuels. The biomass can be converted to methanol ("wood alcohol") through thermochemical conversion or to ethanol ("grain alcohol") through fermentation or, possibly, thermochemical conversion. Ethanol production from sugars and starches is currently commercial technology. Wood-to-methanol plants can be built with existing technology, although none currently exists, and plant-herbage-to-methanol technology needs to be demonstrated.

Most cost calculations indicate that methanol from coal will be less expensive than either alcohol from biomass. Until and unless a domestic liquid fuels surplus develops, however, this cost difference is not likely to exclude the biomass alcohols from the market.

Spark Ignition Engines—Effects From Alcohols and Blends

Alcohols make excellent fuels for spark-ignited engines which are designed for their use. However, when considering alcohol or alcohol-gasoline blend use in gasoline engines, there are four specific factors that are of overriding importance. One factor is the material from which the engine is constructed. Another is the ratio of air to fuel (A/F ratio) in the mixture that is burned in the engine. A third is proper fuel distribution among the cylinders, and a fourth is cold-starting ability.

Some materials in some automobiles are incompatible with alcohols. Contact with alcohols can damage some gaskets, fuel pump diaphragms, and other plastic and rubber parts. If these parts are adversely affected or fail, the engine is likely to malfunction. Furthermore, some electric fuel pumps are mounted in the fuel tank. Electric currents induced in the alcohol fuels by these motors may cause the protective terneplate coating on fuel tanks to dissolve and leave the tank susceptible to corrosion. There may also be a fire hazard associated with electrical shorting. Under certain circumstances, not totally understood at present, the alcohols or blends can also chemically remove the terneplate coating.[1][2][3] Finally, alcohols can cause some deposits in the fuel tanks and lines to loosen and dislodge, leading to a blockage in the fuel filter or carburetor.

Three major classes of automobiles are in use today: pre-1975 cars, oxidation (two-way) catalyst cars (most post-1975 cars in States other than California), and California three-way catalyst cars. The range of A/F ratios* intended for each class of cars is shown in figure 37 together with the effect of this ratio on the engine power, efficiency, and emissions. If the A/F ratio extends beyond the ranges of this figure, most engines will hesitate or stall. (Stratified-charge engines like the Honda CVCC and Ford Proco have somewhat wider ranges.)

Since the pre-1975 cars and oxidation catalyst cars usually have carburetors with fixed fuel metering passageways, the alcohol blend fuels, which require less air per volume of fuel than gasoline, will make the effective fuel mixture leaner (i.e., move the effective A/F ratio to less fuel and more air). California three-way catalyst cars, however, have a sensor that adjusts the fuel delivery system to the A/F value intended by the manufacturer. Nevertheless, exhausts from alcohol fuels "fool" this sensor somewhat, so the compensation is not com-

---

[1]K. R. Stamper, "50,000 Mile Methanol/Gasoline Blend Fleet Study—A Progress Report," in *Proceeding of the Alcohol Fuels Technology Third International Symposium,* Asilomar, Calif. (Bartlesville, Okla.: Bartlesville Energy Technology Center, Department of Energy, May 1979).

[2]S. Gratch, Director of Chemical Science Lab, Ford Motor Co., Dearborn, Mich., private communication, 1979.

[3]J. L. Keller, G. M. Nakagucki, and J. C. Ware, "Methanol Fuel Modification for Highway Vehicle Use," Union Oil Co. of California, Brea, Calif., final report to the Department of Energy, contract No. FY 76-04-3683, July 1978.

*The figure shows the equivalence ratio which is found by dividing the stoichiometric A/F ratio which is exactly sufficient to completely burn all of the fuel by the actual A/F ratio used in the car. Leaner mixtures are to the left and richer to the right.

Figure 37.—Efficiency, Power, and Emissions as a Function of Equivalence Ratio

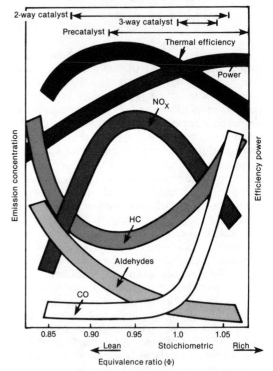

The equivalence ratio is the ratio of the A/F ratio which is exactly sufficient to completely burn all of the fuel to the actual A/F ratio used in the car. Leaner mixtures have an excess of air, while richer ones have an excess of fuel.

SOURCE: H. Adelman, et al., "End Use of Fluids From Biomass as Energy Resources in Both Transportation and Nontransportation Sectors," University of Santa Clara, Santa Clara, Calif., contractor report to OTA, 1979.

plete.[4] Consequently, while the difference between the A/F ratio for alcohol fuels and pure gasoline is less for California three-way catalyst cars than for other cars, the A/F ratio is still somewhat leaner with alcohol fuels compared to gasoline.

For pure alcohols the fuel metering rate must be increased significantly relative to gasoline. This increased rate can result in streaming flow rather than well-disbursed droplets as the fuel enters the air stream when carburetors are retrofitted for alcohol fuel. This change

---

[4]Gratch, private communication, op. cit.

can seriously aggravate the variation in the A/F ratio among the cylinders which in turn can reduce performance and economy and increase emissions. Proper design of the fuel delivery system and intake manifold can avoid this penalty.

The alcohols do not inherently provide good cold engine starting. Below 40° F, special attention must be paid to avoid cold-starting problems with alcohol. Aids such as electric heating in the intake manifold, blending agents such as gasoline, butane, or pentane added to the alcohols, or auxiliary cold-start fuel are providing solutions to this problem in the alcohol vehicle fleets now in operation.

Gasohol

Materials Compatibility

Gasohol is a mixture of 10 percent ethanol and 90 percent unleaded gasoline. The ethanol blended in gasohol must be dry (anhydrous) or the blend will separate into two phases or layers under certain conditions. Typically, gasohol can hold more water than gasoline, but it can contain no more than about 0.3 percent if separation is to be avoided down to −40° F. Various additives have been tried to improve the water tolerance of gasohol, but none have proved satisfactory to date.[5]

Although the use of anhydrous ethanol should minimize water tolerance problems with gasohol, phase separation has been observed to occur in four service stations in Iowa.[6] This phase-separated blend was sold to some customers and caused their vehicles to stall. Both the vehicle tanks and the service station tanks were drained and up to 0.3 percent by volume water was found in the mixture, but the origin of the water contamination is not known.

Gasohol does not appear to significantly affect engine wear as compared to gasoline but more experience with gasohol is needed before a definitive statement can be made. However, an unknown fraction of existing automobiles have specific components that are not compatible with gasohol, which can result in some fuel system failures.

As older cars are replaced with newer ones warranted for gasohol use and as experience develops in handling ethanol-gasoline blends, these problems should gradually disappear.

Thermal Efficiency

The leaning effect of gasohol relative to gasoline will affect the three classes of cars somewhat differently. For precatalyst and oxidation catalyst cars, the thermal efficiency can either increase or decrease with gasohol depending on the original A/F setting (see figure 37). In general, automobiles that operate rich will increase in efficiency, while those that operate lean will decrease in thermal efficiency with gasohol.

The Nebraska "two million mile" test showed a large average mileage increase (7 percent) with gasohol.[7] However, the spread of data points is so large that the uncertainty in this difference is greater than the difference itself.[*] This is a generic problem in trying to deduce small differences in mileage with road tests.

Laboratory tests, however, indicate an increase in thermal efficiency of 1 to 2 percent

[5]H. Adelman, et al., "End Use of Fluids from Biomass as Energy Resources in Both Transportation and Non-Transportation Sectors," University of Santa Clara, Santa Clara, Calif., contractor report to OTA, 1979.

[6]Douglas Snyder, Iowa Development Commission, private communication, 1979.

[7]W. A. Scheller, *Nebraska 2 Million Mile Gasohol Road Test Program, Sixth Progress Report* (Lincoln, Nebr.: University of Nebraska, January 1977).

[*]Taking data from figure 4 of Scheller, OTA has analyzed the uncertainty in the average mileage difference. Using a standard statistical test ("t" test) reveals that the spread in data points (standard deviation) is so large that the mileage difference between gasohol and regular unleaded would have to be more than 30 percent (two times the standard deviation) before OTA would consider that the test had demonstrated a difference in mileage. While more sophisticated statistical tests might indicate that the measured difference in mileage is meaningful, the validity of these statistical methods is predicated on all the errors being strictly random; and the assumption of random errors is suspect unless the number of vehicles in the test fleet is orders of magnitude larger than any tests conducted to date.

with gasohol in precatalyst and oxidation catalyst cars, which is within the measurement errors.[8] The changes in thermal efficiency with three-way catalyst cars will be less and therefore negligible.

Since gasohol contains 3.5 percent less energy per gallon than gasoline, precatalyst and oxidation catalyst cars are expected to experience about a zero- to 4-percent decrease in miles per gallon. Three-way catalyst cars are expected to experience about a 3- to 4-percent decrease in mileage. Probably neither of these decreases would be noticeable by motorists and, as stated above, will depend on the A/F setting of the automobile for all but the three-way catalyst cars.

These conclusions are in complete agreement with the results of an American Petroleum Institute study released in the spring of 1980,[9] in which all available data on gasohol mileage were compared, averaged, and treated statistically to determine the significance of the results.

Drivablility

Post-1970 noncatalyst and oxidation catalyst cars that are set at lean A/F ratios on gasoline can experience drivability problems such as stumbling, surging, hesitation, and stalling with gasohol due to further mixture leaning. While no drivability problems have been reported for precatalyst cars, laboratory tests on 1978 and 1979 oxidation catatlyst cars suggested slight deterioration in drivability.[10] If the percentage of ethanol is increased beyond 10 percent, more and more cars are expected to experience drivability problems due to the leaning effect.

Since three-way catalyst cars largely compensate for the leaning effect of gasohol, no

drivability problems are expected with gasohol, as long as the mixture does not go beyond the capability of the compensation mechanism in these cars.

In most cars there may also be some minor problems with vapor lock, if the vapor pressure of the blend is not adjusted properly by removing some butanes from the gasoline.

Octane

Addition of ethanol to gasoline increases the octane* for the mixture over that of the gasoline. The exact increase will depend on the gasoline, of which there is a great variety. On the average, for the range of 5 to 30 percent ethanol, each percent of ethanol added to one base unleaded regular gasoline (88 octane) raised the octane number by 0.3.[11] However, the octane increases per unit alcohol are larger for lower percentages of ethanol and lower octane gasolines and level off at higher alcohol percentages and gasoline octane. A 10-percent blend would raise the octane by about 2 to 4 using an "average gasoline."

The octane-boosting properties of ethanol can be exploited in either of two ways to save energy: 1) by reducing the oil refinery energy by producing a lower octane gasoline or 2) by increasing the octane of all motor fuels so that automobile manufacturers can increase the compression ratios and thus the efficiency of new cars.

There is considerable uncertainty and variability in the amount of premium fuel that can be saved at refineries by using ethanol as an octane-boosting additive. As shown in table 63, reported or derived values vary from nearly zero to more than 60,000 Btu/gal of ethanol, depending on the average octane of the refinery gasoline pool, the octane boost assumed from ethanol, the type of gasoline and the ratio of gasoline to middle distillates produced by the refinery, the refinery technology used, and other specifics.

[8]R. K. Pefley, et al., "Characterization and Research Investigatlon of Methanol and Methyl Fuels," University of Santa Clara, Santa Clara, Calif., contractor report to the Department of Energy, contract No. FY 76-5-02-1258, 1979.

[9]Mueller Associates, Inc., "A Comparative Assessment of Current Gasohol Fuel Economy Data," commissioned by the American Petroleum Institute, 1980.

[10]R. Lawrence, "Gasohol Test Program," Technology Assessment and Evaluation Branch, Environmental Protection Agency, Ann Arbor, Mich., December 1978.

*Octane here refers to the average of research octane and motor octane.
[11]Keller, Nakagucki, and Ware, op. cit.

Table 63.—Various Estimates of Refinery Energy Savings From Use of Ethanol as Octane-Boosting Additive

Source	10³ Btu saved/gal of ethanol blended 10% in gasoline	Conditions
Energy Research Advisory Board[a]	8	Unknown
Kozinski[b]	16	86 pool octane, reduction in gasoline to distillate ratio, 3 octane number boost by ethanol
OTA[c]	40-45	91 pool octane, reduction in gasoline to distillate ratio, 3 octane number boost by ethanol
Adelman[d]	53	Pool of 91 and 96 research octane, 5 research octane boost by ethanol
Office of Alcohol Fuels[e]	63	Unknown

[a]Energy Research Advisory Board, "Gasohol," Department of Energy, Apr. 29, 1980.
[b]A. A. Kozinski, Amoco Oil Co., Naperville, Ill., private communication, 1980 from data in D. K. Lawrence, et al., "Automotive Fuels—Refinery Energy and Economics," Amoco Oil Co., SAE technical paper series No. 800225, 1980.
[c]OTA from data in Lawrence, op. cit.; and from figure 5 in G. W. Michalski and G. H. Unzelman, "Effective Use of Antiknocks During the 1980's," American Petroleum Institute preprint No. 22-79, from 44th Refinery Midyear Meeting, May 16, 1979.
[d]H. Adelman, et al., "End Use of Fluids From Biomass as Energy Resources in Both Transportation and Nontransportation Sectors," University of Santa Clara, Santa Clara, Calif., contractor report to OTA, 1979.
[e]Office of Alcohol Fuels, Department of Energy, "Comments by the DOE Office of Alcohol Fuels or the Energy Research Advisory Board, April 29, 1980, Gasohol Study Group, Report," June 3, 1980.

SOURCE: Office of Technology Assessment.

Based on published computer simulations of an oil refinery,[12] Kozinski has estimated a savings of about 16,000 Btu/gal of ethanol on the basis of an average gasoline pool octane of 86, and appropriate reduction in the gasoline-to-distillate ratio, which is appropriate for the current situation where the octane of about half the gasoline is raised with tetraethyl lead.[13] In the future, if most of the gasoline produced is unleaded, then the pool will have to increase to at least 89 octane, which is the current average after lead has been added. Moreover, the octane requirements of new cars is increasing,[14] which will induce refiners to increase the pool octane further.

Assuming an average pool octane requirement of 91, which can be reduced to 88 by the addition of 10 percent ethanol, and assuming

an 8-percent reduction in the gasoline-to-distillate ratio from the ethanol, the refinery energy savings from using ethanol as an octane-boosting additive are about 40,000 to 45,000 Btu/gal of ethanol.[15][16] This corresponds to about 0.3 to 0.4 gal of gasoline equivalent per gallon of ethanol.

The refinery energy savings are nonlinear with the pool octane and the greatest savings occur with the first increment of ethanol used. Consequently, since the supply of ethanol will likely be limited to less than a universal 10-percent blend, 0.4 gal of gasoline equivalent per gallon of ethanol is used in the calculations.

If the energy savings from ethanol represent the major economic incentive to the refiner, then refineries with the highest potential for energy savings would be the most likely to use it and savings would be maximized. Some refineries, however, may have additional incentives for using ethanol, including capital savings and greater gasoline yields from reduced reforming requirements, and access to stronger markets with gasohol. These incentives may not coincide with maximum energy savings. Moreover, the widespread use of technically advanced refining methods could reduce the potential for energy savings through ethanol use. Clearly, there are numerous factors which can lower the actual savings below that which is technically possible for the refineries modeled previously.[17][18] Consequently, although 0.4 gal of gasoline equivalent per gallon of ethanol is used as the refinery energy savings, it should be viewed as a potential savings, which probably will not be achieved in practice for all cases. However, too many assumptions about future refinery operations are required in calculating the energy savings to be able to determine a single, correct value; and the actual savings achieved will be very site specific.

[12]D. K. Lawrence, et al., "Automotive Fuels—Refinery Energy and Economics," Amoco Oil Co., SAE technical series No. 800225, 1980.
[13]A. A. Kozinski, Amoco Oil Co., Naperville, Ill., private communication, 1980.
[14]Bob Tippee, "U.S. Refiners Adjusting to Changing Requirements," Oil and Gas Journal, June 23, 1980.

[15]Based on Lawrence, et al., op. cit.
[16]Based on figure 5 of G. W. Michalski and G. H. Unzelman, "Effective Use of Antiknocks During the 1980's," American Petroleum Institute preprint No. 22-79, from 44th Refinery Midyear Meeting, May 16, 1979.
[17]Lawrence, et al., op. cit.
[18]Michalski and Unzelman, op. cit.

The other possibility is to use the ethanol nationwide to gradually increase the octane of motor fuels. Auto manufacturers could use the increased octane to improve engine efficiencies by increasing the compression ratios in automobile engines. The energy savings per gallon of ethanol would be comparable to that calculated above, but there would be little savings before higher octane fuels were readily available and older automobiles had been replaced with the newer, more efficient engines.

Some stratified-charge engines (e.g., Ford Proco) do not require high-octane fuels and the benefits from a high-octane fuel would be substantially less than for conventional engines. If these or other such engines are used extensively, the octane-boosting properties of ethanol would be of little value, but it is likely that large numbers of automobiles will need high-octane fuels well into the 1990's.

Value of Ethanol in Gasohol

The value of ethanol or the price at which it becomes competitive as a fuel additive can be calculated in several ways. Two alternatives are presented here.

At the oil refinery, each gallon of ethanol used as an octane booster saves the refinery the equivalent of 0.4 gal of gasoline by allowing the production of a lower octane gasoline (see section on octane above). In addition, 1 gal of ethanol will displace about 0.8 gal of gasoline when used as a gasohol blend (i.e., gasohol mileage is assumed to be 2 percent less than gasoline mileage). At the refinery gate, unleaded regular costs about 1.6 times the crude oil price. Assuming that the fuels saved by the octane boost, which are of lower value than gasoline, cost about the same as crude oil, the ethanol is valued at about (gasoline saved) × (gasoline price) + (refinery fuel saved) × fuel price = 0.8 × 1.6 + 0.4 × 1.0 = 1.7 times the crude oil price.*

*This is in agreement with the value of 1.6 to 1.8 times the crude oil price that can be calculated using Bonner and Moore's[19] estimates based on $12.25/bbl crude oil.

[19]A Formula for Estimating Refinery Cost Changes Associated With Motor Fuel Reformation (Houston, Tex.: Bonner and Moore Associates, Inc., Jan. 13, 1978).

If the gasoline retailer blends the gasohol, the value of the ethanol is somewhat different. Gasoline retailers bought regular unleaded gasoline for about $0.70/gal in July 1979 and sold gasohol for a rough average of $0.03/gal more than regular unleaded. (The difference between this and the retail price of gasoline is due to taxes and service station markup, which total about $0.29/gal.) One-tenth gallon of ethanol displaces $0.07 worth of gasoline and the mixture sold for $0.03/gal more. Therefore, 0.1 gal of ethanol was valued at $0.10 or $1.00/gal. This is 2.5 times the July 1979 average crude oil price of $0.40/gal.

Both of these estimates are approximate, and changing price relations between crude oil and gasoline could affect them. Moreover, several other factors can change the estimated value of ethanol. If a special, low-octane, low vapor pressure gasoline is sold for blending with ethanol, at low sales volumes the wholesaler might assign a larger overhead charge per gallon sold. Also, the refinery removes relatively inexpensive gasoline components in order to lower the vapor pressure* of the gasoline, and this increases its cost. On the other hand, in areas where gasohol is popular, the large sales volumes lower service station overhead per gallon thus raising ethanol's value. These factors can change the value of ethanol by as much as $0.40/gal in either direction (i.e., $0.04/gal of gasohol) and the pricing policies of oil refiners and distributors will, to a large extent, determine whether ethanol is economically attractive as an octane-boosting additive.

Ethanol

If pure ethanol is used, carburetors have to be modified to accommodate this fuel. New engines designed for ethanol could have

*The more volatile components of gasoline (e.g., butanes) may be removed to decrease evaporative emissions and reduce the possibility of vapor lock. Although these components can be used as fuel, removing them decreases the quantity of gasoline and the octane boost achieved by the ethanol. Consequently, the advantages of having a less volatile gasoline must be weighed against the resultant decrease in the gasoline quantity and the value of the ethanol. This is a matter of business economics in each refinery and there are no simple rules which would be universally applicable.

higher compression ratios (due to the higher octane of ethanol) and burn leaner fuel mixtures which would improve engine efficiency; and laboratory tests indicate the improvement could be 10 to 20 percent.[20][21] Furthermore, the increased compression ratio provides more power, so engine sizes could also be reduced, thereby increasing the efficiency still further.

In cold climates there can be problems starting and during warmup of engines fueled by pure ethanol, due to its low volatility and high heat of vaporization. Consequently, special equipment will be necessary to enable cold starting in vehicles fueled with straight ethanol. Alternatively, it may be possible to blend small quantities of light hydrocarbons in the alcohol to alleviate the cold-start problem, or one could use a combination of these strategies.

Methanol-Gasoline Blends

In general the effects of adding methanol to gasoline are similar to those for ethanol addition, but more extreme. Thus methanol separates from gasoline at lower moisture levels and damages alcohol-susceptible parts and some paints[22] more quickly or extensively. Therefore, there would be some added cost associated with using metals, plastic and rubber parts, or paints that are tolerant to methanol over using those tolerant to ethanol, but the added cost is probably quite small.

As with ethanol, the change in thermal efficiency for methanol-gasoline blends depends on the original gasoline A/F ratio, but would generally be in the range of zero to 4 percent for a 10 percent methanol blend, leading to an estimated 1- to 5-percent drop in mileage (miles per gallon).

The octane boost that can be achieved with methanol is comparable to that for ethanol, or about 3 octane numbers for a 10-percent blend.[23][24]

For cars adjusted lean on gasoline, drivability problems will occur with a 10 percent methanol blend due to additional leaning. However, at lower percentages of methanol, these problems decrease. Indeed methanol is used as a de-icer at concentrations of about 0.5 percent, with no apparent impairment of drivability.[25]

The principal problems with methanol blends are the large increase in vapor pressure when methanol is added and the poor water tolerance of the blends. The higher vapor pressure can lead to increased evaporative emissions and possibly vapor lock. The decreased water tolerance can lead to fuel separation (into layers), which can lead to poorer drivability in automobiles.

The vapor pressure of the blend can be decreased by reducing the gasoline vapor pressure, but this significantly reduces the volume of gasoline blending stock and can result in less total automotive fuel.[26] Newer cars, however, are fitted with charcoal to trap evaporative emissions from fuel tanks,[27] but these filters may have to be replaced yearly in order to maintain their effectiveness.[28] Evaporative emissions from carburetor boiloff increase with alcohol blends. However, charcoal air filters are being used on some 1980 model vehicles to trap the evaporative emissions from the carburetor and may reduce blend evaporative emissions. Moreover, fuel injection systems have less fuel losses. Vapor lock may also be a problem in some cases,[29] but studies indicate

---

[20]H. Menrad, "Recent Progress in Automotive Alcohol Fuel Application," in *Proceedings of the Fourth International Symposium on Automotive Propulsion Systems,* held on Apr. 18-22, 1977, (NATO Committee on the Challenges of Modern Society, February 1978).

[21]Winfried Berhardt, "Possibilities for Cost-Effective Use of Alcohol Fuels in Otto Engine-Powered Vehicles," in *Proceedings of the International Symposium on Alcohol Fuel Technology, Methanol and Ethanol,* West Germany, Nov. 21-23, 1973, english translation by the Department of Energy.

[22]Keller, Nakagucki, and Ware, op. cit.

[23]Adelman, et al., op. cit.

[24]F. W. Cos, "The Physical Properties of Gasoline/Alcohol Automotive Fuels," in *Proceedings of the Third International Symposium on Alcohol Fuel Technology, vol. II,* Asilomar, Calif. May 28-31, 1979.

[25]Adelman, et al., op. cit.

[26]Keller, Nakagucki, and Ware, op. cit.

[27]Ibid.

[28]K. R. Stamper, Bartlesville Energy Technology Center, Department of Energy, Bartlesville, Okla., 1979.

[29]Keller, Nakagucki, and Ware, op. cit.

that proper vehicle design can also eliminate this problem.[30]

The water tolerance problem may require some sort of cosolvent, or additive, which helps to retain methanol in gasoline.* One such cosolvent, t-butanol (another alcohol), is currently being test marketed in t-butanol-methanol-gasoline blends by Sun Oil.[32] The energy cost of producing the t-butanol, however, is not known. Alternatively, hexanol (still another alcohol) has been used successfully to recombine the phases in a separated methanol-gasoline blend.[33]

Each of the problems with methanol blends has numerous solutions, but it is unclear at present which will be the most effective at increasing motor fuel supplies at the least cost to consumers. Additional work is needed to clarify this matter.

Methanol

In order to use methanol, carburetors suitable for methanol have to be installed on the engine or old ones modified. New engines designed for methanol could have higher compression ratios and burn leaner fuel mixtures leading to a potential 20-percent improvement in thermal efficiency.[34][35] As with ethanol, slightly smaller engines could be used because of the greater power associated with the higher compression ratio, which could provide still greater efficiency improvements.

Another possible approach with methanol is to decompose the alcohol into carbon monoxide (CO) and hydrogen in the carburetor. This gas mixture is then used to fuel the engine. Exhaust heat from the engine is used to fuel the decomposition; and the CO-hydrogen mixture contains 20 percent more energy than the methanol from which it came. This offers the possibility of improving the engines thermal efficiency by 20 percent, but it is too soon to know whether this potential increase can be achieved in practice.

Methanol can cause gasoline fuel injection pumps to fail, due to its low lubricity.[36] Other tests indicate that methanol combustion corrodes cast iron piston rings and may affect normal lubricating oils, particularly in very cold weather starting.[37] However, in actual engine and vehicle tests in warm weather conditions, methanol has not been found to cause premature engine wear.[38]

Below about 40° F, methanol-fueled engines can experience starting problems, due to the same factors that affect ethanol-fueled vehicles. As with ethanol, special equipment and/or blending of volatile hydrocarbons in the fuel will be needed to enable cold starting.

Summary

Ethanol-gasoline blends are currently being marked commercially, and methanol blends (with a cosolvent) are being test marketed. In addition, automobiles fueled with straight ethanol are being used in Brazil and extensive tests with methanol-fueled vehicles are underway. Nevertheless, because the alcohols are not fully compatible with the existing liquid fuels delivery system and automobile fleet, some initial difficulties in using alcohol fuels

[30]A. W. Crowley, et al., "Methanol-Gasoline Blends Performance in Laboratory Tests and Vehicles," Inter Industry Emission Control Program-2, Progress Report No. 1 (Society of Automotive Engineers, 1975).
*Nevertheless, one test of automobiles operating on a phase-separated blend showed fewer drivability problems than would have been expected.[31]
[31]Stamper, "50,000 Mile Methanol/Gasoline Blend Fleet Study," op. cit.
[32]B. C. Davis and W. H. Douthut, "The Use of Alcohol Mixtures as Gasoline Additives," Suntech, Inc., Marcus Hook, Pa., presented at 1980 National Petroleum Refiners Association Annual Meeting, March 1980.
[33]Stamper, private communication, op. cit.
[34]W. J. Most and J. P. Longwell, "Single Cylinder Engine Evaluation of Methanol-Improved Energy Economy and Reduced NO$_x$," SAE paper No. 750119, February 1975.
[35]Pefley, et al., op. cit.

[36]J. C. Ingamells and R. H. Lindquist, "Methanol as a Motor Fuel or a Gasoline Blending Component," SAE paper No. 750123, Automotive Engineering Congress and Exposition, Detroit, Mich., February 1975.
[37]E. C. Owens, "Methanol Effects on Lubrication and Engine Wear," in Proceedings of the International Symposium on Alcohol Fuel Technology, Methanol and Ethanol, Wolfsburg, West Germany, Nov. 21-23, 1977.
[38]Pefley, et al., op. cit.

are to be expected. These problems should disappear with time, however, as more experience is gained at handling and using the alcohol fuels and as older automobiles are replaced with vehicles designed for use with these fuels.

For ethanol the preferred use probably is as an octane-boosting additive to gasoline be- cause of the energy saved by allowing refiners to produce a lower octane gasoline. The situation with methanol is less clear because of the greater difficulties associated with methanol blends and the possible need for cosolvents. The use of methanol both in blends and as a straight fuel is currently being pursued.

Diesel Engines

Alcohols have only limited solubility in diesel fuel, making diesel-alcohol blends impractical at present.* If "ignition accelerators"** are dissolved in the alcohols to enable them to ignite in diesel engines, they can be used as a replacement for diesel fuel, but the fuel metering system would have to be modified to provide the full range of power for which the engine was designed and some provisions made for the decreased lubricity of the alcohols.

Alternatively, the alcohols can be used in dual fuel systems, i.e., where the alcohol and diesel fuel are kept in separate fuel tanks and separate fuel metering systems are used. The two main possibilities are fumigation and dual injection. In a fumigation system, the alcohol is passed through a carburetor or injected into the air intake stream and the alcohol-air mixture replaces the intake air. In dual injection, each fuel is injected separately into the combustion chamber.

Most diesel engines are speed governed, i.e., more or less diesel fuel is injected automatically to maintain a constant engine speed for a given accelerator setting. If alcohol is fumigated into the cylinder, the diesel injection decreases automatically when the engine is not at full power to compensate for the additional power from the alcohol. At full power, the alcohol will give the engine additional power. Consequently, once a fumigation system is installed, alcohol usage is optional, since the engine will run normally without the alcohol present, but the higher power at full power can cause additional engine wear if the engine is not designed for this power. Alternatively, the diesel injection can be modified to allow less fuel to be injected, but it would have to be returned to its original state when alcohol is not being used. Dual injection systems also can be designed to run with or without alcohol, but the injection controls would probably be more complicated.

If the fuel systems are separate, alcohol containing up to 20 percent water probably can be used. However, the diesel engines must be modified to accommodate the alcohols. The modifications for alcohol fumigation are relatively simple and can be performed for an estimated $150 if, for example, a farmer does it himself and uses mostly spare parts.[39] Costs could range up to $500 to $1,500 if installed by a mechanic using stainless steel fuel tanks and all new parts.[40] Cost estimates for modifying engines for dual injection are not available, however, but would be more expensive. In either case the long-term effects, such as possibly increased engine wear, are unknown at the present time.

Fumigation systems will generally be limited to 30 to 45 percent ethanol or about 20 percent methanol, because evaporation of the alcohol cools the combustion air and the cooling from higher concentrations is sufficient to prevent the diesel fuel from igniting. However, consid-

---

*Emulsions are possible, but still in the R&D phase.
**Although alkyl nitrates have generally been used as ignition accelerators, sunflower seed oil and other vegetable oils have been suggested for biomass-derived ethanol.

[39]Pefley, et al., op. cit.
[40]Ibid.

erably higher proportions of alcohol can be used with dual injection systems.[41]

In fumigation systems, some tests have shown thermal efficiency increases of up to 30 percent for certain combinations of alcohol and diesel fuel.[42] Other tests[43] showed slight increases (5 percent) in thermal efficiency when the engine is at two-thirds to full load, while there are large decreases (25 percent) in efficiency at one-third full load. Similar am-

biguities exist for dual injection systems.[44] These differences in efficiency are due to differences in engine tuning and design. An accurate determination is not available at present, but it is unlikely that there will be significant differences in the thermal efficiencies of engines optimized for the respective fuels.

Considerable uncertainty exists about the thermal efficiencies that can be obtained in practice if, for example, tractors are converted to alcohol use. Assuming, however, that the thermal efficiency does not change, 1 gal of ethanol would replace 0.61 gal of diesel fuel, and 1 gal of methanol would replace 0.45 gal of diesel fuel.

[41] F. F. Pischinger and C. Havenith, "A New Way of Direct Injection of Methanol in a Diesel Engine," in *Proceedings of the Third International Symposium on Alcohol Fuels Technology,* vol. II, Asilomar, Calif., May 1979.

[42] K. Bro and P. S. Pederson, SAE paper No. 770794, September 1977.

[43] K. D. Barnes, D. B. Kittleson, and T. E. Murphy, SAE paper No. 750469, Automotive Engineering Congress and Exposition, Society of Automotive Engineers, Detroit, Mich., February 1975.

[44] E. Holmer, "Methanol as a Substitute Fuel in the Diesel Engine," in *Proceedings of the International Symposium on Alcohol Fuel Technology, Methanol and Ethanol,* West Germany, November 1977.

Environmental Impacts of Automotive Use of Biomass Fuels

The use of alcohol fuels and gasoline-alcohol blends in automobiles will have a number of environmental impacts associated with changes in automotive emissions as well as differences in the toxicity and handling characteristics of the fuel alternatives. The potential changes in automotive emissions have been identified as the impact of major concern and are treated in the greatest detail in this discussion.

Air Pollution—Spark Ignition Engines

Predictions of emissions changes can be based on a combination of theoretical considerations, laboratory tests, and field measurements. Unfortunately, the results of the emissions tests that have been completed to date are varied and confusing. Difficulties with using these results for predicting emissions changes include:

- Tests are rarely comparable because of different base fuels (gasolines), fuel mixtures, automobiles, state of "tune," driving cycle, etc.

- In some important tests, methodological problems may seriously weaken the derived conclusions. For example, the Environmental Protection Agency's (EPA) 1978 tests of "gasohol" (10 percent ethanol blend) included some vehicles that either operated too "fuel-rich" initially (four vehicles) or exceeded the nitrogen oxide (NO_x) standard on indolene (two vehicles).[45] If these noncompliance vehicles are dropped from the test sample, the changes caused by using gasohol are less than test-to-test variability in exhaust emissions for the same vehicle.[46]

- Test results have generally been obtained from laboratory engines or, in testing alcohol blends, from relatively unmodified automobile engines. A strategy that provided reliable and plentiful supplies of alcohol fuels would presumably be ac-

[45] *Characterization Report: Analyses of Gasohol Fleet Data to Characterize the Impact of Gasohol on Tailpipe and Evaporative Emissions* (Washington, D.C.: Technical Support Branch, Mobile Source Enforcement Division, Environmental Protection Agency, December 1978).

[46] Wiplore K. Juneja, et al., "A Treatise on Exhaust Emission Test Variability," *Society of Automotive Engineers,* vol. 86, paper 770136, 1977.

companied by design changes that would take advantage of the different properties of these fuels. Thus, extrapolations from current test data may be overly pessimistic, at least for the long run.

Aside from test results, emission changes can be explained in great part by the dependence of emissions on the operating conditions of the engine. Emissions of CO, HC, NO_x, and aldehydes are strongly influenced by the "equivalence ratio Φ" (stoichiometric A/F ratio/actual A/F ratio) and the emission control system (none, oxidation catalyst, etc.). Figure 37 shows how CO, HC, NO_x, and aldehydes are likely to vary with Φ.

Both methanol (6.4:1) and ethanol (9:1) have lower stoichiometric A/F ratios than gasoline (14.5:1). Thus, blends of either alcohol fuel result in lower equivalence ratios ("leaner" operation) if no changes are made in the fuel metering devices. Examining figure 37, emissions changes can be predicted qualitatively by observing that adding alcohol pushes the equivalence ratio to the left. For an automobile normally operating "lean," CO may be expected to remain about the same, HC remain the same or increase slightly, and NO_x decrease.

For out-of-tune automobiles, which usually operate in a "fuel rich" mode, CO and HC may be expected to decrease while NO_x increases. Vehicles equipped with three-way catalysts have feedback-controlled systems that operate to maintain a predetermined value of Φ and thus should be less affected by the blend leaning effect of the alcohol fuels. However, this feedback system is usually overridden during cold starting to deliver a fuel-rich mixture; during this time period, HC and CO are more likely to decrease and NO_x to increase with alcohol fuels. Also, catalysts with oxygen sensors can be fooled into adjusting to leaner operation because the exhaust emissions from alcohol blends oxidize faster than gasoline-based exhausts and drive down the oxygen level in the exhaust stream (giving the appearance of overly fuel-rich operation);[47] this should also tend to decrease HC and CO and increase NO_x emissions when alcohol blends are used in vehicles equipped with such catalysts.

Table 64 provides a summary of the type of emissions changes that may be expected by combining knowledge of test results and the

*[47]Gratch, op. cit.

Table 64.—Emission Changes (compared to gasoline) From Use of Alcohol Fuels and Blends

Pollutant/fuel	Methanol	Methanol/gasoline	Ethanol	Ethanol/gasoline "gasohol"
Hydrocarbon or unburned fuels	About the same or slightly higher, but much less photochemically reactive, and virtual elimination of PNAs; can be catalytically controlled	May go up or down in unmodified vehicles, unchanged when Φ remains constant. Composition changes, tho, and PNAs go down. Can be controlled. Higher evaporative emissions	Not very much data, should be about the same or higher but less reactive. Expected reduction in PNA	May go up or down in unmodified vehicles, about the same when Φ remains constant; composition may change, expected reduction in PNAs. Evaporative emissions up
Carbon monoxide	About the same, slightly less for rich mixtures; can be catalytically controlled; primarily a function of Φ	Essentially unchanged if Φ remains constant, lower if leaning is allowed to occur	About the same, can be controlled; primarily a function of Φ	Decrease in unmodified vehicles (i.e., leaning occurs), about the same when Φ remains constant
Nitrogen oxides	1/3 to 2/3 less at same A/F ratio, can be lowered further by going very lean; can be controlled	Mixed; decreases when Φ is held constant, but may increase from fuel "leaning" effect in unmodified vehicles	Lower, but not as low as with methanol; can be controlled	Slight effect, small decrease when Φ is held constant, but may increase or decrease further from fuel "leaning" effect in unmodified vehicles
Oxygenated compounds	Much higher aldehydes, particularly significant with precatalyst vehicles	Aldehydes increase somewhat, most significant in precatalyst vehicles	Much higher aldehydes, particularly significant with precatalyst vehicles	Aldehydes increase, most significant in precatalyst vehicles
Particulates	Virtually none	Little data	Expected to be near zero	Little data, no significant change expected
Other	No sulfur compounds, no HCN or ammonia	Unknown	No sulfur compounds	No data

SOURCE: Office of Technology Assessment.

theoretical model discussed above. The most significant changes, and their environmental implications, are:

- **Substantial reductions in reactive HC and NO$_x$ exhaust emissions with 100 percent (neat) methanol and, to a lesser extent, ethanol.** — Although HC emissions are expected to remain approximately the same with alcohol fuel use at the same Φ,[48][49] the reactivity of these emissions is much lower than that of gasoline-based HC emissions. Reductions in reactive HC and NO$_x$ should reduce photochemical smog formation, although predictions of the magnitude of these effects are difficult.

- **Increase in aldehyde emissions with neat alcohols and blends.** — Use of pure alcohol fuels yields several-fold increases in aldehydes,[50][51] whereas blends increase aldehyde emissions to a lesser extent. Because catalytic converters are effective in removing aldehydes, catalyst-equipped vehicles tend to have low aldehyde emissions whether or not alcohol is used;[52][53] the major problem lies with cars not equipped with catalysts.

  Aldehydes cause eye and respiratory irritations and are photochemically reactive. Despite this, aldehydes are not specifically regulated in automobiles, and

the most abundant aldehyde in automotive emissions—formaldehyde—is not detectable with conventional HC measuring instrumentation. Aldehyde increases may somewhat negate the positive effects of reductions in emissions of HC as well as in NO$_x$ emissions from alcohol use. The magnitude of the potential impacts, however, is not well understood.

- **Substantial reductions in particulate emissions if neat alcohol fuels are used.** — Use of neat alcohol fuels may reduce particulate emissions virtually to zero. This is particularly significant when the fuel substituted for is leaded gasoline; particulate emissions from autos using leaded gasoline are on the order of 0.6 g/mile on the Federal emission test cycle,[54] and most of the particles are toxic (mostly lead by weight, with polynuclear aromatic (PNA) compounds adsorbed on their surfaces) and in the inhalable size range (whereas particulate emissions from autos using unleaded gasoline are on the order of 0.2 g/mile on the same test cycle, are about 90-percent controllable with catalytic converters, and are composed mainly of carbon particles.[55]

- **Substantial reductions in PNA compounds with neat alcohols and blends.** — PNA compounds emitted in small quantities in automobile exhausts are toxic and carcinogenic.[56] Methanol and methanol blends appear to provide substantial reductions in these emissions (methanol exhaust contains only about 1 percent of the PNA compounds observed in gasoline exhaust),[57] which may be of some significance in reducing the cancer hazards of urban air pollution. Ethanol and ethanol blends may be expected to provide similar effects, but this has not yet been verified.

[48]David L. Hilden and Fred B. Parks, "A Single Cylinder Engine Study of Methanol Fuel—Emphasis on Organic Emissions," Society of Automotive Engineers paper No. 760378, presented at the Automotive Engineering Congress, Dearborn, Mich., Feb. 23-27, 1976.

[49]Samuel O. Lowry and R. S. Devoto, "Exhaust Emissions From a Single-Cylinder Engine Fueled With Gasoline, Methanol, and Ethanol," Combustion Science and Technology, vol. 12, Nos. 4, 5, and 6, 1976, pp. 177-82.

[50]W. Lee and W. Geffers, "Engine Performance and Exhaust Emission Characteristics of Spark Ignited Engines Burning Methanol and Methanol-Gasoline Mixtures," Volkswagon Research and Development Division, Wolfsburg, West Germany, presented at AICLE meeting, Boston, Mass., September 1975.

[51]Comparative Automotive Engine Operation When Fueled With Ethanol and Methanol (Washington, D.C.: Alcohol Fuels Program, Alternative Fuels Utilization Program, Department of Energy, May 1978).

[52]J. R. Allsup, "Experimental Results Using Methanol and Methanol/Gasoline Blends as Automotive Engine Fuel," Bartlesville Energy Technology Center, No. B9RC/R1-76/15, January 1977.

[53]J. R. Allsup and D. B. Eccleston, "Ethanol/Gasoline Blends as Automotive Fuels," Bartlesville Energy Technology Center, draft No. 4.

[54]R. E. Sampson and G. S. Springer, "Effects of Exhaust Gas Temperature and Fuel Composition on Particulate Emission From Spark Ignited Engines," Environmental Science and Technology, vol. 7, No. 1, January 1973.

[55]Ibid.

[56]American Petroleum Institute, "API Toxicological Review: Gasoline," 1967.

[57]On the Trail of New Fuels—Alternative Fuels for Motor Vehicles (Bonn, West Germany: Federal Ministry for Research and Technology, 1974), translated by Addis Translation International.

Air Pollution—Diesel Engines

Very little data is available to allow the prediction of emission changes from the use of alcohol fuels and blends in diesel engines. Predictions of some limited reliability may be made from the small number of tests, extrapolation from spark ignition tests, and knowledge of diesel characteristics.

The major environmental reason why alcohol fuels appear to be attractive for diesels is their ability to burn without producing particulate emissions. Domestic manufacturers are having problems meeting the proposed EPA particulate standard of 0.6 g/mile. Particulate emissions from diesel engines are 50 to 100 times those from gasoline engines[58] and may contain more PNA; particulate reductions thus appear to be especially attractive environmentally.

HC emissions from diesels are more photochemically reactive than automobile HC emissions. Although a switch to alcohol fuels by itself will have an uncertain effect on uncontrolled emissions, the elimination of particulate emissions may allow the use of oxidation catalysts to improve HC control (because particulates otherwise would plug up the catalyst).[59]

If alcohol fuels behave in diesels in a manner similar to their behavior in spark ignition engines, they should cause NO_x emissions to decrease and aldehydes to increase. CO levels have been observed in tests to double their originally low values when shifting from diesel fuel to alcohol;[60] however, this is thought to be a correctable problem with the fuel injection systems.

Alcohol fuels have poor ignition capabilities when injected into the compressed and heated air in a diesel engine. To counteract this difficulty, ignition accelerating agents containing nitrates can be added to the fuel to provide an ignition source for the alcohol. There appears to be some potential for the formation of hydrogen cyanide or ammonia in the combustion process when these additives are used. Laboratory testing will be necessary to verify the existence of this effect.

Emission characteristics of mixed fuel operation with alcohol and diesel fuels depend on the method of introducing the alcohol into the combustion chamber.

When the alcohol is mixed with the intake air (fumigation), the following changes have been observed to occur:[61]

- increase in HC emissions and aldehydes,
- little change in CO,
- increase or decrease in NO_x, and
- decrease in particulates.

The emissions effects of other fuel systems are poorly understood.

Occupational Exposures to Fuels and Emissions

In general, the effects of gasoline and gasoline-based emissions are more acute, in an occupational setting, than those of methanol and ethanol. For example:

- Short-term exposure to gasoline is considered more poisonous, tissue disruptive, and irritative than methanol when effects of eye contact, inhalation, skin penetration, skin irritation, or ingestion are considered.[62] Effects of the more severe (ingestion) exposures to methanol are generally reversible, although in some extreme cases there can be irreversible effects on the central nervous system, optic nerve end, and heart.[63]

[58]K. J. Springer and T. M. Baines, "Emissions From Diesel Versions of Production Passenger Cars," *Society of Automotive Engineers Tranlations,* vol. 86, sec. 4, paper No. 770818, 1977.

[59]M. Amano, et al., "Approaches to Low Emission Levels for Light-Duty Diesel Vehicles," Society of Automotive Engineers paper No. 760211, February 1976.

[60]W. F. Marshall, *Experiments With Novel Fuels for Diesel Engines* (Bartlesville, Okla.: Bartlesville Energy Technology Center, Department of Energy, February 1978), BERCITPR-7718.

[61]B. S. Murthy and L. G. Pless, "Effectiveness of Fuel Cetane Number for Combustion Control in Bi-Fuel Diesel Engine," *Journal of the Institution of Engineers (India),* vol. 45, No. 7, pt. ME 4, March 1965.

[62]N. V. Steer, ed., *Handbook of Laboratory Safety* (Cleveland, Ohio: Chemical Rubber Co., 1971).

[63]M. N. Gleason, et al., *Clinical Toxicology of Commercial Poisons,* Williams and Williams.

- The effects of both acute and chronic exposure to ethanol are considered to be much less disruptive than methanol and, therefore, gasoline.
- The automotive exhaust emissions that are most dangerous in an enclosed space — such as a garage without adequate ventilation — are CO from gasoline and CO and formaldehyde from methanol. If a fleet of methanol-powered cars is compared directly to a fleet of gasoline-powered cars, CO will be the most dangerous pollutant (and equally dangerous for both fleets, because methanol should not substantially change CO emissions)—so long as three-way catalysts are used. Without catalysts, formaldehyde emissions could be more dangerous than CO in the methanol-powered fleet.

It is interesting to observe that, for the catalyst-equipped methanol fleet, formaldehyde will act as a "tracer" for CO; if eye and respiratory irritation from formaldehyde becomes acute, this will be an almost sure sign that CO is at dangerous levels.

Safety Hazards

The risks of fire and explosion appear to be lower with alcohol fuels than with gasoline, although evidence is mixed:

- gasoline has a lower flash point and ignition temperature and is more flammable and explosive in open air than either ethanol or methanol,[64]
- alcohols are the greater hazard in closed areas,[65]
- higher electrical conductivity of alcohol lessens danger of spark ignition,
- high solubility in water makes alcohol fires easier to fight than gasoline fires, and
- alcohol fires are virtually invisible, adding to their danger (but addition of trace materials could overcome this drawback).

Alcohol blends will be similar to gasoline but they may be more ignitable in open spaces and less ignitable in closed containers when the blends have higher evaporation rates than the pure gasoline. Diesel fuels and diesel alcohol emulsions are considered to be safer than gasoline or alcohol fuels.[66]

Environmental Effects of Spills

To the extent that domestic alcohol production substitutes for significant quantities of imported oil, a reduction in fuel transportation and a consequent reduction in spills can be expected. If alcohol is shipped by coastal tanker, the possibility of large alcohol spills is a realistic one, and the effects of such spills should be compared to the effects of oilspills.

Alcohol fuels appear to be less toxic than oil in the initial acute phase of the spill and have fewer long-term effects. Except in areas where alcohol concentrations reach or exceed 1 percent, the immediate effects of a spill should be minimal. For example, a concentration of about 1 percent methanol in seawater is tolerated by many common components of intertidal, mud-flat, and estuarine ecosystems unless the alcohol is contaminated with heavy metals.[67] In contrast, crude oil contains several highly toxic water soluble components that can be damaging at low concentrations. Furthermore, alcohols are extremely biodegradable—toxic effects may be eliminated in hours—whereas the effects of heavy fuel oils can last for years.

Hazards to the Public

The widespread distribution and use of alcohol fuels will result in the public facing the same potential dangers as exist in the occupational environment. A true assessment of ethanol and methanol public health risks must incorporate an analysis of probable exposure, however, and such an analysis is likely to show that both alcohol fuels may have considerably

[64]CRC Handbook of Laboratory Safety, op. cit.
[65]Ibid.
[66]M. E. LePera, "Fine Safe Diesel Emulsions,"Conference on Transportation Synfuels, sponsored by the Department of Energy, San Antonio, Tex., November 1978.
[67]P. N. D'Eliscu, "Biological Effects of Methanol Spills Into Marine, Estuarine, and Freshwater Habitats," in Proceedings of the International Symposium on Alcohol Fuel Technology, Methanol and Ethanol, Wolfsburg, West Germany, Nov. 21-23, 1977.

greater potential than gasoline to harm the public. For example, although methanol and gasoline are comparably toxic upon ingestion, methanol has a long history of improper ingestion and gasoline does not. Ethanol is even more likely to be improperly used, and the ethanol used for motor fuel blending will be contaminated with dangerous toxic chemicals. Although vile tasting and smelling denaturants may be added to fuel ethanol to discourage improper use, enterprising individuals are likely to try to filter out these additives. Also, fuel ethanol may be diverted to consumption be-

fore these denaturants are added. The probability of such diversion will be especially high if small, onfarm ethanol stills are widely used.

A careful risk assessment of ethanol and methanol fuels and blends could identify and quantify these types of risks and would be invaluable both in setting priorities for research and in devising risk mitigation strategies that must accompany promotion of alcohol fuel use. Such an assessment has not as yet been conducted by DOE.

Gas Turbines

Alcohol fuels can be used readily in gas turbine generators used to generate peakload electric power. The fuel metering system has to be modified to meter the larger volumes of alcohols necessary to maintain the same power output and to accommodate the lower lubricity of alcohols relative to light fuel oil. These modifications are minor. In some cases, however, the alcohols may attack the turbine blades or other metal parts and the modifications needed to use alcohol fuels would be considerably more expensive.

If alcohol fuels are used, care must be taken to ensure that no salts are dissolved in the alcohols by, for example, contamination with seawater during barge transport. The salts could greatly reduce the life of the turbines.

The thermal efficiency of a gas turbine is determined by the ratio of the pressures at the turbine inlet and outlet. This ratio is limited by the combustion temperature. The alcohols have slightly lower combustion temperatures and should allow higher efficiencies than with light fuel oil in redesigned turbines. In unmodified turbines, the thermal efficiency of the alcohol fuels is about the same (within ± 2 percent) as for light fuel oils.[68] [69] Thus, 1 gal of

ethanol would replace 0.67 gal of light oil and 1 gal of methanol would replace 0.48 gal of light fuel oil.

Currently about 0.25 Quad/yr of oil and 0.2 Quad/yr of natural gas are consumed for peakload electric generation.[70] This represents about 6 percent of the electricity generated in the United States. Use of alcohol fuels here would save about 0.2 trillion ft³ of natural gas per year and about 130,000 bbl/d of light distillate oil.

Air Pollution Effects of Alcohol Fuel Use in Gas Turbines

Although alcohol fuels have been tested in gas turbines and the resulting emissions levels have been measured, there is some doubt as to whether those levels represent true indicators of emissions to be expected from an optimized system. For example, methanol use in an automotive gas turbine produced a tenfold increase in HC emissions in one test,[71] but it is quite possible that more optimal design of the fuel injection nozzles could lower these values considerably.

[68]L. W. Huellmantel, S. G. Teddle, and D. C. Hammond, Jr., "Combustion of Methanol in an Automotive Gas Turbine," in *Future Automotive Fuels*, J. M. Colucci and N. E. Gallapoulous, ed. (New York: Plenum Press, 1977).

[69]P. M. Jarvis, "Methanol as Gas Turbine Fuel," presented at the 1974 Engineering Foundation Conference, *Methanol as an Alternate Fuel*, Henneker, N.H., July 1974.

[70]Adelman, et al., op. cit.

[71]C. W. Lapointe and W. L. Schultz, "Comparison of Emission Indices Within a Turbine Combustor Operated on Diesel Fuel or Methanol," Society of Automotive Engineers paper No. 730669, June 1973.

The most significant emission change should be a substantial drop in NO_x emissions, which are typically quite high in gas turbines. Methanol has achieved 76-percent reductions in NO_x emissions in large turbines because it has a significantly lower combustion temperature than distillate fuels.[72] Ethanol, which has a combustion temperature intermediate between methanol and distillates, should achieve somewhat smaller reductions.

_____
[72]Jarvis, op. cit.

Chapter 11
ENERGY BALANCES FOR
ALCOHOL FUELS

Chapter 11.—ENERGY BALANCES FOR ALCOHOL FUELS

ENERGY BALANCES FOR ALCOHOL FUELS

The energy objective of using alcohol fuels from biomass is the displacement of foreign oil and gas with domestic synthetic fuels. The effectiveness of a fuel alcohol program depends on the energy consumed in growing and harvesting the feedstock and converting it into alcohol, the type of fuel used in the conversion process, and the use of the alcohol.

The major sources of biomass alcohol fuels are grains, sugar crops, wood, grasses, and crop residues. Ethanol from grains and sugar crops is considered first, including a comparison of various feedstocks and end uses. Methanol and ethanol from the other feedstocks are then considered, and the use of these feedstocks directly as fuels is compared with the production of alcohols from them. Finally, some general considerations about the energy balance of these fuels are given.

Ethanol From Grains and Sugar Crops

Corn is currently a principal feedstock for ethanol production, but other grains and sugar crops could also be used. The energy balance for gasohol from corn is discussed in detail below, followed by a summary of the energy balance for various possible feedstocks and for use of the ethanol either as an octane booster or as a standalone fuel.

For each gallon of ethanol derived from corn, farming and grain drying consume, on the average, the energy equivalent of 0.29 gal of gasoline* in the form of oil (for fuel and petrochemicals) and natural gas (for nitrogen fertilizers). (See ch. 3 in pt. I.) The exact amount will vary with farming practices (e.g., irrigation) and yields. In general, however, the farming energy input per gallon of ethanol produced will increase when the farmland is of poorer quality (e.g., setaside acreage) and/or in dryer or colder climates (i.e., most of the western half of the country, excluding Hawaii).

The type of fuel used in the distillation process is perhaps the most important factor in determining the displacement potential of etha-

nol. Even under the most favorable circumstances, distillery energy consumption is significant. The distillery producing most of the fuel ethanol used today reportedly consumes 0.25 gal of gasoline equivalent (0.24 in the form of natural gas) per gallon of ethanol produced.[1] This number, however, involves some arbitrary decisions about what energy inputs should be attributed to the facility's food-processing operations. Total processing energy inputs in this plant amount to about 0.55 gal of gasoline equivalent per gallon of ethanol (see ch. 7).

Energy-efficient standalone fuel ethanol distilleries would consume the equivalent of about 0.45 gal of gasoline per gallon of ethanol produced (see ch. 7). Because the energy consumption of distilleries is not likely to be insignificant in relation to the alcohol produced in the foreseeable future, it is essential that distilleries use abundant or renewable domestic energy sources such as coal, biomass, and/or solar heat or obtain their heat from sources that would otherwise be wasted. Reliance on these fuels would reduce the total use of oil and gas at the distillery to insignificant levels.

---

*Some authors have included the energy used to manufacture farming equipment and the materials from which they are made as part of the farm energy inputs. However, for consistency one should also include, as a credit, the energy used in manufacturing the goods that would have been exported to pay for importing the oil displaced by the ethanol. Because of the uncertainty in these factors, and the fact that they are relatively small, they are not included in the energy balance calculations.

---

[1]Archer Daniels Midland Co., Decatur, Ill., "Update of Domestic Crude Oil Entitlements, Application for Petroleum Substitutes," ERA-03, submitted to the Department of Energy, May 17, 1979.

The amount of petroleum displaced by ethanol fuel also depends on the manner in which it is used. As a standalone fuel, each gallon of ethanol displaces about 0.65 gal of gasoline equivalent. As an additive in gasohol, each gallon of ethanol displaces about 0.8 ($\pm$0.2) gal of gasoline.* (See ch. 10.) If the oil refinery produces a lower grade of gasoline to take advantage of the octane-boosting properties of ethanol, up to 0.4 gal of gasoline energy equivalent can be saved in refinery processing energy (see ch. 10) for each gallon of ethanol used.

Additional energy savings are achieved by using the byproduct distillers' grain as an animal feed. To the extent that crop production is displaced by this animal feed substitute, the energy required to grow the feed crop is displaced.

Table 65 summarizes the oil and natural gas used and displaced for the entire gasohol fuel cycle. The energy is expressed as gallons of gasoline energy equivalent for each gallon of ethanol produced and used in gasohol (i.e., 1.0 in the table represents 117,000 Btu/gal of ethanol, 0.5 represents 58,500 Btu/gal of ethanol, etc.) The three cases presented correspond to: 1) two ways to calculate the present situation, 2) future production of ethanol from the less productive land that can be brought into crop production and using coal as a distillery fuel,

*The greater displacement results from the alcohol's leaning effect.

and 3) the same as (2) except that the octane of the gasoline is lowered to exactly compensate for ethanol's octane-boosting properties. They result in net displacements of: 1) from zero to one-third gal, 2) about one-half gal, and 3) slightly less than 1 gal of gasoline and natural gas equivalent per gallon of ethanol used.

In all, the total displacement of premium fuels (oil and natural gas) achieved per gallon of ethanol can be nearly 1 gal of gasoline equivalent per gallon of ethanol if petroleum and natural gas are not used to fuel ethanol distilleries and 2) lower octane gasoline is used in gasohol blends. Failure to take these steps, however, can result in the fuel cycle consuming slightly more oil and natural gas than it displaces leading to a net increase in oil and gas consumption with ethanol production and use. This is the situation that is alluded to in most debates over gasohol's energy balance, but it is a situation that can be avoided with appropriate legislation.

Nevertheless in the most favorable case (case 3) and with an energy-efficient distillery, the ratio of total energy displaced to total energy consumed is 1.5 ($\pm$0.4), i.e., the energy balance is positive (a ratio greater than 1). And if the feedstocks are derived from more productive farmland, or local conditions allow energy savings at the distillery, e.g., not having to dry the distillers' grain, then the balance is even more favorable. Alternatively, an energy

Table 65.—Energy Balance of Gasohol From Corn: Oil and Natural Gas Used (+) and Displaced (–)
(in gallons of gasoline equivalent per gallon of ethanol produced and useda)

| | Present | | Set-aside and potential cropland | | |
| | | | Coal-fired distillery | Coal-fired distillery and lowering of gasoline octane | Uncertainty |
	Entire plant	Ethanol only			
Farming	0.3b	0.3b	0.4c	0.4c	±0.15
Distillery	0.55	0.24	0e	0e	—
Distillery byproduct	−0.09d	0	−0.09d	−0.09d	±0.03
Automobile	−0.8	−0.8	−0.8	−0.8	±0.2
Oil refinery	—	—	—	−0.4	±0.2
Total	−0.0	−0.3	−0.5	−0.9	±0.3

aLower heat content of gasoline and ethanol taken to be 117,000 Btu/gal and 76,000 Btu/gal, respectively.
b0.16 as nitrogen fertilizer (from natural gas) and 0.13 mostly as petroleum products.
cEstimated uncertainty of ±0.15; assumes 75% of the yield achievable on average cropland.
dBased on soybean cultivation and crushing energy. The byproduct of 1 gal of ethanol from corn displaces 12 lb of crushed soybeans, which requires 0.09 gal of gasoline equivalent to produce. Private communication with R. Thomas, Van Arsdall, National Council of Farmer Cooperatives.
e55,000 Btu of coal per gallon of ethanol.

SOURCE: Office of Technology Assessment.

credit could be taken for the crop residues, which would also improve the calculated balance. This general approach to the energy balance, however, does not consider the different values of liquid versus solid fuels.

The uncertainty factor in table 65 of ±0.3 gal of gasoline per gallon of ethanol is due primarily to inherent differences in farming practices and yields, errors in fuel efficiency measurements, uncertainties in oil refinery savings, and the magnifying effect on these errors of the low (10 percent) ethanol content of gasohol. These factors make more precise estimates unlikely in the near term.

Not only does the farming energy used for grain or sugar crop production vary considerably from State to State, but also the average energy usage displays some differences between the various feedstocks. A more significant difference arises, however, between use of the ethanol as an octane-boosting additive to gasoline and as a standalone fuel, e.g., in diesel tractors or for grain drying. As an octane-boosting additive, each gallon of ethanol displaces up to 1.2 gal of gasoline energy equivalent in the automobile and at the refinery (see table 65).* As a standalone fuel, however, the displacement at the end use is only 0.65 to 0.8 gal of gasoline energy equivalent per gallon of ethanol.**

Table 66 summarizes the net displacement of premium fuels (oil and natural gas) for various feedstocks and the two end uses. In each case it was assumed that the feedstocks would be grown on marginal cropland with yields that are 75 percent of those obtained on average U.S. cropland.

The striking feature displayed in table 66 is that use of ethanol as a standalone fuel is considerably less efficient in displacing premium fuels than use of it as an octane-boosting additive. In some cases, e.g., with grain sorghum and in areas with poor yields of the other

*0.4 gal of gasoline equivalent is due to the octane-boosting properties of ethanol and 0.15 gal is due to the leaning effect of the alcohol.

**Used as a standalone fuel in spark-ignition engines, alcohol-fueled engines can have a 20 percent higher thermal efficiency than their gasoline-fueled counterparts (see ch. 10).

Table 66.—Net Displacement of Premium Fuels (oil and natural gas) From Various Feedstocks and Two End Uses
(energy expressed as gallons of gasoline equivalent per gallon of ethanol produced and used[a])

Feedstock	Ethanol used as an octane-boosting additive to gasoline[b]	Ethanol used as a standalone fuel[c]
Corn	0.9	0.4
Grain sorghum	0.7	0.1
Spring wheat	1.0	0.5
Oats	1.0	0.5
Barley	1.0	0.4
Sugarcane	0.9	0.3

[a]Assuming lower heat content of gasoline and ethanol to be 117,000 Btu/gal, respectively; crops grown on marginal cropland with yields of 75 percent of average cropland yields; distillers' grain energy credits as in table 65 for all grains and no credit for sugars; distillers fueled with nonpremium fuels; national average energy inputs. S. Barber, et al., "The Potential of Producing Energy From Agriculture," contractor report to OTA.
[b]Uncertainty ±0.3.
[c]Uncertainty ±0.2.

SOURCE: Office of Technology Assessment.

grains, ethanol produced from grains and used as a standalone fuel (e.g., onfarm as a diesel fuel substitute) may actually lead to an increased use of premium fuels, even if nonpremium fuels are used in the distillery. Consequently, caution should be exercised if onfarm ethanol production and use are encouraged as a means of reducing the U.S. dependence on imported fuels.

Furthermore, the agricultural system is so complex and interconnected that it is virtually impossible to ensure that large levels of grain production for standalone ethanol fuel would not lead to a net increase in premium fuel consumption. Two examples illustrate this point. If grain sorghum from Nebraska is used as an ethanol feedstock (to produce a standalone fuel), the net displacement of premium fuels per gallon of ethanol is similar to the national average for corn. A secondary effect of this, however, could be an increase in grain sorghum production on marginal cropland in Texas, and the increased energy required to grow this grain sorghum could more than negate the fuel displaced by the Nebraska sorghum. Similarly, ethanol production from corn could raise corn prices and lead to some shift from corn to grain sorghum as an animal feed. Depending on where the shifts occurred, U.S. premium fuel consumption could either increase or decrease as a result.

The crucial point is that the energy usage in agriculture is an important consideration in determining the effectiveness of a fuel ethanol program. Because of this, there can be situations where energy from agriculture does not result in a net displacement of premium fuels.

In order to avoid this situation, care should be taken to ensure that ethanol derived from grains and sugar crops be used in the most energy-efficient manner possible, i.e., as an octane-boosting additive.

Methanol and Ethanol From Wood, Grasses, and Crop Residues

Methanol, like ethanol, can be used as an octane-boosting additive and the oil refinery energy saved per gallon of methanol is roughly equivalent to that of ethanol. The lower energy content (per gallon) of methanol, however, leads to a smaller displacement of gasoline in the automobile per gallon of alcohol (0.6 gal of gasoline equivalent per gallon of methanol versus 0.8 for ethanol; see ch. 10). On the other hand, the energy used to grow, collect, and transport wood and plant herbage for methanol production is less than for ethanol feedstocks such as grain and sugars. There are, however, considerable local variations and where, for example, crop residues are collected on lands with poor yields, the energy consumed in collection could be comparable to that needed to produce some grains and sugar crops.

Table 67 presents a summary of the net displacement per gallon of alcohol for the various lignocellulosic feedstocks and two end uses. The net displacement per gallon of methanol is comparable to that obtained for ethanol from grains and sugar crops, because the lower energy content of methanol (as compared to ethanol) is largely compensated for by the lower energy required to obtain methanol feedstocks.

Another aspect of the energy balance for the lignocellulosic feedstocks is the net displacement of premium fuels per ton of feedstock. In table 68, direct combustion, airblown gasification, and alcohol fuels production are compared with wood as the feedstock. Similar results can also be derived for crop residues and grasses.

Table 67.—Net Displacement of Premium Fuels (oil and natural gas) With Alcohol Production From Various Feedstocks and Two End Uses (energy expressed as gallons of gasoline equivalent per gallon of alcohol produced and used[a])

Feedstock	Fuel	Used as an octane-boosting additive to gasoline	Used as a standalone fuel
Wood	Methanol	0.9[b]	0.4[c]
Grasses or crop residues	Methanol	0.8[b]	0.3[c]
Wood	Ethanol	1.1[d]	0.6[d]
Grasses or crop residues	Ethanol	1.0[d]	0.5[d]

[a]Assumes: 1) lower heating values of 57,000, 76,000, and 117,000 Btu/gal for methanol, ethanol, and gasoline, respectively; 2) cultivation (grasses) collection and transport (all feedstocks) energy of 0.75 million Btu/dry ton for wood and 2 million Btu/dry ton for grasses and crop residues (including fertilizers for grasses and fertilizer replacements needed when crop residues are collected); 3) methanol yields of 120 gal/ton for wood and 100 gal/ton for grasses and crop residues (50% energy conversion efficiency); 4) ethanol yields are 100 gal/ton of feedstock fermented, but additional feedstock amounting to 25,000 Btu/gal of ethanol is required for distillery energy over and above that obtained from burning the byproduct lignin (based on G. H. Emert and R. Katzen, "Chemicals From Biomass by Improved Enzyme Technology," presented at the *Symposium on Biomass as a Non-Fossil Fuel Source*, ACS/CST Joint Chemical Congress, Honolulu, Hawaii, Apr. 1-6, 1979); resulting in net yields of 86 and 84 gal/ton of feedstock for wood and grasses/crop residues, respectively; 5) methanol and ethanol displace 1.0 and 1.2 gal, respectively, of gasoline energy equivalent (per gallon of alcohol) at the refinery and in the automobile when used as octane-boosting additives to gasoline; 6) they replace 0.48 and 0.65 gal of gasoline equivalent (per gallon of alcohol) at the end use when used as standalone fuels.
[b]Uncertainty ±0.3.
[c]Uncertainty ±0.1.
[d]Uncertainty large, since future processes for producing ethanol from these feedstocks are not fully defined (see footnote a).
SOURCE: Office of Technology Assessment.

Table 68.—Net Displacement of Premium Fuel (oil and natural gas) per Dry Ton of Wood for Various Uses

Use	Net displacement of premium fuel (10⁶ Btu/dry ton of feedstock)	(% of feedstock energy content)
Direct combustion (68% efficiency)	12[ab]	75
Air gasification and combustion of fuel gas (85% overall efficiency)	15[ab]	95
Methanol (used as octane-boosting additive)	13[c]	80
Ethanol (used as octane-boosting additive)	11[c]	70
Methanol (standalone fuel)	6[c]	40
Ethanol (standalone fuel)	6[c]	40

[a]Assuming 16 million Btu/dry ton; 0.75 million Btu/ton required for collection and transport.
[b]Assuming it replaces oil burned with 85% efficiency.
[c]Based on table 67.
SOURCE: Office of Technology Assessment.

Care should be exercised when interpreting table 68. The ethanol yields (per ton of wood) and the energy that will be required by wood-to-ethanol distilleries are still highly uncertain. Nevertheless, this table does display the general feature that alcohol fuels used as octane-boosting additives can be nearly as efficient in displacing premium fuels as the direct combustion or airblown gasification of wood. On the other hand, if the alcohols are used as standalone fuels, the premium fuels displacement is considerably smaller.

General Considerations

The results presented in tables 65 through 68 are based on OTA's estimates of average values for the energy consumed and displaced by the various feedstocks. These figures, however, cannot be taken too literally since local variations and changing circumstances can influence the results. Two of the more important factors which influence the results—the energy required to obtain the feedstock and the end use of the fuel—are discussed below.

The energy needed to grow, harvest, and transport the feedstocks varies considerably, depending on a number of site-specific factors such as quantity of available biomass per acre, terrain, soil productivity, plant type, harvesting techniques, etc. Generally, however, factors that increase the energy requirements also increase the costs. For example, where the quantity of collectable crop residues per acre is small both the energy used and the cost (per ton of residue) will be higher than the average. The economics will therefore usually dictate that—locally, at least—the more energy-efficient source of a given feedstock be used.

As the use of bioenergy increases, however, the tendency will be to move to less energy-efficient sources of feedstocks, and large Government incentives could lead to the use of bioenergy that actually increases domestic consumption of premium fuels. The danger of this is minimal with wood, but somewhat greater for grasses and crop residues due to the larger amount of energy needed to grow and/or collect them. The danger is even greater when grain or sugar feedstocks are used for the production of standalone fuel ethanol.

Another important factor in the energy balance is the end use of the alcohol fuel. As has been emphasized above, there is a significant increase in the displacement of premium fuels when the alcohol is used as an octane-boosting additive. In the 1980's, however, there could be an increased use of automobile engines that do not require high-octane fuels and that have automatic carburetor adjustment to maintain the proper air to fuel ratio (see ch. 10). With these engines, the octane-boosting properties of the alcohols are essentially irrelevant. Consequently, if the automobile fleet is gradually converted in this way, there will be a gradual reduction in the fuel displacement per gallon of alcohol, until the energy balances derived for standalone fuels pertain. The same conclusion would hold if oil refineries convert to more energy-efficient processes for producing high-octane gasoline.

Another consequence of these possible changes would be to increase the importance of the energy required to obtain the feedstock. For example, if ethanol only displaces as much premium fuel as indicated when used as a standalone fuel, then, as mentioned above, cultivating and harvesting the grains or sugar crops used as feedstocks may require more premium fuel than is displaced by the ethanol. The danger of this is considerably less for grasses and crop residues and virtually nonexistent for forest wood used as feedstock for alcohol production.

Chapter 12
CHEMICALS FROM BIOMASS

Chapter 12.—CHEMICALS FROM BIOMASS

Chapter 12
CHEMICALS FROM BIOMASS

Introduction

Biomass is used as a source of several industrial chemicals, including dimethyl sulphoxide, rayons, vanillin, tall oil, paint solvents, tannins, and specialty chemicals such as alkaloids and essential oils. Biomass is also the source of furfural which is used to produce resins and adhesives and can be used in the production of nylon.[1] Aside from paper production, biomass currently is the source of cellulose acetates and nitrates and other cellulose derivatives (4 billion lb annually). Other chemicals include tall oil resin and fat acid, lignosulfonate chemicals, Kraft lignin, bark chemicals, various seed oils, and many more. Every petroleum-derived chemical currently being used could be produced from biomass and nonpetroleum minerals, but some (e.g., carbon disulphide) would require rather circuitous synthesis routes.

In the future biomass-derived chemicals could play an increasing role in the petrochemical industries. The economic decisions to use or not to use biomass will be based on an assessment of the overall process from feedstock to end product and it will probably

[1] I. S. Goldstein, "Potential for Converting Wood Into Plastics," *Science*, vol. 189, p. 847, 1975.

involve consideration of various alternative synthesis routes in most cases. At present, however, too little information is available about the relative merits of biomass-versus coal-derived chemicals to expect widespread, new industrial commitments to biomass chemicals in the near future. This uncertainty depends as much on uncertainties surrounding the costs and possibilities of coal syntheses as on those surrounding biomass chemicals. Continued research into both options is needed to resolve the problem and it is likely (as has been the case in the past) that a mix of feedstocks will result.

Biomass-derived chemicals can be divided into two major areas: 1) those in which the plant has performed a major part of the synthesis and 2) those in which chemical industry feedstocks are derived by chemical synthesis from the more abundant biomass resources such as wood, grasses, and crop residues. Some examples of each type are given below. The possibilities are so enormous, only an incomplete sampling can be given here. A thorough analysis of the options for chemicals from biomass is beyond the scope of this study.

Chemicals Synthesized by Plants

Several plant species produce relatively large quantities of chemicals that can be used to produce plastics, plasticizers, lubricants, coating products (e.g., paints), and various chemicals that can serve as intermediates in the syntheses used for numerous industrial products.[2]

The biologically synthesized chemicals that are most easily used in the chemical industries are those that are either: 1) identical to existing feedstock or intermediate chemicals, or 2)

[2] L. H. Princen, "Potential Wealth Is New Crops: Research and Development," *Crop Resources* (New York: Academic Press, 1977).

have properly placed chemical groups which are susceptible to chemical attack so that they can be readily converted to the needed industrial chemicals. There is also the possibility of using biologically derived chemicals to produce products (such as plastics) which would be expected to have similar properties to the products currently produced. For example, nylon could be made from acids and amines other than the six carbon acids and amines currently used for nylon synthesis.

Some plant species producing various classes of chemicals are shown in tables 69

227

through 72. (Note that these lists are incomplete and used only to illustrate some possibilities.) Included are the following types:

- long-chain fatty acids which might be used for the production of polymers, lubricants, and plasticizers;
- hydroxy fatty acids which could displace the imported castor oil currently used as a supply of these fatty acids;
- epoxy fatty acids which may be useful in plastics and coating materials; and

Table 69.—Species With Long-Chain Fatty Acids in Seed Oil

Common name	Species	Component in triglyceride oil
Crambe	Crambe abyssinica	60% C_{22}
Money plant	Lunaria annua	40% C_{22}, 20% C_{24}
Meadowfoam	Limnanthes alba	95% C_{22} + C_{20}
Selenia	Selenia grandis	58% C_{22}
—	Leavenworthia alabamica	50 % C_{22}
Marshallia	Marshallia caespitosa	44% C_{22}

SOURCE: L. H. Princen, "Potential Wealth in New Crops: Research and Development," *Crop Resources* (New York: Academic Press, Inc.), 1977.

Table 70.—Species With Hydroxy and Keto Fatty Acids

Common name	Species	Component in triglyceride oil
Bladderpod Consessi	Lesquerella gracilis	14-OH-C_{20} (70%)
Holarrhena	Holarrhena antidysenterica	9-OH-C_{18} (70%)
Bittercress	Cardamine impatiens	Dihydroxy C_{22} and C_{24} (23%)
Thistle	Chamaepeuce afra	Trihydroxy C_{18} (35%)
Bladderpod	Lesquerella densipila	12-OH-C_{18} diene (50%)
Blueeyed Capemarigold	Dimorphotheca sinuata	9-OH-C_{18} conj. diene (67%)
Myrtle Coriaria	Coriaria myrtifolia	13-OH-C_{18} conj. diene (65%)
—	Cuspedaria pterocarpa	Keto acids (25%)

SOURCE: L. H. Princen, "Potential Wealth in New Crops: Research and Development," *Crop Resources* (New York: Academic Press, Inc.), 1977.

Table 71.—Potential Sources of Epoxy Fatty Acids

Common name	Species	Epoxy acid content, %
Kinkaoil ironweed	Vernonia anthelmintica	68-75%
Euphorbia	Euphorbia lagascae	60-70
Stokesia	Stokesia laevis	75
—	Cephalocroton pueschellii	67
—	Erlangea tomentosa	50
Hartleaf Christmasbush	Alchornea cordifolia	50 (C_{20})
—	Schlectendalia luzulaefolia	45

SOURCE: L. H. Princen, "Potential Wealth in New Crops: Research and Development," *Crop Resources* (New York: Academic Press, Inc.), 1977.

Table 72.—Sources of Conjugated Unsaturates

Common name	Species	Type of saturation
Common valeriana	Valeriana officinalis	40% 9,11,13
Potmarigold colendula	Calendula officinalis	55% 8,10,12
Spurvalerian centrathus	Centranthus macrosiphon	65% 9,11,13
Snapweed	Impatiens edgeworthii	60% 9,11,13,15
Blueeyed Capemarigold	Dimorphotheca sinuata	60% 10,12 (+ hydroxy)
Myrtle Coriaria	Coriaria myrtifolia	65% 9,11 (+ hydroxy)

SOURCE: L. H. Princen, "Potential Wealth in New Crops: Research and Development," *Crop Resources* (New York: Academic Press, Inc.), 1977.

- conjugated unsaturates potentially useful as intermediates in the synthesis of various industrial products. (These can also be obtained from structural modification of soybean and linseed fatty acids.[3]).

Other possible new sources of chemicals and materials include natural rubber from guayule (*Parthenium argentatum*),[4] and possibly jojoba (*Simmondia chinensis*) and paper pulp from kenaf (*Hibiscus cannabinus*).[5] Some of these plants have also received considerable attention because it may be possible to grow them on marginal lands or land where the irrigation water is insufficient to support conventional crops (see ch. 4). It is risky, however, to extrapolate unambiguous conclusions about their economic viability from incomplete data on the cultivation. In many cases they would also compete with food production for the available farmland. Nevertheless, continued screening of plant species together with cultivation tests should provide numerous additional options for the cultivation of crops yielding chemicals for industrial use.

Another type of chemical synthesis involves the use of specific bacteria, molds, or yeasts to synthesize the desired chemicals or substance. Commercial production of alcohol beverages by fermentation is one example. Mutant bacteria designed to produce insulin or other

---

[3]W. J. De Jarlais, L. E. Gast, and J. C. Cowan, *J. Am. Oil Chem Soc.*, vol. 50, p. 18, 1973.

[4]K. E. Foster, "A Sociotechnical Survey of Buyule Rubber Commercialization," report to the National Science Foundation, Division of Policy Research and Analysis, grant No. PRA 78-11632, April 1979.

[5]M. O. Bagby, "Kenaf: A Practical Fiber Resource," *TAPPI Press Report: Non-Wood Plant Fiber Pulping Process Report*, No. 8, p. 175, Atlanta, Ga., 1977.

drugs is another.[6][7] Furthermore, other basic biochemical processes such as the reduction of nitrates to ammonia may also be used eventually.[8]

[6]Pearce Wright, "Time for Bug Valley," *New Scientist,* p. 27, July 5, 1979.
[7]"Where Genetic Engineering Will Change Industry," *Business Week,* p. 160, Oct. 22, 1979.
[8]P. Candan, C. Manzano, and M. Losada, "Bioconversion of Light Into Chemical Energy Through Reduction With Water of Nitrate to Ammonia," *Nature,* vol. 262, p. 715, 1976.

Further study into the details of photosynthesis, the biochemistry of plants, and molecular genetics could lead to the development of other plants or micro-organisms that could synthesize specific, predetermined chemicals. The options seem enormous at this stage of development, but considerable additional R&D is needed before the full potential of this approach can be evaluated.

Chemical Synthesis From Lignocellulose

The second major area of chemicals from biomass involves using the abundant biomass resources of wood, grasses, and crop residues (lignocellulosic material) to synthesize large-volume chemical feedstocks, which are converted in the chemical industry to a wide variety of more complex chemicals and materials. The large (polymer) molecules in lignocellulosic materials are converted to the desired chemical feedstocks either: 1) by chemical means or 2) with heat or microwaves. The distinction between these two approaches, however, is not always clear cut.

The chemical means include treatment with acids, alkaline chemicals, and various bacterial processes. Pretreatments also often involve some heating and mechanical grinding (see ch. 8). The three basic polymers—lignin, cellulose, and hemicellulose—are reduced to sugars and various benzene-based (so called aromatic) chemicals, which can be used to synthesize the chemical feedstocks by rather direct and efficient chemical synthesis or fermentation (see figure 38). Some ot the major petrochemical feedstocks that can be produced in this way are shown in table 73, together with the quantities of these chemicals (derived mostly from petroleum) which were used by the chemical industries in 1974.

The quantities of wood that would be required to satisfy the 1974 U.S. demand for plastics, synthetic fibers, and synthetic rubber from the above chemical feedstocks are shown in table 74 for the various types of products.

These estimates were derived by Goldstein[9] using optimistic assumptions about the yields of the sugars- and benzene-based chemicals from wood. Obtaining these sugars- and benzene-based compounds from wood is currently the subject of considerable R&D. (See ch. 8.) The yields for the other chemical reactions were based on established experimental and industrial data.

About 95 percent of these synthetic polymers (plastics, synthetic fibers, and synthetic rubbers) can be derived from wood or other lignocellulosic materials, although the circuitous synthesis route required for some of them might make such processes uneconomic at this time. In all, slightly less than 60 million dry tons (about 1 Quad) of wood per year could supply 95 percent of these synthetic polymer needs; and the ratio of cellulose to lignin required (2:1) would be about the same as their natural abundance in wood. This quantity of wood is relatively modest in comparison to OTA estimates of the quantities that can be made available, and in all cases it serves as a direct substitute for chemicals derived from fossil fuels (mostly oil and natural gas). About three to five times as much wood would be needed to supply all petrochemical needs using more or less established chemical synthesis routes,[10] and again there appears to be no technical barrier to supplying these quantities of wood. In both cases, however, addi-

[9]Goldstein, op. cit.
[10]I. S. Goldstein, Department of Wood and Paper Science, North Carolina State University, Raleigh, N.C., private communication, 1980.

Figure 38.—Synthesis Routes for Converting Lignocellulose Into Select Chemical Feedstocks

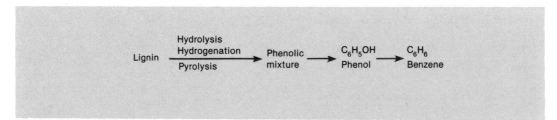

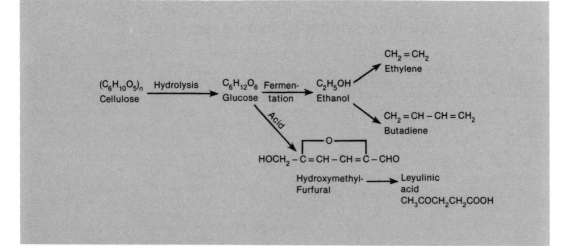

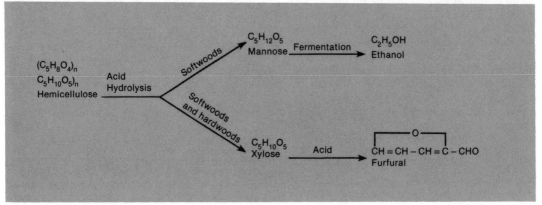

SOURCE: I. S. Goldstein, "Chemicals From Lignocellulose," *Biotechnol. and Bioen. Symp. No. 6,* (New York: John Wiley and Sons, Inc., 1976), p. 293.

Table 73.—Major Petrochemicals That Can Be Synthesized From Lignocellulose

	1974 U.S. production (in billions of pounds)
Total lignocellulose	
Ammonia	31.4
Methanol	6.9
Hemicellulose	
Ethanol	2.0
Cellulose	
Ethanol	2.0
Ethylene	23.5
Butadiene	3.7
Lignin	
Phenol	2.3
Benzene	11.1

SOURCE: From I. S. Goldstein, ''Chemicals From Lignocellulose,'' in *Biotechnology and Bioenergy Symposium No. 6* (John Wiley & Sons, Inc., p. 293, 1976).

tional energy would be needed to provide heat for the syntheses; and, in some cases, this is more than the energy content of the chemical feedstock.[11] As of 1976, the petrochemical industry consumed 1.2 Quads/yr of oil and natural gas for fuel and 2.3 Quads/yr for feedstocks.[12]

Another approach to chemicals from lignocellulose involves heat, partial combustion, or the use of microwave radiation to break the natural polymers into smaller molecules suitable for the synthesis. Biogas derived from the anaerobic digestion of biomass could also be used in some of these processes, but the yields are likely to be lower than for the more direct processes. Some possible synthetic routes are shown in figure 39.

[11]''Big Future for Synthetics,'' *Science,* vol. 208, p. 576, May 9, 1980.

[12]G. B. Hegeman, *Report to the Petrochemical Energy Group on 1976 Petrochemical Industry Profile* (Cambridge, Mass.: Arthur D. Little, Inc., June 28, 1977).

Table 74.—1974 Production of Plastics, Synthetic Fibers, and Rubber, and Estimated Lignocellulose Raw Material Base Required

Material	Production (10^3 tons)	Lignocellulose required[a] (10^3 tons)
Plastics		
Thermosetting resins		
Epoxies	125	355 (L)
Polyesters	455	1,220 (L)
Urea	420	—
Melamine	80	—
Phenolic and other tar-acid resins	670	1,915 (L)
Thermoplastic resins		
Polyamide	100	285 (L)
Polyethylene		
Low density	2,985	11,940 (C)
High density	1,420	5,680 (C)
Polyprophylene and copolymers .	1,125	4,500 (C)
Styrene and copolymers	2,505	7,445 (L)
Polyvinyl chloride	2,425	4,225 (C)
Other vinyl resins	175	440 (C)
Total plastics	12,485	
Synthetic fibers		
Cellulosic		
Rayon	410	—
Acetate	190	—
Noncellulosic		
Nylon	1,065	3,045 (L)
Acrylic	320	640 (C)
Polyester	1,500	4,020 (L)
Olefin	230	920 (C)
Total noncellulosic fibers	3,115	
Synthetic rubber		
Styrene-butadiene	1,615	5,700 (C)
		1,920 (L)
Butyl	180	1,060 (C)
Nitrile	95	190 (C)
Polybutadiene	360	2,120 (C)
Polyisoprene	100	—
Ethylene-propylene	140	825 (C)
Neoprene and others	280	—
Total synthetic rubber	2,770	
Total plastics, noncellulosic fibers, and rubber	18,370	
Obtainable from lignocellulose	17,490	58,445
Cellulose drived (C)		38,240 (C)
Lignin derived (L)		20,205 (L)

[a]Estimated from optimistic approximate yields of monomers obtainable. (C) cellulose derived; (L) lignin derived.

SOURCE: I. S. Goldstein, ''Potential for Converting Wood Into Plastics,'' *Science,* vol. 189, p. 847; 1975.

Figure 39.—Chemical Synthesis Involving Thermal Processes and Microwaves

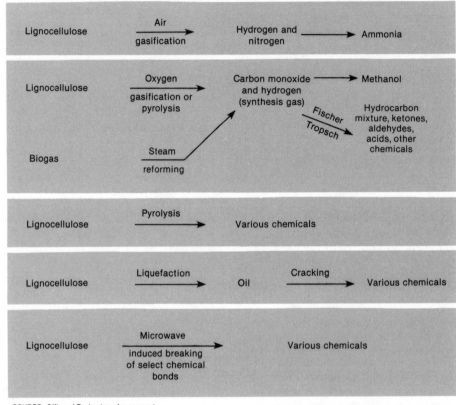

SOURCE: Office of Technology Assessment.

The production of ammonia[13] and methanol from wood can be accomplished with commercial technology (see ch. 7 for further details of the methanol synthesis). The Fischer Tropsch process is commercial in South Africa (although the source of the synthesis gas is coal rather than biomss). The economics of the processes other than methanol synthesis have not been assessed by OTA for this report.

The other processes yielding various chemicals are considerably less developed. The yields of some chemicals that have been produced in laboratory experiments using rapid heating and gasifications (pyrolysis) of various

types of biomass are shown in table 75. Presumably by learning more about pyrolysis, the yields of select chemicals would be increased to a level where it could be economical to extract that chemical from the gas. An example might be the conceptual equation:

$$C_{10}H_{14}O_6 \longrightarrow 3CO_2 + 3.5\,C_2H_4$$

Wood	Carbon dioxide	Ethylene
(solid)	(gas)	(gas)

where 43 weight percent of the dry wood is converted to ethylene (which is by far the largest volume petrochemical used for chemical synthesis in the world). If it becomes practical to achieve relatively high (e.g., 30 weight percent) yields of ethylene, then this process could be competitive with petroleum-derived

[13]R. W. Rutherford and K. Ruschin, "Production of Ammonia Synthesis Gas From Wood Fuel in India," presented at a meeting of the Institute of Chemical Engineers, London, Oct. 11, 1949.

Table 75.—Product Results in Fast Pyrolysis of Biomass and Its Constituents

The final 13 columns (Char through Other) fall under the heading **Approximate weight percent of organics**.

Reference	Type and form of biomass	Rate of heating	Maximum temperature of pyrolysis/gasification	Residence time at temperature	Environment	Char	Tar	Gases	H_2	CH_4	C_2H_4	CO	CO_2	C_2H_6	C_3H_8	C_3H_6	H_2O	Other
Antal (1979)	Whitman filter paper (cellulose) 0.125 g	100°C/min to 500°C/min	700°C	3.5 s	Hot steam or argon	10	2	88	1.2	7.2	5.5	47.2	11.5		1.0	0.15		0.15
Berkowitz-Mattuck and Noguchi (1963)	Cotton cellulose cloth	Carbon-arc radiant at 5 cal/cm²-s, 25 cal/cm²-s	NA	1 s irradiation	Cold helium	~20 / ~8					0 / 0.86	5 / 7.3	0					(a) / (b)
Martin (1965)	α-cellulose + 2% carbon black	Carbon-arc radiant at cal/cm²-s 4.4, 11.6 early, 11.6 late, 10²-10³	NA	0.4-8 s irradiation / ~1 s / ~8 s	Cold helium	20-35 / 4 / 4 / 0	80 / 55		0.02 / 0.5	0.02 / 0.9	0.02 / 0.9	3 / 13	4 / 10				10 / 15	(c) / (d)
Lincoln (1965)	α-cellulose + 2% carbon black	Carbon-arc & Xenon lamp 1.5 cal/cm²-s, 11.0 cal/cm²-s, 3,000 cal/cm²-s	~300°C, ≥600°C, ≥600°C	10 s, 4 s, ½ ms	Cold helium	33 / 3 / 1	19 / 51 / 0	48 / 46 / 99	NA	NA	NA	3 / 13 / 37	9 / 11 / 4				32 / 16 / 28	Volatile organics 3 / 6 / 30
Lewellen, et al. (1976)	Cellulose filter paper 0.75 x 2.5 x 0.01 cm strip	Electrically heated mesh 400°-1,000°C/s	250°-1,000°C	0.2-75,000 s 400°C/s	Vacuum to 1 atm cold He	No char		No gas analysis. Weight loss versus time only										
Hileman, et al. (1976)	Douglas fir 1.5 mg samples	Pyroprobe at ~200°C/s	550°C	3-4 s	Cold argon or air					2.0	1.0	21	5.7		0.3	1.6	15.3	Propane 1.6
Prahacs (1967)	Na, Ca, Mg-base spent liquors	Atomized spray into hot reactor	600°-900°C 5-45 psig	11-100 s	Self-generated steam	Down to 3%			Up to 11%		Illuminants up to 7%					0.6		
Prahacs, et al. (1971)	Bark, slash wood, and spent liquors / Bark	N.A. various reactors up to pilot scale Batch fed	600°-1,000°C 0-25 psig / 810°C 0 psig	3-60 s / 3.6 s	N_2 or self generated			89.5			6.2							
Rensfelt, et al. (1978)	Poplar wood	~1,000°C/s	400°-1,000°C	~1 s	Steam, H_2					5	5 (optimized)							
Brink and Massoudi (1978)	White fir particles 20-40 mesh	~1,000°C/s	316°-871°C / 843°C	3-5 s / 3.0 s	N_2	2.5	7.1		~1	~10	~5	~62	~13					
Diebold and Smith (1979)	Ecofuel II, 200 μm	10⁴°-10⁵°C/s	500°-900°C	50-100 ms	Steam and CO_2	19			1	4	Unsaturates 24	36	16					
Stern, et al. (1965)	Sawdust 20-30 mesh	Fast	1,000°C	Long (steel wool) (Alundum)	Self-generated	22.2 / 14.0	4.4 / 0.4	71.0 / 84.4	4.6 / 4.0	0.3 / 7.1	0 / 0.25	65.5 / 68.4	0.6 / 4.8			2.3 / 2.2		
Brink (1976)	Wood, MSW, and kraft black liquor	Probably fast	(475°-1,125°C) 850°C	Uncertain	6.5% moisture wood / 52.5% moisture wood				1.8 / 3.8	11.2 / 7.2	7.0 / 4.2	73.0 / 47.7	13.2 / 35.2			45		

aTar fraction, mainly levoglucosan. bVariety of polar organics boiling below 187°C. cAcetaldehyde 0.45; acrolein 0.15; acetone, furan, methanol — 0.7. dAcetaldehyde 1.3; acrolein, furan, methanol 0.25; methanol, furan 0.15.

NOTE: References to table 75 appear on p. 234.

SOURCE: T. Milne, "Pyrolysis—The Thermal Behavior Below 600°C," A Survey of Biomass Gasification (vol. 2: Golden, Colo.: Solar Energy Research Institute, 1979).

References to Table 75

Antal, M. J., W. E. Edwards, H. C. Friedman, and F. E. Rogers, ''A Study of the Steam Gasification of Organic Wastes,'' Environmental Protection Agency University grant No. R 804836010, final report, 1979.
Berkowitz-Mattuck, J, B. and T. Noguchi, ''Pyrolysis of Untreated and APO-THPC Treated Cotton Cellulose During 1-Sec. Exposure to Radiation Flux Levels of 5-25 cal/mc² sec.1,'' *J. Applied Polymer Science*, vol. 7, p. 709, 1963.
Brink, D. L., ''Pyrolysis—Gasification—Combustion: A Process for Utilization of Plant Material,'' *Applied Polymer Symposium No. 28*, p. 1377, 1976.
Brink, D. L. and M. S. Massoudi, ''A Flow Reactor Technique for the Study of Wood Pyrolysis. I. Experimental, *J. Fire and Flammability T*, p. 176, 1978.
Diebold, J. P. and G. D. Smith, ''Noncatalytic Conversion of Biomass to Gasoline,'' ASME paper No. 70-Sol-29, 1979.
Hileman, F. D., L. H. Wojeik, J. H. Futrell, and I. N. Einhorn, ''Comparison of the Thermal Degradation Products of Cellulose and Douglas Fir Under Inert and Oxidative Environments,'' *Thermal Uses and Properties of Carbohydrates and Lignins Symposium*, Shafizadek, Sarkenen, and Tillman, ed. (Academic Press, 1976), p. 49.
Lewellen, P. C., W. A. Peters, and J. B. Howard, ''Cellulose Pyrolysis Kinetics and Char Formation Mechanism,'' *Sixteenth Symposium (International on Combustion* (The Combustion Institute, 1976), p. 1471.
Lincoln, K. A., ''Flash Vaporization of Solid Materials for Mass Spectrometry by Intense Thermal Radiation,'' *Analytical Chemistry*, vol 37, p. 541, 1965.
Martin, S., ''Diffusion—Controled Ignition of Cellulosic Materials by Intense Radiant Energy,'' *Tenth Symposium (International) on Combustion* (The Combustion Institute, 1965), p. 877.
Prahacs, S., ''Pyrolytic Gasification of Na-, Ca-, and Mg- Base Spent Pulping Liquors in an AST Reactor,'' *Advances in Chemistry Series*, vol. 69, p. 230, 1967.
Prahacs, S., H. G. Barclay, and S. P. Bhaba, ''A Study of the Possibilities of Producing Synthetic Tonnage Chemicals From Lignocellulosic Residues,'' *Pulp and Paper Magazine of Canada*, vol. 72, p. 69, 1971.
Rensfelt, E., G. Blomkvist, C. Ekstrom, S. Engstrom, B. G. Espenas, and L. Liinanki, ''Basic Gasification Studies for Development of Biomass Medium-Btu Gasification Processes.''
Stern, E. W., A. S. Logindice, and H. Heinerman, ''Approach to Direct Gasification of Cellulosics,'' *Industrial Engineering Chemistry Process Design and Development*, vol. 4, p. 171, 1965.

ethylene. The ethylene could be converted to ethanol, and overall processing costs (wood to ethanol) may be considerably lower than those projected for fermentation processes (see chs. 7 and 8).

The liquefaction process for producing a pyrolytic oil (see ch. 7) might also be carried further by cracking the oil in a way that is analogous to current oil refinery technology. In addition, microwave energy (or other electromagnetic radiation) could possibly be used to break specific predetermined chemical bonds in order to guide and control the decomposition of the biomass.

Rapid pyrolysis, cracking, and microwave processes are still at the research stage and considerable work is required to determine their feasibility. Efforts along these directions, might lead to significant advances in the use of biomass for chemicals and fuels.

RETURN TO: CHEMISTRY LIBRARY
100 Hildebrand Hall • 510-642-3753

LOAN PERIOD	1	2	1-MONTH USE 3
4		5	6

ALL BOOKS MAY BE RECALLED AFTER 7 DAYS.

Renewals may be requested by phone or, using GLADIS, type inv followed by your patron ID number.

DUE AS STAMPED BELOW.

FORM NO. DD 10
3M 7-08

UNIVERSITY OF CALIFORNIA, BERKELEY
Berkeley, California 94720–6000